Road Vehicle Dynamics

GROUND VEHICLE ENGINEERING
Dr. Vladimir V. Vantsevich

Professor and Director
Program of Master of Science in Mechatronic Systems Engineering
Lawrence Technological University, Michigan

Road Vehicle Dynamics

Fundamentals and Modeling with MATLAB®, Second Edition

Georg Rill and Abel Arrieta Castro

Driveline Systems of Ground Vehicles

Theory and Design

Alexandr F. Andreev, Viachaslau Kabanau, Vladimir Vantsevich

Dynamics of Wheel-Soil Systems

A Soil Stress and Deformation-Based Approach

Jaroslaw A. Pytka

Design and Simulation of Heavy Haul Locomotives and Trains

Maksym Spiryagin, Peter Wolfs, Colin Cole, Valentyn Spiryagin, Yan Quan Sun, Tim McSweeney

Automotive Accident Reconstruction: Practices and Principles

Donald E. Struble

Design and Simulation of Rail Vehicles

Maksym Spiryagin, Colin Cole, Yan Quan Sun, Mitchell McClanachan, Valentyn Spiryagin, Tim McSweeney

For more information about this series, please visit: https://www.crcpress.com/Ground-Vehicle-Engineering/book-series/CRCGROVEHENG

Road Vehicle Dynamics

Fundamentals and Modeling with MATLAB®

Second Edition

Georg Rill

Abel Arrieta Castro

CRC Press
Taylor & Francis Group
Boca Raton London New York

CRC Press is an imprint of the
Taylor & Francis Group, an **informa** business

Second edition published 2020
by CRC Press
6000 Broken Sound Parkway NW, Suite 300, Boca Raton, FL 33487-2742

and by CRC Press
2 Park Square, Milton Park, Abingdon, Oxon, OX14 4RN

© 2020 Taylor & Francis Group, LLC

[First edition published by CRC 2011]

CRC Press is an imprint of Taylor & Francis Group, LLC

Library of Congress Control Number: 2020932629

ISBN: 978-0-367-19973-9 (hbk)
ISBN: 978-0-367-49398-1 (pbk)
ISBN: 978-0-429-24447-6 (ebk)

Visit the eResources: https://www.crcpress.com/9780367199739

**Visit the Taylor & Francis Web site at
http://www.taylorandfrancis.com**

**and the CRC Press Web site at
http://www.crcpress.com**

Contents

Series Preface

Ground vehicle engineering took shape as an engineering discipline in the 20th century, and became the foundation for significant advancements and achievements, from personal transportation and agriculture machinery to lunar and planetary exploration. As we step into the 21st century with global economic challenges, there is a need for developing fundamentally novel vehicle engineering technologies, and effectively training future generations of engineers. The Ground Vehicle Engineering series will unite high-caliber professionals from the industry and academia to produce top-quality professional/reference books and graduate-level textbooks on the engineering of various types of vehicles, including conventional and autonomous mobile machines, terrain and highway vehicles, and ground vehicles with novel concepts of motion.

The Ground Vehicle Engineering series concentrates on conceptually new methodologies of vehicle dynamics and operation performance analysis and control; advanced vehicle and system design; experimental research and testing; and manufacturing technologies. Applications include, but are not limited to, heavy-duty, multi-link, and pickup trucks, farm tractors and agriculture machinery, earth-moving machines, passenger cars, human-assist robotic vehicles, planetary rovers, military conventional and unmanned wheeled and track vehicles, and reconnaissance vehicles.

Dr. Vladimir V. Vantsevich
Series Editor

Preface

The basic objective of this textbook is to provide fundamental knowledge of the dynamics of road vehicles and to impart insight into and details on modeling aspects. Road vehicles have been developed and built for more than 125 years. From the beginning to now, the main goal in vehicle dynamics has been to achieve optimal safety and ride comfort. Today, computer simulations have become an essential tool to develop new and enhance existing concepts for road vehicles. Electronically controlled components provide new options. Usually they are developed and tested in software-in-the-loop (SIL) or hardware-in-the-loop (HIL) environments. Today, an automotive engineer requires a basic knowledge of the fundamentals and the skills to build basic simulation models, to handle sophisticated ones, or to operate simulation tools properly. Yet, only books on the fundamentals of vehicle dynamics or on modeling aspects are available currently. Both subjects are combined for the first time in this textbook. A first course in dynamics and a basic knowledge of a programming language are prerequisites.

Lecture notes for an undergraduate course in vehicle dynamics form the basis of this textbook. The notes were then extended to serve as the basis for a graduate course, first delivered at the State University in Campinas (UNICAMP) by the author in 1992 and repeated as a short course in vehicle dynamics several times since then. Part of the notes combined with additional material are also used for in-house seminars at different automotive suppliers and in several workshops at conferences on different topics of vehicle dynamics. A truly motivating personal contact with Vladimir V. Vantsevich, the editor of the Ground Vehicle Engineering series, at the *21st International Symposium on Dynamics of Vehicles on Roads and Tracks* in 2009 resulted in a textbook proposal to Chapman & Hall/CRC. The lecture notes and the material for the seminars formed the skeleton of the first edition of this textbook.

The second edition retains the basic structure of the textbook and provides additional topics, examples, and exercises. Due to a slightly enlarged page layout, the number of pages has increased only slightly.

As far as possible, the symbols are adjusted to the common use in vehicle dynamics now. A newly added section illustrates the primary meaning of the symbols. In addition, a short glossary at the end of the textbook explains phrases typical for vehicle dynamics.

Exercises at the end of each chapter serve to repeat the contents and check the knowledge of the reader. Many programming examples, which are integrated into this textbook and available for download on the website of the publisher, deepen the insight on vehicle dynamics and make a serious self-study possible. Due to its ease of use and popularity, MATLAB® was used as the programming language.

This textbook may be used in a graduate course on vehicle dynamics for classroom teaching and self-study. Skipping the modeling aspects, it will suit an "Introduction to Vehicle Dynamics" at the senior undergraduate level. The modeling aspects may even be integrated into a course on multibody dynamics. In addition, this textbook will also help practicing engineers and scientists in the field of vehicle dynamics by providing a full range of vehicle models that may be used for basic studies, parameter variations, optimization, as well as the simulation of different driving maneuvers.

Modern SI units are used throughout this textbook.

The introduction in Chapter 1 provides an overview of units and quantities, discusses the terminology in vehicle dynamics, deals with definitions, gives an overview of multibody dynamics, and derives the equations of motion for a first vehicle model.

Chapter 2 is dedicated to the road. At first, a sophisticated road model is discussed, then deterministic and stochastic road models are described in detail. The quarter car model with a trailing arm suspension, presented in Chapter 1, is used to demonstrate the impact of a rough road to the wheel and to the chassis.

The handling tire model TMeasy is discussed in Chapter 3. It includes a sophisticated contact calculation and provides all forces and torques. The longitudinal and lateral forces as well as the torque about the vertical axis are modeled in a first-order dynamic approach. Besides that, measuring techniques and modeling aspects are also discussed.

Chapter 4 focuses on the drive train. Components and concepts of different drive trains are discussed first. Then, the dynamics of wheel and tire are studied in detail. A simple vehicle wheel tire model including a lockable brake torque model is presented and investigated by a corresponding MATLAB simulation. A short description of the layout and modeling aspects of differentials, standard drive trains, the transmission, the clutch, and different power sources completes this chapter.

The purpose and components of suspension systems are discussed in Chapter 5. A dynamic model of rack-and-pinion steering as well as a kinematic model of a double wishbone suspension system are presented and analyzed via the simulation results. The newly included design kinematics make it possible to model arbitrary suspension systems or to match kinematic and compliance tests smoothly and efficiently. The design kinematics is even capable of approximating a statically overdetermined suspension system in a quasi-static manner. This specific feature is demonstrated on the twist beam suspension system, which is a cheap and robust solution for cars with front-wheel drive or the axles of semi-trailers. Chapter 5 ends with a model of a typical race car suspension system.

Chapter 6 is dedicated to force elements. A practical air spring model is added here. Besides the standard force elements like springs, anti-roll bars, and dampers, dynamic force elements, including hydro-mounts, are discussed in detail. A sophisticated model for a sweep sine excitation, as well as models for different dynamic forces, are provided here.

The fundamentals of vehicle dynamics in the vertical, longitudinal, and lateral directions are provided in Chapters 7 to 9. Each chapter starts with a simple model approach. Then, the effects of nonlinear, dynamic, or even active force elements, as well as the influence of suspension kinematics, are studied. Section 7.2 and Section 7.7

are completely revised and extended. Practical aspects and applications complete these chapters. Chapter 8 contains a newly added planar vehicle model that demonstrates the basic principle of an anti-lock system. Section 9.4 is supplemented by the discussion of vehicles with additional rear-wheel steering.

Finally, Chapter 10 presents the idea of a full three-dimensional vehicle model and shows and discusses typical results of standard driving maneuvers. A newly added section presents the impact of mechatronic systems on the dynamics of vehicles.

The results of the exercises and all MATLAB examples as well as some additional M files are available for download on the website of the publisher. Executable MATLAB scripts are indicated by names of type app_no_name.m where app stands for application, no is a number referring to the corresponding chapter in the textbook, and name specifies the specific topic. Most of the scripts refer to functions, which have names in the form of fun_no_name.m or uty_name.m. The former refers to specific functions (fun) and the latter to utility functions (uty) of general nature. Additional MATLAB scripts named like xtra_no_name.m are not listed in the textbook. The MATLAB files provided on the website of the publisher represent a set of virtual test rigs which can be used in courses or in self-study. The MATLAB applications make it possible to reflect and to deepen the insight into different problems in vehicle dynamics by studying the influence of the model parameter on the results or by modifying the simulation input.

The first author is very grateful to Vladimir V. Vantsevich, who encouraged him to write such a textbook. The authors are also thankful to the publisher, namely, Senior Editor Jonathan W. Plant and Editorial Assistant Bhavna Saxena for all their support, which allowed us to focus on the contents and not to bother with the layout. The former provided the necessary support for the first edition and initiated the second one, while the latter took care of the administrative tasks of this textbook.

Georg Rill
Abel Arrieta Castro

MATLAB® is a registered trademark of the MathWorks, Inc.
For product information, please contact:
The MathWorks, Inc.
3 Apple Hill Drive
Natick, MA 01760-2098 USA
Tel.: 508 647 7000
Fax.: 508 647 7001
E-mail: info@mathworks.com
Web.: www.mathworks.com

Visit the eResources: https://www.crcpress.com/9780367199739

About the Authors

Georg Rill has been a researcher and an educator for more than thirty years. He is professor at OTH Regensburg, Germany. For nearly five years he was a technical specialist at the Daimler-Benz Company in Stuttgart, Germany. There, he developed a full vehicle model for trucks that was used in the Daimler-Benz driving simulator for many years. He has been a visiting professor at the State University in Campinas, Brazil. He spent one of his three sabbaticals at Ford Motor Company in Detroit, Michigan. He is the author of numerous journal publications on vehicle dynamics. His first book, *Simulation von Kraftfahrzeugen*, was published in 1994. Recently, he contributed "Multibody Systems and Simulation Techniques" to the book *Vehicle Dynamics of Modern Passenger Cars*. He developed the handling tire model TMeasy, which for several years has been part of the SIMPACK® automotive package.

Abel Arrieta Castro got his bachelor's in mechatronic engineering at National University of Engineering (UNI) in Lima, Peru and received his master's degree in applied mechanics at the PUC-Rio in Rio de Janeiro, Brazil. In his PhD thesis, he developed a robust and fault-tolerant integrated control system to improve the stability of road vehicles in critical driving scenarios. During his PhD studies and as a post-doc he spent several months at OTH Regensburg, where he deepened his knowledge of vehicle dynamics and tire modeling. Since January 2019 he has been a research engineer at the Elektronische Fahrwerksysteme (EFS) GmbH, a major supplier of the automotive industry. His focus is on multibody dynamics, vehicle dynamics, real-time applications, and control strategies.

Primary Meaning of Symbols

In vehicle dynamics Greek and Latin symbols are used. Some are pre-defined by DIN-ISO-directives. These definitions are respected in this textbook as much as possible. However, the number of Greek and Latin letters is limited. That is why some symbols have to be used with different meanings in separate contexts. The primary meanings used in this textbook are listed below.

Greek Symbols

α	Slip angle, rotation angle about x-axis, angular acceleration	μ	Friction coefficient
		π	Number pi
β	Sideslip angle, rotation angle about y-axis	ϱ	Air density, radius of curvature
γ	Camber angle, rotation angle about z-axis	σ	Standard deviation, effective value
δ	Steering angle	φ	Rotation angle, wheel rotation
ζ	Viscous damping		
Θ	Inertia	ϕ	Roll angle of vehicle
θ	Pitch angle of vehicle	ψ	Yaw angle of vehicle
κ	Polytropic exponent	Ω, ω	Angular frequency, angular velocity
λ	Eigenvalue		

Latin Symbols

A	Area	e_{yR}	Unit vector normal to the rim center plane defining the wheel rotation axis
A_{ij}	Rotation matrix which transforms vectors from frame j to frame i		
a	Acceleration, wheel base	F	Force
a_x	Longitudinal acceleration	F_x	Longitudinal force
a_y	Lateral acceleration	F_y	Lateral force
B	Width	F_z	Vertical force, wheel load
b	Width of vehicle	f	Frequency
		G	Weight
c	Spring rate	g	Constant of gravity
c_S	Suspension stiffness	H, h	Height
d	Damping parameter	k_S	Cornering stiffness
d_S	Suspension damping	L, ℓ	Length

M, m	Mass	s_y	Lateral slip
n	Pneumatic trail	t	Time
p	Pressure	T	Torque
q	Generalized forces	u	Displacement
r, R	Radius	v	Velocity
r_0	Unloaded wheel radius	w	Waviness
r_S	Static wheel radius	x, y, z	Coordinates and axis of coordi-
r_D	Dynamic wheel radius		nate systems and frames
r_{ij}	Position vector pointing from i to j	x	State vector
S	Power spectral density (psd)	y	Generalized coordinates
s	Spatial coordinate, track width	z	Generalized velocities,
s_x	Longitudinal slip		road height

1

Introduction

CONTENTS

1.1 Units and Quantities

1.1.1 SI System

The International System of Units[1] is the world's most widely used system of measurement. The International Organization for Standardization (ISO) laid it out in ISO 31 and ISO 1000. The modern SI system will be used throughout this textbook. Common quantities, their variations, as well as conversion tables may be found in [3]. The most important physical quantities, their standardized units, as well as some old but still important units are collected in Table 1.1. The unit rad can be replaced by the

TABLE 1.1
Physical quantities, common symbols, and units

Quantity	Name	Unit	SI	Relationship
Length ℓ	meter	m	yes	
	inch	in	no	$1\,\text{in} = 0.0254\,\text{m}$
Area A	square meter	m^2	yes	
Volume V	cubic meter	m^3	yes	
Angle φ	radian	rad	yes	
	degree	deg or °	no	$1° = 0.01745\,\text{rad}$
Mass m	kilogram	kg	yes	
	pound	lb	no	$1\,\text{lb} = 0.45359\,\text{kg}$
Time t	second	s	yes	
Frequency f	hertz	Hz	no	$1\,\text{Hz} = 1/\text{s}$
Angular freq. ω		s^{-1}	yes	$\omega = 2\pi f$
Velocity v	meters / second	m/s	yes	
	kilometers / hour	km/h	no	$1\,\text{km/h} = 1/3.6\,\text{m/s}$
	miles / hour	mph	no	$1\,\text{mph} = 1.609\,\text{km/h}$
Angular velocity v	radians / second	rad/s	yes	
Rotational speed n	rpm	1/min	no	$1\,\text{rpm} = (1/60)\,\text{s}^{-1}$
Acceleration a		m/s^2	yes	
Angular accel.		rad/s^2	yes	
Force F	newton	N	yes	$1\,\text{N} = 1\,\text{kg m s}^{-2}$
Torque T, Moment M	newtonmeter	Nm	yes	$1\,\text{Nm} = 1\,\text{kgm}^2/s^2$
Pressure p	pascal	Pa	yes	$1\,\text{Pa} = 1\,\text{N/m}^2$
	bar	bar	no	$1\,\text{bar} = 10^5\,\text{Pa}$
	pound / square inch	psi	no	$1\,\text{psi} = 6\,894.8\,\text{Pa}$
Temperature T	kelvin	K	yes	

numeral 1 in calculations. With the exception of hours (1 h = 3 600 s) and minutes (1 min = 60 s), decimal multiples and submultiples will form additional legal units.

---

[1]Abbreviated SI from the French *le Système International d'Unités*.

1.1.2 Tire Codes

Although the SI unit system has been nearly globally adopted, the alphanumeric code molded into the sidewall of the tire is still a mixture of modern and old units. To take an example, P215/60R16 signifies by the letter P a passenger car tire with a 215-millimeter tire width. The height of the sidewall of the tire is 60% of the width, and the letter R states that the cord threads of the tire casing run in the radial direction. Finally, the tire is designed to fit rims of 16 inches. So, the height of this tire amounts to 215*0.60 = 129 *mm*, and the conversion from inches to meters and millimeters, given in Table 1.1, will produce a rim diameter of 16*0.0254 = 0.4064 *m* = 406.4 *mm*. Adding twice the tire height to the rim diameter will then result in an overall tire diameter of 2*129 *mm* + 406.4 *mm* = 664.4 *mm*, which is equivalent to a tire radius of 332.2 *mm*. By adding the load index (a number code, from 0 to 279), along with the speed rating symbol (a letter code, from A to Z), we finally end up in the ISO Metric Sizing System.

Some light truck tires follow the Light Truck High Flotation System, which for example, may read as 37x12.5R17LT. Here, the code starts with a two-digit number separated by the character x from a three- or four-digit number, indicating the diameter of the tire and the section width (cross section) of the tire. The following letter B for belted bias, D for diagonal bias, or R for radial tire names the construction type. The following two-digit number gives the diameter in inches of the wheel rim that this tire is designed to fit. Finally, the letters LT indicate that this is a Light Truck tire. Load index and speed rating are not required for this type of tire but may be provided by the manufacturer.

1.2 Terminology

1.2.1 Vehicle Dynamics

Vehicle dynamics are a part of engineering primarily based on classical mechanics but may also involve control theory, physics, electrical engineering, chemistry, communications, psychology, etc. Here, the focus is laid on ground vehicles supported by wheels and tires. Vehicle dynamics encompass the interaction of the following:

- Driver
- Vehicle
- Load
- Environment

Vehicle dynamics mainly deals with:

- Improvement in active safety and driving comfort
- Reduction in road destruction

Vehicle dynamics employs:

- Computer calculations
- Test rig measurements
- Field tests

In the following, the interactions between the single systems and the problems with computer calculations and/or measurements are discussed.

1.2.2 Driver

By various means, the driver can interfere with the vehicle:

$$\text{Driver} \left\{ \begin{array}{l} \text{Steering wheel} \quad\quad \text{Lateral dynamics} \\ \text{Accelerator pedal} \\ \text{Brake pedal} \quad\quad\quad\quad \text{Longitudinal dynamics} \\ \text{Clutch} \\ \text{Gear shift} \end{array} \right\} \longrightarrow \text{Vehicle}$$

The vehicle provides the driver with the following information:

$$\text{Vehicle} \left\{ \begin{array}{ll} \text{Vibrations:} & \text{longitudinal, lateral, vertical} \\ \text{Sounds:} & \text{motor, aerodynamics, tires} \\ \text{Instruments:} & \text{velocity, external temperature, ...} \end{array} \right\} \longrightarrow \text{Driver}$$

The environment also influences the driver:

$$\text{Environment} \left\{ \begin{array}{l} \text{Climate} \\ \text{Traffic density} \\ \text{Track} \end{array} \right\} \longrightarrow \text{Driver}$$

The driver's reaction is very complex. To achieve objective results, an "ideal" driver is used in computer simulations; and in driving experiments, automated drivers (e.g., steering machines) are employed. Transferring results to normal drivers is often difficult if field tests are made with test drivers. Field tests with normal drivers must be evaluated statistically. Of course, the driver's security must have absolute priority in all tests. Driving simulators provide an excellent means of analyzing the behavior of drivers even in limit situations without danger. Many have tried to analyze the interaction between driver and vehicle with complex driver models for some years.

1.2.3 Vehicle

The following vehicles are listed in the ISO 3833 directive:

- Motorcycles
- Passenger cars
- Buses

- Trucks
- Agricultural tractors
- Passenger cars with trailer
- Truck trailer/semitrailer
- Gigaliners and road trains

For computer calculations these vehicles must be depicted in mathematically describable substitute systems. The generation of the equations of motion, the numeric solution, as well as the acquisition of data require great expense. In these days of PCs and workstations, computing costs hardly matter anymore. At an early stage of development, often only prototypes are available for field and/or laboratory tests. Results can be falsified by safety devices, e.g., jockey wheels on trucks.

1.2.4 Load

Trucks are conceived for taking up load. Thus, their driving behavior changes.

$$\text{Load} \left\{ \begin{array}{l} \text{Mass, Inertia, Center of gravity} \\ \text{Dynamic behavior (liquid load)} \end{array} \right\} \longrightarrow \text{Vehicle}$$

In computer calculations, problems occur when determining the inertias and the modeling of liquid loads. Even the loading and unloading process of experimental vehicles takes some effort. When carrying out experiments with tank trucks, flammable liquids must be substituted with water. Thus, the results achieved cannot be simply transferred to real loads.

1.2.5 Environment

The environment influences primarily the vehicle:

$$\text{Environment} \left\{ \begin{array}{ll} \text{Road:} & \text{irregularities, coefficient of friction} \\ \text{Air:} & \text{resistance, cross wind} \end{array} \right\} \longrightarrow \text{Vehicle,}$$

but also affects the driver:

$$\text{Environment} \left\{ \begin{array}{l} \text{Climate} \\ \text{Visibility} \end{array} \right\} \longrightarrow \text{Driver}$$

Through the interactions between vehicle and road, roads can quickly be destroyed. The greatest difficulty with field tests and laboratory experiments is the virtual impossibility of reproducing environmental influences. The main problems with computer simulation are the description of random road irregularities, the interaction of tires and road, as well as the calculation of aerodynamic forces and torques.

1.3 Definitions

1.3.1 Coordinate Systems

If the chassis is supposed to be a rigid body, one coordinate system fixed to the vehicle, which in general is located in the center C, will be sufficient to describe the overall motions of the vehicle body, Figure 1.1. The earth-fixed system 0 with the axis x_0, y_0,

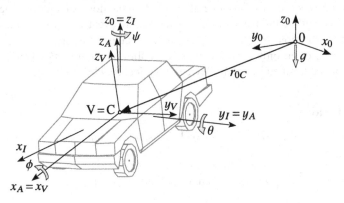

FIGURE 1.1
Position and orientation of the vehicle body.

z_0 serves as the inertial reference frame. Its origin 0 lies in a reference ground plane. Throughout this textbook the z_0-axis will point in the opposite direction of the gravity vector g. Within the vehicle-fixed system V, the x_V-axis points forward, the y_V-axis to the left, and the z_V-axis upward, which will correspond with the definitions in the ISO 8855 directive.

The orientation of the vehicle-fixed axis system V with respect to the inertial frame 0 may be defined by the Cardan or Bryant angles ψ, θ, and ϕ, which represent the yaw, the pitch, and the roll motion of the vehicle body. The first rotation about the $z_0=z_I$-axis defines the intermediate axis system with x_I and y_I parallel to the horizontal ground.

In complex vehicle models it is often more convenient to attach the vehicle-fixed axis system V to a representative chassis point rather than to the center of gravity of the vehicle C, because the latter will change with different loading conditions.

The wheel consists of the tire and the rim. Handling tire models approximate the contact patch by a local road plane, which is represented by the contact point P and the unit vector e_n perpendicular to this plane. The contact geometry is discussed in detail in Chapter 3. The rim is mounted at the wheel carrier or knuckle. The suspension system, which attaches the wheel carrier to the chassis, is extensively described in Chapter 5. Depending on the type of suspension system, the wheel carrier and the attached wheel can perform a hub motion z and optionally a steering motion δ, Figure 1.2. To describe the position and orientation of the wheel carrier and

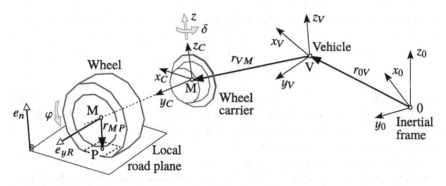

FIGURE 1.2
Wheel position and orientation.

the wheel, a reference frame with the axes x_C, y_C, z_C is fixed to the wheel carrier. The origin of this axis system is supposed to coincide with the wheel center M. The position and the orientation of the wheel carrier depend on the hub motion z and optionally on the steer motion δ. In the design position, the corresponding axes of the frames C and F are supposed to be parallel. The wheel itself rotates with the angle φ about an axis that is determined by the unit vector e_{yR}, Figure 1.2.

1.3.2 Design Position of Wheel Center

The design position of the wheel center M_0 is roughly[2] determined by the wheelbase $a = a_1 + a_2$ and the track widths s_1, s_2 at the front and rear axles, Figure 1.3. If

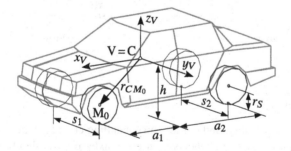

FIGURE 1.3
Design position of wheels.

---

[2]Note: The track width is defined as the distance of the contact points at an axle. On cambered wheels, the distance of the wheel centers is slightly different.

left/right symmetry is assumed, then the position vector

$$r_{CM_0,V} = \begin{bmatrix} a_1 \\ s_1/2 \\ -h + r_S \end{bmatrix} \tag{1.1}$$

denoted in the vehicle-fixed reference frame V will define the design position of the front left wheel center M_0 relative to the vehicle center C. Here, h denotes the height of the vehicle center C above the ground and r_S names the static tire radius, which takes the tire deflection, caused by the weight of the vehicle, into account.

By changing the sign in the second component ($s_1/2 \rightarrow -s_1/2$), Equation (1.1) applies for the right wheel too. Finally, the design position of rear wheels is obtained by replacing a_1 and s_1 by $-a_2$ and s_2, respectively.

1.3.3 Toe-In, Toe-Out

Wheel toe-in is an angle formed by the center line of the wheel and the longitudinal axis of the vehicle, looking at the vehicle from above, Figure 1.4. When the extensions of the wheel center lines tend to meet in front of the direction of travel of the vehicle, this is known as toe-in. If, however, the lines tend to meet behind the direction of travel of the vehicle, this is known as toe-out. The amount of toe can be expressed in degrees as the angle δ to which the wheels are out of parallel, or, as the difference between the track widths as measured at the leading and trailing edges of the tires or wheels. Note that at the left wheel the sign of toe angle δ does not correspond to a positive rotation around the z-axis. Toe settings affect three major areas of performance: tire

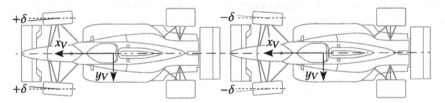

FIGURE 1.4
Toe-in and toe-out.

wear, straight-line stability, and corner entry handling characteristics. For minimum tire wear and power loss, the wheels on a given axle of a car should point directly ahead when the car is running in a straight line. Excessive toe-in or toe-out causes the tires to scrub, as they are always turned relative to the direction of travel. Toe-in improves the directional stability of a car and reduces the tendency of the wheels to shimmy.

1.3.4 Wheel Camber

Wheel camber is the angle of the wheel relative to vertical, as viewed from the front or the rear of the car, Figure 1.5. If the wheel leans away from the car, it has

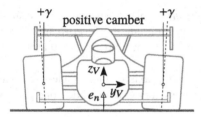 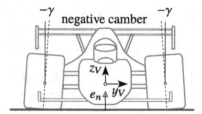

FIGURE 1.5
Positive and negative camber angles.

positive camber; if it leans in toward the chassis, it has negative camber. Again, at the left wheel, the sign of camber angle γ does not correspond to a positive rotation around the x-axis. The wheel camber angle must not be mixed up with the tire camber angle, which is defined as the angle between the wheel center plane and the local track normal e_n. Excessive camber angles cause nonsymmetric tire wear. A tire can generate maximum lateral force during cornering if it is operated with a slightly negative tire camber angle. As the chassis rolls when cornering, the suspension must be designed such that the wheels perform camber changes as the suspension moves up and down. An ideal suspension will generate an increasingly negative wheel camber as the suspension deflects upward. The kinematic property of a camber compensation is realized in general in high-performance cars [58].

1.3.5 Design Position of the Wheel Rotation Axis

Usually, the wheel rotation axis, which is described by the unit vector e_{yR}, will not coincide with the y_C-axis, which is part of the corresponding axis system located in the wheel center and fixed to the wheel carrier, Figure 1.6. The orientation of the unit

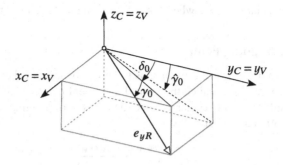

FIGURE 1.6
Design position of wheel rotation axis.

vector e_{yR} can be defined either by the angles δ_0 and γ_0, or by δ_0 and $\hat{\gamma}_0$, where δ_0 is the angle between the y_C-axis and the projection line of the wheel rotation axis into the $x_C\,y_C$-plane. The angle $\hat{\gamma}_0$ describes the angle between the y_C-axis and the projection line of the wheel rotation axis into the $y_C z_C$-plane, whereas γ_0 is the angle between the wheel rotation axis e_{yR} and its projection into the $x_C y_C$-plane. Toe-in and a positive camber angle are indicated by $\delta_0 >$ and $\gamma_0 > 0$ or $\hat{\gamma}_0 > 0$ at the left wheel.

In the design position, where the corresponding axis of the vehicle-fixed axis system V and the wheel carrier-fixed coordinate system C are parallel, one gets by inspecting Figure 1.6,

$$e_{yR,V} = e_{yR,C} = \frac{1}{\sqrt{\tan^2 \delta_0 + 1 + \tan^2 \hat{\gamma}_0}} \begin{bmatrix} \tan \delta_0 \\ 1 \\ - \tan \hat{\gamma}_0 \end{bmatrix} \qquad (1.2)$$

On the other hand, applying a series of elementary rotations results in

$$e_{yR,V} = e_{yR,C} = \begin{bmatrix} \sin \delta_0 \cos \gamma_0 \\ \cos \delta_0 \cos \gamma_0 \\ - \sin \gamma_0 \end{bmatrix} \qquad (1.3)$$

On a flat and horizontal road where the track normal e_n points in the direction of the vertical axis $z_C = z_F$, the angles δ_0 and γ_0 correspond to the toe angle and the camber angle, respectively. To specify the difference between γ_0 and $\hat{\gamma}_0$, the ratio between the third and second component of the unit vector e_{yR} is considered now. Equations (1.2) and (1.3) deliver

$$\frac{- \tan \hat{\gamma}_0}{1} = \frac{- \sin \gamma_0}{\cos \delta_0 \cos \gamma_0} \qquad \text{or} \qquad \tan \hat{\gamma}_0 = \frac{\tan \gamma_0}{\cos \delta_0} \qquad (1.4)$$

Hence, for small angles $\delta_0 \ll 1$, the difference between the angles γ_0 and $\hat{\gamma}_0$ is hardly noticeable. Kinematics and Compliance (KnC) test machines usually measure the angle $\hat{\gamma}_0$. That is why the automotive industry mostly uses this one instead of γ_0 to determine the orientation of the wheel rotation axis in the design position.

1.3.6   Wheel Aligning Point

Often the position of the wheel rotation axis is defined by the wheel center W and an additional point D which is called the "wheel alignment point," Figure 1.7. If point D is located inside the wheel center W, then the unit vector pointing in the direction of the wheel rotation axis is given by

$$e_{yR,V} = \frac{r_{VW,V} - r_{VD,V}}{\left| r_{VW,V} - r_{VD,V} \right|} \qquad (1.5)$$

where the vectors $r_{VW,V}$ and $r_{VD,V}$ describe the design position of the wheel center W and the wheel alignment point D relative to the origin V of the vehicle-fixed reference frame. This approach is commonly used in complex vehicle models because it avoids any misunderstandings.

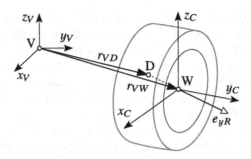

FIGURE 1.7
Wheel alignment point.

1.4 Active Safety Systems

Today many passenger cars as well as commercial vehicles are equipped with several control systems, which assist or even counteract the driver in order to improve the maneuverability or to maintain the stability of the vehicle.

The anti-lock brake system (ABS) used in road vehicles was developed in 1970 in a joint venture partnership between Robert Bosch and Mercedes-Benz. Simply put, ABS mainly copies the earlier technique of skilled drivers that performed cadence braking to avoid the permanent lockup of a braked wheel on emergency stops. The ABS control unit processes the measured angular wheel velocities and operates a braked wheel at or close to the limit of adhesion, regardless of the actual road friction properties. ABS also generates an optimal braking force distribution between front and rear axle which maintains the stability of the vehicle when braking in a turn and minimizes the stopping distance at different loading conditions. The latter is demonstrated in Section 8.4 with a planar vehicle model and a simple three-point brake torque controller.

The anti-lock system makes it possible to brake each wheel individually. The electronic stability program (ESP), discussed in more detail in Section 9.4, makes use of this specific feature.

The driving torque applied to an axle is in general distributed equally to both wheels. Controlled differentials or separate electric motors at each wheel make it possible to transmit more torque to the outer wheel when cornering. This solves the traction problem, because the wheel load transfer from the inner to the outer wheels reduces the capacity at the inner wheel to put adequate longitudinal and lateral tire forces to the road. This "torque vectoring" automatically generates a turn torque which reduces the steering effort and makes, in particular, larger vehicles more agile. Torque vectoring may even reduce the cornering resistance [54].

The anti-roll bar is part of the suspension system. It reduces the roll angle of a vehicle during cornering, but it also reduces the comfort by coupling the up and down motions of the wheels at one axle when driving straight ahead on an uneven road. That is why premium-segment cars use electronically controlled anti-roll bars to remove this conflict of objectives. Anti-foll bars also affect the stability of cars by influencing the wheel load transfer at the front and rear axle during cornering.

In most road vehicles the wheels at the front axle(s) are steered. Controlled rear-wheel steering (RWS) can increase the maneuverability and the stability of a vehicle. The term all-wheel steering (AWS) is more general and incorporates active or semi-active control at all wheels.

A complex obstacle-avoiding maneuver presented in Section 10.6 will demonstrate the benefits of integrated control by coordinating the control strategies of steered rear wheels and active anti-roll bars.

Today, many driver-assistance systems are available in addition. Advanced driver-assistance systems (ADAS) provide aid to the driver in nearly any driving situations. Steer-by-wire and drive-by-wire systems finally result in automated driving, where in a final step the vehicle itself performs all driving tasks and does not require a human driver anymore.

Automated driving is currently in full development. The complexity of the interaction of sensors, controllers, environmental conditions, and different vehicles, requires many tests. Some of them can be performed securely and very efficiently in virtual environments. That is why this textbook puts special emphasis on the modeling aspect which is underlined by MATLAB® examples included throughout the text.

1.5   Multibody Dynamics Tailored to Ground Vehicles

1.5.1   Modeling Aspects

For dynamic simulation, the vehicles are usually modeled by multibody systems (MBS) [21]. Typically, the overall vehicle model is separated into different subsystems [37]. Figure 1.8 shows the components of a passenger car model that can be used to investigate handling and ride properties. The vehicle model consists of the vehicle framework and subsystems for the steering system and the drive train.

The vehicle framework represents the kernel of the model. It at least includes the module chassis and modules for the wheel/axle suspension systems. The vehicle framework is supplemented by modules for the load, an elastically suspended engine, and passenger/seat models. A simple load module just takes the mass and inertia properties of the load into account. To describe the sloshing effects of liquid loads, dynamic load models are needed [49]. The subsystems, elastically suspended engine, passenger/seat, and in heavy truck models a suspended driver's cabin, can all be handled by a generic free-body model [44].

For standard vehicle dynamics analysis, the chassis can be modeled by one rigid body [2]. However, in the case of heavy trucks, the compliance of the frame and the elastically suspended driver's cab will require at least a lumped mass model approach or the embedding of an enhanced finite element structure. Most wheel/axle suspension

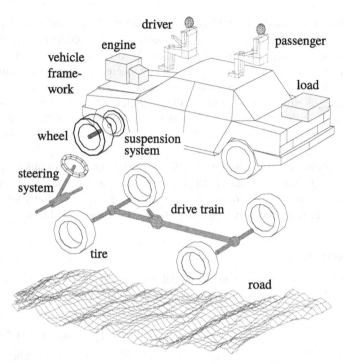

FIGURE 1.8
Vehicle model structure.

systems can be described by typical MBS elements such as rigid bodies, links, joints, and force elements [42]. Due to their robustness, leaf springs are still a popular choice for solid axles. They combine guidance and suspension properties, which causes many problems in modeling [12]. A lumped mass approach can overcome these problems [51].

The steering system at least consists of the steering wheel, a flexible steering shaft, and the steering box, which may also be power-assisted. A very sophisticated model of the steering system that includes compliances, dry friction, and clearance can be found in [29].

Tire forces and torques have a dominant influence on vehicle dynamics. Usually, semi-empirical tire models are used for vehicle handling analysis. They combine a reasonable computer runtime performance with sufficient model accuracy. Complex tire models are valid even for high frequencies and on really rough roads. But, they are computer time consuming and therefore used in special investigations only. The "Tyre Model Performance Test (TMPT)" provides information about the efficiency and problems of tire modeling and parameterization as well as the integration in standard multibody system program codes [25]. In this textbook the tire model "TMeasy" is discussed in detail. This semi-empirical tire model meets the requirements of both user friendliness and sufficient model accuracy [17, 52].

The drive train model in [44] takes lockable differentials into account, and it combines front-wheel, rear-wheel, and all-wheel drive. A simplified model for a rear-wheel drive will be presented in this textbook. The drive train is supplemented by a module describing the engine torque. It may be modeled, as done here, quite simply by a first-order differential equation or by enhanced engine torque modules.

Road irregularities and variations in the coefficient of friction present significant impacts on the vehicle. Simple road models are discussed in Chapter 2. A more enhanced model approach for generating two-dimensional reproducible random profiles is presented in [43].

This textbook restricts itself to the fundamentals of the dynamics of ground vehicles and will therefore focus on simple comfort and handling models for passenger cars. Besides some planar models, a simple three-dimensional model with a rigid chassis and four independently suspended wheels will be provided too.

1.5.2   Kinematics

A simple three-dimensional vehicle model consists of at least five rigid bodies, Figure 1.9. The position and orientation of each model body $i = 1, 2, \ldots$ will be

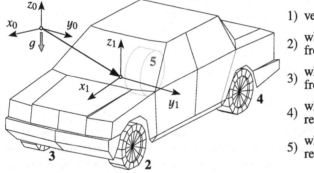

1) vehicle body (chassis)

2) wheel and wheel carrier front left

3) wheel and wheel carrier front right

4) wheel and wheel carrier rear left

5) wheel and wheel carrier rear right

FIGURE 1.9
Bodies of a vehicle model.

described relative to the earth-fixed coordinate system 0 that serves here as an inertial reference frame. Throughout this textbook the z_0-axis of the earth-fixed reference frame will point in the opposite direction of the gravity vector g.

The movements of each wheel and wheel carrier relative to the vehicle body (chassis) are restricted by links, joints, or other guiding elements. The resulting constraint forces and torques may either be eliminated from the equations of motion by appropriate algorithms or taken into account via Lagrange multipliers. The first method is cumbersome but will result in a minimized number of Ordinary Differential Equations (ODEs), whereas the latter will lead to Differential Algebraic Equations (DAEs). A more detailed description of multibody system used for vehicle modeling is given in [53].

A right-handed Cartesian coordinate system is fixed to each body in its center. Then, the position and the orientation of body i with respect to the inertial reference frame 0 is determined by the position vector

$$r_{0i,0} = r_{0i,0}(y) \tag{1.6}$$

and the rotation matrix

$$A_{0i} = A_{0i}(y) \tag{1.7}$$

where the generalized coordinates $y_1, y_2, \ldots y_f$ needed to describe the motions of the multibody system are collected in the vector y and the comma-separated subscript 0 indicates that the coordinates of the position vector r_{0i} are measured in the reference frame 0.

The columns of the rotation matrix A_{0i} are orthogonal unit vectors measured in the inertial frame 0 and pointing in the direction of the axis of the body-fixed coordinate system i. For such kinds of orthonormal matrices

$$A_{0i}^T A_{0i} = A_{0i} A_{0i}^T = I \quad \text{or} \quad A_{0i}^{-1} = A_{0i}^T \tag{1.8}$$

will hold, where I denotes the corresponding matrix of identity. If the components of the vector r are measured in the body-fixed coordinate system i, multiplication with the rotation matrix A_{0i} transforms this vector via

$$r_{,0} = A_{0i} r_{,i} \tag{1.9}$$

to the earth-fixed axis system 0, and

$$r_{,i} = A_{0i}^T r_{,0} \tag{1.10}$$

quite simply defines the inverse transformation.

The velocity with which body i is moving relative to the inertial system 0 is determined by the time derivative of the position vector defined in Equation (1.6)

$$v_{0i,0} = \frac{d}{dt} r_{0i,0}(y) = \dot{r}_{0i,0}(y) = \sum_{m=1}^{f} \frac{\partial r_{0i,0}(y)}{\partial y_m} \dot{y}_m = v_{0i,0}(y, \dot{y}) \tag{1.11}$$

The time derivative of the rotation matrix multiplied by its transposed results in a skew-symmetric matrix

$$\tilde{\omega}_{0i,0} = \frac{d}{dt}\left(A_{0i,0}(y)\right) A_{0i,0}^T(y) = \sum_{m=1}^{f} \frac{\partial A_{0i}(y)}{\partial y_m} \dot{y}_m A_{0i,0}^T(y) = \tilde{\omega}_{0i,0}(y, \dot{y}) \tag{1.12}$$

Its essential components

$$\tilde{\omega}_{0i,0} = \begin{bmatrix} 0 & -\omega_{0i,0}(3) & \omega_{0i,0}(2) \\ \omega_{0i,0}(3) & 0 & -\omega_{0i,0}(1) \\ -\omega_{0i,0}(2) & \omega_{0i,0}(1) & 0 \end{bmatrix} \tag{1.13}$$

define the vector of the angular velocity $\omega_{0i,0} = \left[\omega_{0i,0}(1),\ \omega_{0i,0}(2),\ \omega_{0i,0}(3)\right]^T$ with which the body-fixed axis system i rotates relative to the earth-fixed axis system 0. A direct calculation shows that

$$\widetilde{\omega}_{0i,0}\, r_{,0} \;=\; \omega_{0i,0} \times r_{,0} \tag{1.14}$$

holds for any vector $r_{,0}$, which means that multiplication of the skew-symmetric matrix of the angular velocities can be replaced by the corresponding vector- or cross-product.

Depending on the kind of constraints, the algebraic representation of the velocity vector $v_{0i,0} = v_{0i,0}(y, \dot{y})$ and the vector of the angular velocities $\omega_{0i,0} = \omega_{0i,0}(y, \dot{y})$ may become very complex. However, significant simplifications are possible if the time derivative of the vector of the generalized coordinates $\dot{y}$ is replaced via

$$z \;=\; K(y)\,\dot{y} \tag{1.15}$$

by a corresponding vector of generalized velocities $z = z(y, \dot{y})$. Then, the algebraic representation for the resulting velocities and angular velocities

$$v_{0i,0}(y, \dot{y}) \;\Longrightarrow\; v_{0i,0}(y, z) \quad \text{and} \quad \omega_{0i,0}(y, \dot{y}) \;\Longrightarrow\; \omega_{0i,0}(y, z) \tag{1.16}$$

will be less complicated. In many cases a simple inspection of the resulting velocity terms already leads to appropriate generalized velocities [22]. The trivial choice

$$z \;=\; \dot{y} \tag{1.17}$$

is always possible. Here, the kinematical matrix $K = K(y)$ simplifies to the corresponding matrix of identity.

The time derivatives of the velocities and the angular velocities finally result in the corresponding accelerations

$$\begin{aligned}
a_{0i,0} &= \frac{d}{dt}\, v_{0i,0}(y, z) \;=\; \sum_{m=1}^{f} \frac{\partial v_{0i,0}(y, z)}{\partial y_m}\, \dot{y}_m + \sum_{m=1}^{f} \frac{\partial v_{0i,0}(y, z)}{\partial z_m}\, \dot{z}_m \\
\alpha_{0i,0} &= \frac{d}{dt}\, \omega_{0i,0}(y, z) \;=\; \sum_{m=1}^{f} \frac{\partial \omega_{0i,0}(y, z)}{\partial y_m}\, \dot{y}_m + \sum_{m=1}^{f} \frac{\partial \omega_{0i,0}(y, z)}{\partial z_m}\, \dot{z}_m
\end{aligned} \tag{1.18}$$

1.5.3   Equations of Motion

The motions of one rigid body are governed by Newton's law

$$m_i\, a_{0i,0} \;=\; F_{i,0} \tag{1.19}$$

and Euler's equation

$$\Theta_{i,0}\, \alpha_{0i,0} + \omega_{0i,0} \times \Theta_{i,0}\, \omega_{0i,0} \;=\; T_{i,0} \tag{1.20}$$

where m_i is the mass of body i, and $\Theta_{i,0}$ denotes the inertia tensor of body i defined with respect to its center of mass and measured in the inertial frame. If the body is exposed to kinematical constraints, then the forces and torques acting on body i can be split into two parts

$$F_{i,0} = F_{i,0}^c + F_{i,0}^a \quad \text{and} \quad T_{i,0} = T_{i,0}^c + T_{i,0}^a \tag{1.21}$$

where $F_{i,0}^c$, $T_{i,0}^c$ represent the forces and torques provided by the constraints and $F_{i,0}^a$, $T_{i,0}^a$ collect all other forces and torques applied to body i.

Similar to D' Alembert's principle of virtual work, Jourdain postulated that the virtual power of all constraint forces and torques must vanish. For a system with k rigid bodies, we get

$$\sum_{i=1}^{k} \left\{ \delta v_{0i,0}^T F_{i,0}^c + \delta \omega_{0i,0}^T T_{i,0}^c \right\} \tag{1.22}$$

The virtual velocity and the virtual angular velocity of body i are defined by

$$\delta v_{0i,0} = \frac{\partial v_{0i,0}}{\partial z} \delta z \quad \text{and} \quad \delta \omega_{0i,0} = \frac{\partial \omega_{0i,0}}{\partial z} \delta z \tag{1.23}$$

where the $f \times 1$-vector δz collects the variations of the generalized velocities δz_1, $\delta z_2, \dots \delta z_f$ and the partial derivatives simply named as partial velocities and partial angular velocities are arranged in the $3 \times f$-Jacobian matrices of translation and rotation

$$\frac{\partial v_{0i,0}}{\partial z} = \left[\frac{\partial v_{0i,0}(y,z)}{\partial z_1}, \frac{\partial v_{0i,0}(y,z)}{\partial z_2} \cdots \frac{\partial v_{0i,0}(y,z)}{\partial z_f} \right] \tag{1.24}$$

$$\frac{\partial \omega_{0i,0}}{\partial z} = \left[\frac{\partial \omega_{0i,0}(y,z)}{\partial z_1}, \frac{\partial \omega_{0i,0}(y,z)}{\partial z_2} \cdots \frac{\partial \omega_{0i,0}(y,z)}{\partial z_f} \right] \tag{1.25}$$

Using the Jacobian matrices, the accelerations provided by Equation (1.18) can be written as

$$a_{0i,0} = \frac{\partial v_{0i,0}}{\partial z} \dot{z} + a_{0i,0}^R \quad \text{and} \quad \alpha_{0i,0} = \frac{\partial \omega_{0i,0}}{\partial z} \dot{z} + \alpha_{0i,0}^R \tag{1.26}$$

where $\dot{z}$ is the time derivative of the vector of generalized velocities and

$$a_{0i,0}^R = \sum_{m=1}^{f} \frac{\partial v_{0i,0}(y,z)}{\partial y_m} \dot{y}_m \quad \text{and} \quad \alpha_{0i,0}^R = \sum_{m=1}^{f} \frac{\partial \omega_{0i,0}(y,z)}{\partial y_m} \dot{y}_m \tag{1.27}$$

abbreviates the remaining terms in the accelerations. By combining Equation (1.21) with Equations (1.19) and (1.20), one is able to put the constraint forces and torques down to dynamic terms and the applied forces and torques. Using the notation in Equations (1.26) and (1.27), Jourdain's principle reads as

$$\sum_{i=1}^{k} \left\{ \frac{\partial v_{0i,0}^T}{\partial z} \left[m_i \frac{\partial v_{0i,0}}{\partial z} \dot{z} + m_i a_{0i}^R - F_{i,0}^a \right] \right.$$
$$\left. + \frac{\partial \omega_{0i,0}^T}{\partial z} \left[\Theta_{i,0} \frac{\partial \omega_{0i,0}}{\partial z} \dot{z} + \Theta_{i,0} \alpha_{0i}^R + \omega_{0i,0} \times \Theta_{i,0} \omega_{0i,0} - T_{i,0}^a \right] \right\} \delta z = 0 \tag{1.28}$$

The variations of the generalized velocities δz are arbitrary. Hence, the expression in the braces must vanish. The resulting first-order differential equation can be written as

$$M(y)\dot{z} = q(y, z) \tag{1.29}$$

where the $f \times f$-mass-matrix is defined by

$$M(y) = \sum_{i=1}^{k} \left[\frac{\partial v_{0i,0}^T}{\partial z} m_i \frac{\partial v_{0i,0}}{\partial z} + \frac{\partial \omega_{0i,0}^T}{\partial z} \Theta_{i,0} \frac{\partial \omega_{0i,0}^T}{\partial z} \right] \tag{1.30}$$

and the $f \times 1$-vector of generalized forces

$$q(y, z) = \sum_{i=1}^{k} \left[\frac{\partial v_{0i,0}^T}{\partial z} \left(F_{i,0}^a - m_i a_{0i,0}^R \right) \right.$$
$$\left. + \frac{\partial \omega_{0i,0}^T}{\partial z} \left(T_{i,0}^a - \Theta_{i,0} \alpha_{0i,0}^R - \omega_{0i,0} \times \Theta_{i,0} \omega_{0i,0} \right) \right] \tag{1.31}$$

combines the inertia and gyroscopic forces and torques with the applied forces and torques. The equations of motion result in a set of two first-order differential equations. The definition of generalized velocities which is done by Equation (1.15) or in the trivial form by Equation (1.17) represent the first set and the dynamic equation defined in Equation (1.29) the second one.

1.6   A Quarter Car Model

1.6.1   Modeling Details

The quarter car model shown in Figure 1.10 consists of the chassis, a trailing arm that is rigidly attached to the knuckle, and the wheel. The model represents a quarter car on a hydropulse test rig. That is why the chassis will here perform vertical motions z only. A revolute joint in B connects the trailing arm with the chassis. The rotation of the trailing arm and the knuckle described by the angle β is affected by a torsional spring damper arrangement. The simple torsional damper will be replaced by a point-to-point damper element with a nonlinear characteristic in Chapter 6. The angle φ characterizes the wheel rotation about the y_0-axis. The position of the actuator that supports the wheel is controlled to follow a prescribed displacement time history, $u = u(t)$.

A quarter car model is quite a good, but surely limited, approximation of real vehicle dynamics. So, the simplification that the wheel center W, the center of the knuckle and trailing arm K, and the joint in B are arranged on a straight line will correspond to the overall model quality. In addition, the wheel is supposed to roll, and a simple vertical spring acting between the contact point Q and the wheel center W models the compliance of the tire.

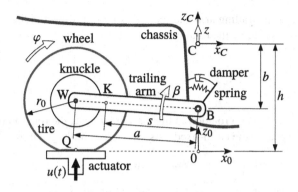

FIGURE 1.10
Quarter car model with trailing arm suspension.

1.6.2 Kinematics

The quarter car model consists of $n = 3$ rigid bodies, Figure 1.11. Their position and orientation are determined by $f = 3$ generalized coordinates $y(1) = z$, $y(2) = \beta$, and $y(3) = \varphi$. Position, velocity, and acceleration of the chassis center C are simply

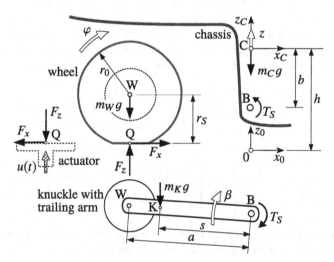

FIGURE 1.11
Quarter car model: Bodies and applied forces and torques.

defined by

$$
r_{0C,0} = \begin{bmatrix} 0 \\ 0 \\ h+z \end{bmatrix}, \quad v_{0C,0} = \begin{bmatrix} 0 \\ 0 \\ \dot{z} \end{bmatrix}, \quad a_{0C,0} = \begin{bmatrix} 0 \\ 0 \\ \ddot{z} \end{bmatrix} \quad (1.32)
$$

The orientation of the trailing arm and the knuckle with respect to the knuckle-fixed axis system is described by the rotation matrix

$$A_{CK} = \begin{bmatrix} \cos\beta & 0 & \sin\beta \\ 0 & 1 & 0 \\ -\sin\beta & 0 & \cos\beta \end{bmatrix} \tag{1.33}$$

As the chassis performs no rotation in this particular case, the rotation matrix of the knuckle with respect to the earth-fixed axis system is simply given by

$$A_{0K} = A_{CK} \tag{1.34}$$

and the angular velocity and the angular acceleration are obtained as

$$\omega_{0K,0} = \begin{bmatrix} 0 \\ 1 \\ 0 \end{bmatrix} \dot{\beta} \quad \text{and} \quad \alpha_{0K,0} = \begin{bmatrix} 0 \\ 1 \\ 0 \end{bmatrix} \ddot{\beta} \tag{1.35}$$

The position vector

$$r_{0K,0} = \begin{bmatrix} 0 \\ 0 \\ h+z \end{bmatrix} + \begin{bmatrix} 0 \\ 0 \\ -b \end{bmatrix} + \begin{bmatrix} -s\cos\beta \\ 0 \\ s\sin\beta \end{bmatrix} \tag{1.36}$$

describes the momentary position of the center of the knuckle trailing arm combination K, where the first two parts characterize the momentary position of the revolute joint B. The time derivatives yield the velocity

$$v_{0K,0} = \begin{bmatrix} 0 \\ 0 \\ 1 \end{bmatrix} \dot{z} + \begin{bmatrix} s\sin\beta \\ 0 \\ s\cos\beta \end{bmatrix} \dot{\beta} \tag{1.37}$$

and the acceleration

$$a_{0K,0} = \begin{bmatrix} 0 \\ 0 \\ 1 \end{bmatrix} \ddot{z} + \begin{bmatrix} s\sin\beta \\ 0 \\ s\cos\beta \end{bmatrix} \ddot{\beta} + \begin{bmatrix} s\cos\beta \\ 0 \\ -s\sin\beta \end{bmatrix} \dot{\beta}^2 \tag{1.38}$$

The orientation of the wheel with respect to the earth-fixed axis system 0 and the momentary position of the wheel center W are determined by

$$A_{0W} = \begin{bmatrix} \cos\varphi & 0 & \sin\varphi \\ 0 & 1 & 0 \\ -\sin\varphi & 0 & \cos\varphi \end{bmatrix} \quad \text{and} \quad r_{0W,0} = \begin{bmatrix} 0 \\ 0 \\ h+z-b \end{bmatrix} + \begin{bmatrix} -a\cos\beta \\ 0 \\ a\sin\beta \end{bmatrix} \tag{1.39}$$

Then, the velocity and acceleration state of the wheel and its center are given by

$$\omega_{0W,0} = \begin{bmatrix} 0 \\ 1 \\ 0 \end{bmatrix} \dot{\varphi}, \quad \alpha_{0K,0} = \begin{bmatrix} 0 \\ 1 \\ 0 \end{bmatrix} \ddot{\varphi} \tag{1.40}$$

and

$$v_{0W,0} = \begin{bmatrix} 0 \\ 0 \\ 1 \end{bmatrix} \dot{z} + \begin{bmatrix} a \sin\beta \\ 0 \\ a \cos\beta \end{bmatrix} \dot{\beta} \qquad (1.41)$$

$$a_{0W,0} = \begin{bmatrix} 0 \\ 0 \\ 1 \end{bmatrix} \ddot{z} + \begin{bmatrix} a \sin\beta \\ 0 \\ a \cos\beta \end{bmatrix} \ddot{\beta} + \begin{bmatrix} a \cos\beta \\ 0 \\ -a \sin\beta \end{bmatrix} \dot{\beta}^2 \qquad (1.42)$$

Using the trivial choice of generalized velocities, the partial velocities which according to Equations (1.26) and (1.27) form the Jacobians of translation and rotation, can easily be depicted from the corresponding velocities. The results are collected in the Tables 1.2 and 1.3.

TABLE 1.2
Partial velocities, applied forces, and remaining acceleration terms

Body name mass	Partial Velocities $\partial v_{0i}/\partial z_j$			Applied Forces $-F_i$	Remaining Terms a_{0i}^R
	$z_1 = \dot{z}$	$z_2 = \dot{\beta}$	$z_3 = \dot{\varphi}$		
Chassis m_C	0 0 1	0 0 0	0 0 0	0 0 $-m_C g$	0 0 0
Knuckle and trailing arm m_K	0 0 1	$s \sin\beta$ 0 $s \cos\beta$	0 0 0	0 0 $-m_K g$	$s\dot{\beta}^2 \cos\beta$ 0 $-s\dot{\beta}^2 \sin\beta$
Wheel m_W	0 0 1	$a \sin\beta$ 0 $a \cos\beta$	0 0 0	F_x 0 $F_z - m_W g$	$a\dot{\beta}^2 \cos\beta$ 0 $-a\dot{\beta}^2 \sin\beta$

TABLE 1.3
Partial angular velocities, applied torques, and remaining angular acceleration terms

Body name inertia	Partial Angular Velocities $\partial \omega_{0i}/\partial z_j$			Applied Torques T_i	Remaining Terms α_{0i}^R
	$z_1 = \dot{z}$	$z_2 = \dot{\beta}$	$z_3 = \dot{\varphi}$		
Chassis no rotation	0 0 0	0 0 0	0 0 0	0 $-T_S$ 0	0 0 0
Knuckle and trailing arm Θ_K	0 0 0	0 1 0	0 0 0	0 T_S 0	0 0 0
Wheel Θ_W	0 0 0	0 0 0	0 1 0	0 $-r_S F_x$ 0	0 0 0

The chassis performs translational motions only. That is why the corresponding rows in Table 1.3 will all vanish. There are also no remaining terms in the angular accelerations within this simple model.

1.6.3 Applied Forces and Torques

Assuming linear characteristics, the torsional spring damper combination acting at the revolute joint in B will generate the torque

$$T_S = -\left(T_S^0 + c_S \beta + d_S \dot\beta\right) \tag{1.43}$$

where T_S^0 describes the preload in the steady design position defined by $\beta = 0$ and $\dot\beta = 0$, c_S denotes the torsional spring rate, and d_S denotes the torsional damper constant.

The constants c_x and c_z characterize the compliance of the tire in longitudinal and vertical directions. As long as the tire is in contact with the actuator, the vertical tire force is defined by

$$F_z = -c_z (r_0 - r_S) \tag{1.44}$$

where

$$r_S = h + z - b + a \sin\beta - u \tag{1.45}$$

denotes the static tire radius. Assuming adhesion in the contact area,

$$F_x = -c_x \left(a(1-\cos\beta) - r_S\, \varphi\right) - d_x \left(a \sin\beta\dot\beta - r_S\, \dot\varphi\right) \tag{1.46}$$

will model the longitudinal tire force. The constant d_x models the tire damping in the longitudinal direction. The suspension damping here provided by the torque T_s affects the translational motions of the wheel too. However, the wheel rotation is determined by the longitudinal tire force only. That is why damping must be provided via the constant d_x to avoid undamped wheel oscillations.

A detailed model for all tire forces and torques, which also includes nonlinearities and lift-off, is developed in Chapter 3.

1.6.4 Equations of Motion

Jourdain's principle results in a set of two first-order differential equations. As trivial generalized velocities were chosen here, they read as

$$\dot y = z \quad \text{and} \quad M\dot z = q \tag{1.47}$$

where y denotes the vector of generalized coordinates, the vector z defines trivial generalized velocities, the mass matrix M is defined by Equation (1.30), and Equation (1.31) generates the vector q, which contains the generalized forces and torques. Using the notations in Tables 1.2 and 1.3, we get

$$M = \begin{bmatrix} m_C + m_K + m_W & (s\, m_K + a\, m_W)\cos\beta & 0 \\ (s\, m_K + a\, m_W)\cos\beta & \Theta_K + s^2 m_K + a^2 m_W & 0 \\ 0 & 0 & \Theta_W \end{bmatrix} \tag{1.48}$$

and

$$q = \begin{bmatrix} F_z - (m_C + m_K + m_W)\, g + (s\, m_K + a\, m_W)\sin\beta\, \dot\beta^2 \\ T_S - (s\, m_K + a\, m_W)\cos\beta\, g + a\,(F_x \sin\beta + F_z \cos\beta) \\ -r_S\, F_x \end{bmatrix} \tag{1.49}$$

Inspecting the elements of the mass matrix, we recognize that $m_C + m_K + m_W$ represents the overall mass of the quarter car model and $\Theta_K + s^2 m_K + a^2 m_W$ denotes the inertia of the wheel mass, the knuckle, and the trailing arm with respect to the revolute joint in B. The inertia of the wheel Θ_W will not show up here because the wheel rotation φ is measured not relative to the knuckle but with respect to the earth-fixed system 0.

In steady state, the time derivatives of the generalized coordinates will vanish. Then, the suspension torque provided by Equation (1.43) simplifies to

$$T_S \longrightarrow T_S^{st} = -\left(T_S^0 + c_S \beta^{st}\right) \tag{1.50}$$

and the balance of generalized forces expressed by $q = 0$ will yield

$$F_z^{st} - (m_C + m_K + m_W)\, g = 0 \tag{1.51}$$

$$-\left(T_S^0 + c_S \beta^{st}\right) - s_S m_S \cos \beta^{st} g + a\left(F_x^{st} \sin \beta^{st} + F_z^{st} \cos \beta^{st}\right) = 0 \tag{1.52}$$

$$-r_S F_x^{st} = 0 \tag{1.53}$$

where the term $s m_K + a m_W$ is replaced by $s_S m_S$ by introducing the suspension mass $m_S = m_K + m_W$ (knuckle with trailing arm) and s_S as the position of the corresponding mass center. The first and the third equations deliver the steady-state tire forces

$$F_z^{st} = (m_C + m_K + m_W)\, g \quad \text{and} \quad F_x^{st} = 0 \tag{1.54}$$

where the vertical force F_z^{st} just equals the weight $G = (m_C + m_K + m_W)\, g$ of the quarter car.

1.6.5 Simulation

For any given vehicle data, including the preload of the torsional suspension spring T_S^0, the steady-state rotation angle $\beta = \beta_{st}$ of the trailing arm can be obtained from Equation (1.52). The MATLAB-Script in Listing 1.1 will do the calculation where the MATLAB-Function uty_par_qcmta given in Listing 1.2, generates a structure that provides the basic data of the quarter car model.

Listing 1.1
Script app_01_qcm_steady_state.m: steady-state position of quarter car model

```
1 p = uty_par_qcmta(); % vehicle parameter
2
3 % compute steady state tire forces
4 Fx = 0;
5 Fz = ( p.mC + p.mK + p.mW ) * p.g ;
6
7 % mass and CoG of knuckle and trailing arm
8 mS = p.mK + p.mW;    sS = ( p.s*p.mK + p.a*p.mW ) / mS;
9
10 % solve non-linear equation by using matlab-function fzero
11 be_st = fzero(@(be) ...
12    -(p.Ts0+p.cs*be)-sS*mS*cos(be)*p.g+p.a*(Fx*sin(be)+Fz*cos(be)),0);
13 disp(['be_st = ',num2str(be_st*180/pi),' in degree'])
```

Listing 1.2

Function `uty_par_qcmta`: define parameter of quarter car model with trailing arm

```
1 function p = uty_par_qcmta()
2
3 % model parameter for quarter car model with trailing arm
4 p.g = 9.810;  % [m/s^2] gravity
5 p.s = 0.250;  % [m]     distance joint B knuckle/trailing arm center
6 p.a = 0.400;  % [m]     distance joint B wheel center
7 p.b = 0.310;  % [m]     vertical distance chassis center to joint B
8 p.h = 0.600;  % [m]     height of chassis center
9 p.r0= 0.305;  % [m]     tire radius
10
11 p.mC= 200;    % [kg]      corresponding chassis mass (quarter car)
12 p.mK=  35;    % [kg]      mass of knuckle/trailing arm
13 p.mW=  15;    % [kg]      mass of wheel (rim and tire)
14
15 p.ThetaK=0.6; % [kgm^2]   inertia of knuckle/trailing arm
16 p.ThetaW=0.8; % [kgm^2]   inertia of wheel
17
18 p.Ts0=1200;   % [Nm]      preload in torsional spring
19 p.cs=10000;   % [Nm/rad]  torsional spring rate
20 p.ds=  800;   % [Nms]     torsional damping
21 p.cx=180000;  % [N/m]     longitudinal tire stiffness
22 p.cz=220000;  % [N/m]     vertical tire stiffness
23 p.dx=  150;   % [Ns/m]    longitudinal tire damping
24
25 end
```

The piston movements u which represent the road profile z_R in this test rig arrangement as well as the vertical wheel load F_z are stored in the structure out which is defined as an additional output of the function `dyn_01_qcm_step.m` but are not used in this introductory example. Finally, the MATLAB-Script given in Listing 1.3 performs a time simulation in the interval from $t_0 = 0\,s$ to $t_E = 1.5\,s$.

Listing 1.3

Script `app_01_qcm_step.m`: quarter car model with step input

```
1 % define step input
2 tstep=0.75; % [s]  step at @ t=tstep
3 ustep=0.05; % [m]  actuator step value
4
5 p = uty_par_qcmta(); % assign model parameter to structure
6
7 % initial states
8 x0 = [ 0; 0; 0; 0; 0; 0];
9
10 % time simulation and plots of results
11 t0=0; tE=1.5;
12 [to,xo]= ode45(@(t,x) fun_01_qcm_step(t,x,p,tstep,ustep),[t0,tE],x0);
13
14 subplot(3,1,1)
15 plot(to,xo(:,1)), grid on, xlabel('t/s'), ylabel('z/m')
16 subplot(3,1,2)
17 plot(to,xo(:,2)*180/pi),grid on,xlabel('t/s'),ylabel('\beta/deg')
18 subplot(3,1,3)
19 plot(to,xo(:,3)*180/pi),grid on,xlabel('t/s'),ylabel('\phi/deg')
```

The MATLAB-Function ode45[3] solves the first-order ordinary differential equations (ODEs) provided by the function given in Listing 1.4 and plots the time histories of

---

[3]ode45 is based on an explicit Runge-Kutta (4,5) formula, the Dormand-Prince pair.

the chassis displacement z, the rotation of the knuckle and trailing arm β, as well as the wheel rotation φ versus time t, Figure 1.12.

Listing 1.4

Function `fun_01_qcm_step`: quarter car model dynamics with step excitation

```
1 function [xdot,out] = fun_01_qcm_step(t,x,p,tstep,ustep)
2 % dynamics of quarter car model with trailing arm & step excitation
3
4 % step input to actuator @ t = tstep
5 if t < tstep, u = 0; else, u = ustep; end;  out.zr=u;
6
7 % state variables
8 z  = x(1);  be  = x(2);   phi  = x(3);
9 zd = x(4);  bed = x(5);   phid = x(6);
10
11 % torque in revolute joint
12 Ts = - ( p.Ts0 + p.cs*be + p.ds*bed );
13
14 % tire deflection (static tire radius)
15 rS = p.h + z - p.b + p.a*sin(be) - u ;
16
17 % longitudinal tire force (adhesion assumed)
18 Fx = - p.cx *( p.a*(1-cos(be)) - rS*phi  ) ...
19      - p.dx *( p.a*sin(be)*bed - rS*phid ) ;
20
21 % vertical tire force (contact assumed)
22 Fz =  p.cz *( p.r0 - rS );   out.Fz=Fz;
23
24 % mass matrix
25 Massma=[ p.mC+p.mK+p.mW             (p.s*p.mK+p.a*p.mW)*cos(be)    0 ; ...
26   (p.s*p.mK+p.a*p.mW)*cos(be) p.ThetaK+p.s^2*p.mK+p.a^2*p.mW  0 ; ...
27            0                             0              p.ThetaW ];
28
29 % vector of generalized forces and torques
30 q=[Fz-(p.mC+p.mK+p.mW)*p.g+(p.s*p.mK+p.a*p.mW)*sin(be)*bed^2 ; ...
31    Ts-(p.s*p.mK+p.a*p.mW)*cos(be)*p.g+p.a*(Fx*sin(be)+Fz*cos(be)); ...
32    -rS*Fx ];
33
34 % state derivatives
35 xdot = [ zd; bed; phid;  Massma\q ];
36
37 end
```

During the first time interval $0\,\text{s} \leq t \leq 0.75\,\text{s}$, where the actuator displacement is kept to $u = 0\,\text{m}$, the quarter car model performs the transition from the initial values to steady state. After a short time the rotation angle of knuckle and trailing arm adjusts to the value $\beta = -2.087°$ which is in conformity with the steady-state value be0 calculated and printed by the MATLAB-Script in Listing 1.1. At $t = 0.75\,\text{s}$ the actuator is moved in an instant (step input) to the value of $u = 0.05\,\text{m}$. The resulting oscillations decay in a short time indicating an appropriate suspension damping. The rotation of the trailing arm $\beta = \beta(t)$ causes the wheel center M to move back and forth. As a consequence, the wheel which is in contact with the strictly vertical moving actuator is forced to rotate too, $\varphi = \varphi(t)$.

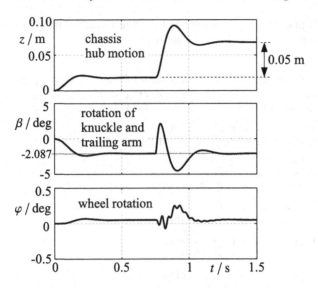

FIGURE 1.12
Simulation results to step input from $u = 0$ to $u = 0.05\ m$ at $t = 0.75\ s$.

Exercises

1.1 Given the tire codes P245/65R17 and 32x10.5R15LT, calculate the radius and width for both tires.

1.2 Measurements on a K&C (Kinematics and Compliance) test rig provide the toe angle $\delta_0 = 0.7°$ and the camber angle $\hat{\gamma}_0 = -1.0°$ in design position. Calculate the components of the unit vector e_{yR} pointing in the direction of the wheel rotation axis.

1.3 Use the unit vector e_{yR} calculated in Exercise 1.2 and generate the coordinates of the wheel alignment point D when its distance to the wheel center W is given by $\overline{DW} = 0.1$ m and the position of the wheel center relative to the origin C of the vehicle-fixed reference frame F is defined by the vector $r_{CW,F} = [\,0.00\ 0.76\ 0.00\,]^T$.

1.4 The steady-state position of the trailing arm depends on the pre-load in the torsional suspension spring. Find the value for T_S^0 which will keep the trailing arm in steady state in a horizontal position, $\beta^{st} = 0°$. Check the result with the MATLAB-Script provided by Listing 1.1. Do not forget to adjust the value for the torsional spring preload Ts0 in Listing 1.2.

1.5 Use the MATLAB-Scripts in Section 1.6.5 and perform simulations with different values of the suspension damping (double or halve the given value).

2

Road

CONTENTS

2.1 Modeling Aspects

Besides single obstacles or track grooves, the irregularities of a road are stochastic in nature. A vehicle driving over a random road profile mainly performs hub, pitch, and roll motions. The local inclination of the road profile also induces longitudinal and lateral motions as well as yaw motions. On normal roads the latter motions will have less influence on ride comfort and ride safety. To limit the effort of the stochastic description, usually simpler road models are used.

Sophisticated three-dimensional road models provide not only the road height $z = z(x, y)$ but also the local friction coefficient $\mu = \mu(x, y)$ at each point x, y, Figure 2.1. In addition, simple road models will often generate the local road normal and the local curvature. Within this general approach the tire model is responsible for calculating the local road inclination and curvature. By separating the horizontal course description from the vertical layout and the surface properties of the roadway almost arbitrary road layouts are possible [5]. Today, high-resolution measurements of road surfaces are performed in moving traffic by measuring vans and generate a huge amount of data. Therefore, a compact but still accurate representation of measured data will be essential for straightforward applications in simulation environments. The open-source project OpenCRG provides a three-dimensional road model where a curved regular grid (CRG) is used to achieve a compact and accurate representation of measurements [62].

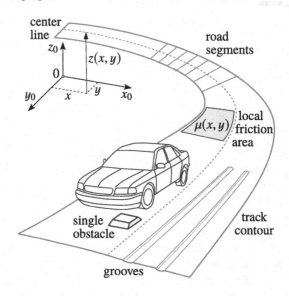

FIGURE 2.1
Sophisticated road model.

If the vehicle drives along a given path, its momentary position can be described by the path variable $s = s(t)$. Hence, a fully three-dimensional road model can be reduced to a parallel track model, Figure 2.2. Now, the road heights on the left and

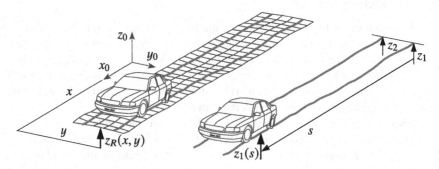

FIGURE 2.2
Parallel track road model.

right track are provided by two one-dimensional functions $z_1 = z_1(s)$ and $z_2 = z_2(s)$. Within the parallel track model, no information about the local lateral road inclination is available. If this information is not provided by additional functions, the impact of a local lateral road inclination on vehicle motions is not taken into account.

For basic studies, the irregularities at the left and the right track can be considered approximately the same, $z_1(s) \approx z_2(s)$. Then, a single track road model with $z_R(s) = z_1(x) = z_2(x)$ can be used. In this case, the roll excitation of the vehicle is neglected too.

2.2 Deterministic Profiles

2.2.1 Bumps and Potholes

Bumps and potholes on the road are single obstacles of nearly arbitrary shape. Already with simple rectangular cleats, the dynamic reaction of a vehicle or a single tire to a sudden impact can be investigated. If the shape of the obstacle is approximated by a smooth function, like a cosine wave, then discontinuities will be avoided. Usually, deterministic obstacles are described in local coordinate systems, Figure 2.3.

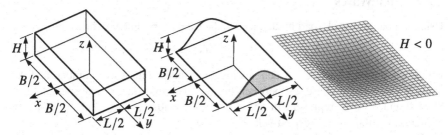

FIGURE 2.3
Rectangular cleat, cosine-shaped bump, and rounded obstacle or pothole respectively.

Then, the rectangular cleat is simply defined by

$$z(x, y) = \begin{cases} H & \text{if} & -\frac{L}{2} < x < \frac{L}{2} \ \& \ -\frac{B}{2} < y < \frac{B}{2} \\ 0 & \text{else} \end{cases} \qquad (2.1)$$

and the cosine-shaped bump is given by

$$z(x, y) = \begin{cases} \frac{1}{2}H\left(1+\cos\left(2\pi\frac{x}{L}\right)\right) & \text{if} & -\frac{L}{2} < x < \frac{L}{2} \ \& \ -\frac{B}{2} < y < \frac{B}{2} \\ 0 & \text{else,} \end{cases} \qquad (2.2)$$

where L, B, and H denote the length, width, and height of the obstacle. Potholes are obtained if negative values for the height ($H < 0$) are used. The function in Listing 2.1 computes the rectangular cleat when the switch parameter OT is set to 1 or with OT=2 the cosine-shaped bump. Case OT=3 generates a rounded bump with a cosine-like transition to the default road height $z = 0$ in the x-direction and a sharp but still smooth blending in the y-direction. The function also returns the road friction property μ which can be used to define an area with low friction.

Listing 2.1

Function uty_obstacle: deterministic obstacles

```
1 function [z,mu] = uty_obstacle(x,y,OT,L,B,H)
2 % set default road height & friction property outside obstacle
3    z = 0; mu=1;
4 % check if x,y-coordinates are inside obstacle
5    if x > -L/2 && x < +L/2 && y > -B/2 && y < +B/2
6       switch OT  % select type of obstacle
7          case 1  %    cleat
8             z = H;
9          case 2  %    cosine shaped obstacle
10            z = H/2 * ( 1 + cos(2*pi*x/L) );
11         case 3  %    rounded obstacle
12            zx=(1-(x/(L/2))^2)^2; zy=(1-(y/(B/2))^4)^2; z=H*zx*zy;
13      end
14   end
15 end
```

In a similar way, track grooves can be modeled too [66]. By appropriate coordinate transformations, the obstacles can then be integrated into the global road description.

2.2.2 Sine Waves

Using the parallel track road model, a periodic excitation can be realized by

$$z_1(s) = A \sin(\Omega s) \quad \text{and} \quad z_2(s) = A \sin(\Omega s - \Psi), \tag{2.3}$$

where s is the path variable, A denotes the amplitude, Ω the wave number, and the angle Ψ describes a phase lag between the left and right tracks. The special cases $\Psi = 0$ and $\Psi = \pi$ represent in-phase excitation with $z_1 = z_2$ and out-of-phase excitation with $z_1 = -z_2$, respectively.

If the vehicle runs with constant velocity $ds/dt = v_0 = const.$, then the momentary position of the vehicle is simply determined by $s = v_0 t$, where the initial position $s = 0$ at $t = 0$ was assumed. By introducing the wavelength

$$L = \frac{2\pi}{\Omega} \tag{2.4}$$

the term Ωs in Equation (2.3) can be written as

$$\Omega s = \frac{2\pi}{L} s = \frac{2\pi}{L} v_0 t = 2\pi \frac{v_0}{L} t = \omega t \tag{2.5}$$

Hence, in the time domain, the excitation frequency is given by

$$f = \omega/(2\pi) = \frac{2\pi \frac{v_0}{L}}{2\pi} = \frac{v_0}{L} \tag{2.6}$$

For most vehicles, the rigid body vibrations are between 0.5 Hz and 15 Hz. This range is covered by waves that satisfy the conditions $v_0/L \geq 0.5$ Hz and $v_0/L \leq 15$ Hz.

For a given wavelength, let's say $L = 4$ m, the rigid body vibrations of a vehicle are excited if the velocity of the vehicle will be varied from $v_0^{min} = 0.5$ Hz $* 4$ m $= 2$ m/s $= 7.2$ km/h to $v_0^{max} = 15$ Hz $* 4$ m $= 60$ m/s $= 216$ km/h. Hence, to achieve an excitation in the whole frequency range with moderate vehicle velocities, road profiles with varying wavelengths are needed.

2.3 Random Profiles

2.3.1 Statistical Properties

Road profiles fit the category of stationary Gaussian random processes [8]. Hence, the irregularities of a road can be described either by the profile itself $z_R = z_R(s)$ or by its statistical properties, Figure 2.4.

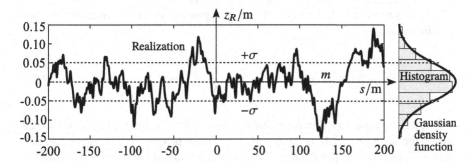

FIGURE 2.4
Road profile and statistical properties.

By choosing an appropriate reference frame, a vanishing mean value

$$m = E\{z_R(s)\} = \lim_{X \to \infty} \frac{1}{X} \int_{-X/2}^{X/2} z_R(s)\, ds = 0 \qquad (2.7)$$

can be achieved, where $E\{\}$ denotes the expectation operator. Then, the Gaussian density function that corresponds with the histogram of a normally distributed stationary random process is given by

$$p(z_R) = \frac{1}{\sigma \sqrt{2\pi}}\, e^{-\frac{z_R^2}{2\sigma^2}} \qquad (2.8)$$

where a vanishing mean value ($m = 0$) was taken for granted. The standard deviation or the effective value σ is obtained from the variance of the process $z_R = z_R(s)$

$$\sigma^2 = E\left\{z_R^2(s)\right\} = \lim_{X \to \infty} \frac{1}{X} \int_{-X/2}^{X/2} z_R(s)^2\, ds \qquad (2.9)$$

Alteration of σ affects the shape of the density function. In particular, the points of inflection occur at $\pm\sigma$. The probability of a value $|z| < \zeta$ is given by

$$P(\pm\zeta) = \frac{1}{\sigma\sqrt{2\pi}} \int_{-\zeta}^{+\zeta} e^{-\frac{z^2}{2\sigma^2}} dz \tag{2.10}$$

In particular, one gets the values $P(\pm\sigma) = 0.683$, $P(\pm2\sigma) = 0.955$, and $P(\pm3\sigma) = 0.997$. Hence, the probability of a value $|z| \geq 3\sigma$ is 0.3%.

As an extension to the variance of a random process, the autocorrelation function is defined by

$$R(\xi) = E\{z_R(s)\,z_R(s+\xi)\} = \lim_{X\to\infty} \frac{1}{X} \int_{-X/2}^{X/2} z_R(s)\,z_R(s+\xi)\,ds \tag{2.11}$$

The autocorrelation function is symmetric, $R(\xi) = R(-\xi)$, and it plays an important part in the stochastic analysis. In any normal random process, as ξ increases, the link between $z_R(s)$ and $z_R(s+\xi)$ diminishes. For large values of ξ, the two values are practically unrelated. Hence, $R(\xi \to \infty)$ will tend to 0. In fact, $R(\xi)$ is always less than $R(0)$, which coincides with the variance σ^2 of the process. If a periodic term is present in the process, it will show up in $R(\xi)$.

Usually, road profiles are characterized in the frequency domain. Here, the autocorrelation function $R(\xi)$ is replaced by the power spectral density (psd) $S(\Omega)$. In general, $R(\xi)$ and $S(\Omega)$ are related to each other by the Fourier transformation

$$S(\Omega) = \frac{1}{2\pi} \int_{-\infty}^{\infty} R(\xi)\,e^{-i\Omega\xi}\,d\xi \quad \text{and} \quad R(\xi) = \int_{-\infty}^{\infty} S(\Omega)\,e^{i\Omega\xi}\,d\Omega \tag{2.12}$$

where i is the imaginary unit, and Ω in rad/m denotes the wavenumber. To avoid negative wavenumbers, usually a one-sided psd is defined. With

$$\Phi(\Omega) = 2\,S(\Omega), \quad \text{if} \quad \Omega \geq 0 \quad \text{and} \quad \Phi(\Omega) = 0, \quad \text{if} \quad \Omega < 0 \tag{2.13}$$

the relationship $e^{\pm i\Omega\xi} = \cos(\Omega\xi) \pm i\,\sin(\Omega\xi)$, and the symmetry property $R(\xi) = R(-\xi)$ Equation (2.12) results in

$$\Phi(\Omega) = \frac{2}{\pi} \int_{0}^{\infty} R(\xi)\,\cos(\Omega\xi)\,d\xi \quad \text{and} \quad R(\xi) = \int_{0}^{\infty} \Phi(\Omega)\,\cos(\Omega\xi)\,d\Omega \tag{2.14}$$

Now, the variance is obtained from

$$\sigma^2 = R(\xi=0) = \int_{0}^{\infty} \Phi(\Omega)\,d\Omega \tag{2.15}$$

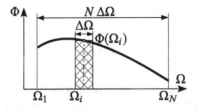

FIGURE 2.5
Power spectral density in a finite interval.

In reality, the psd $\Phi(\Omega)$ will be given in a finite interval $\Omega_1 \leq \Omega \leq \Omega_N$, Figure 2.5. Then, Equation (2.15) can be approximated by a sum, which for N equal intervals will result in

$$\sigma^2 \approx \sum_{i=1}^{N} \Phi(\Omega_i)\,\Delta\Omega \quad \text{with} \quad \Delta\Omega = \frac{\Omega_N - \Omega_1}{N} \tag{2.16}$$

2.3.2 Classification of Random Road Profiles

Road elevation profiles can be measured point by point or by high-speed profilometers. The power spectral densities of roads show a characteristic drop in magnitude with the wavenumber, Figure 2.6a. This simply reflects the fact that the irregularities of the road may amount to several meters over the length of hundreds of meters, whereas those measured over the length of one or fewer meters are normally only some centimeters in amplitude.

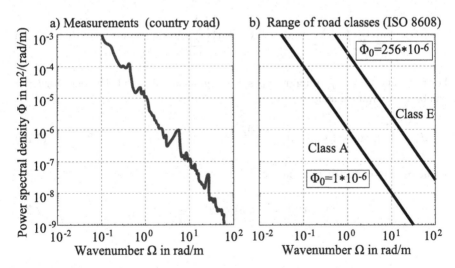

FIGURE 2.6
Road power spectral densities: (a) measurements [4], (b) classification.

Random road profiles can be approximated by a psd in the form of

$$\Phi(\Omega) = \Phi(\Omega_0) \left(\frac{\Omega}{\Omega_0}\right)^{-w} \tag{2.17}$$

where $\Omega = 2\pi/L$ in rad/m denotes the wavenumber and $\Phi_0 = \Phi(\Omega_0)$ in $\mathrm{m^2/(rad/m)}$ or $\mathrm{m^3/}$ describes the value of the psd at the reference wavenumber $\Omega_0 = 1\,\mathrm{rad/m}$. The drop in magnitude is modeled by the waviness w.

According to the international directive ISO 8608 [20], typical road profiles can be grouped into classes from A to E. By setting the waviness to $w = 2$, each class is simply defined by its reference value Φ_0. Class A with $\Phi_0 = 1*10^{-6}\,\mathrm{m^2/(rad/m)}$ characterizes very smooth highways, whereas Class E with $\Phi_0 = 256*10^{-6}\,\mathrm{m^2/(rad/m)}$ represents rather rough roads, Figure 2.6b.

2.3.3   Sinusoidal Approximation

A random profile of a single track can be approximated by a superposition of $N \to \infty$ sine waves

$$z_R(s) = \sum_{i=1}^{N} A_i \sin(\Omega_i\, s - \Psi_i) \tag{2.18}$$

where each sine wave is determined by its amplitude A_i and its wavenumber Ω_i. By different sets of uniformly distributed phase angles Ψ_i, $i = 1(1)N$ in the range between 0 and 2π different profiles can be generated that are similar in general appearance but different in details.

The variance of the sinusoidal representation is then given by

$$\sigma^2 = \lim_{X \to \infty} \frac{1}{X} \int_{-X/2}^{X/2} \left(\sum_{i=1}^{N} A_i \sin(\Omega_i\, s - \Psi_i)\right)\left(\sum_{j=1}^{N} A_j \sin(\Omega_j\, s - \Psi_j)\right) ds \tag{2.19}$$

For $i = j$ and for $i \neq j$, different types of integrals are obtained. The ones for $i = j$ can be solved immediately:

$$J_{ii} = \int A_i^2 \sin^2(\Omega_i s - \Psi_i)\, ds = \frac{A_i^2}{2\Omega_i}\left[\Omega_i s - \Psi_i - \frac{1}{2}\sin(2(\Omega_i s - \Psi_i))\right] \tag{2.20}$$

Using the trigonometric relationship

$$\sin x\, \sin y = \frac{1}{2}\cos(x-y) - \frac{1}{2}\cos(x+y) \tag{2.21}$$

the integrals for $i \neq j$ can be solved also

$$
\begin{aligned}
J_{ij} &= \int A_i \sin\left(\Omega_i s - \Psi_i\right) A_j \sin\left(\Omega_j s - \Psi_j\right) ds \\
&= \frac{1}{2} A_i A_j \int \cos\left(\Omega_{i-j}\, s - \Psi_{i-j}\right) ds - \frac{1}{2} A_i A_j \int \cos\left(\Omega_{i+j}\, s - \Psi_{i+j}\right) ds \\
&= -\frac{1}{2} \frac{A_i A_j}{\Omega_{i-j}} \sin\left(\Omega_{i-j}\, s - \Psi_{i-j}\right) + \frac{1}{2} \frac{A_i A_j}{\Omega_{i+j}} \sin\left(\Omega_{i+j}\, s - \Psi_{i+j}\right)
\end{aligned}
$$

(2.22)

where the abbreviations $\Omega_{i \pm j} = \Omega_i \pm \Omega_j$ and $\Psi_{i \pm j} = \Psi_i \pm \Psi_j$ were used. The sine and cosine terms in Equations (2.20) and (2.22) are limited to values of ± 1. Hence, Equation (2.19) simply results in

$$
\sigma^2 = \underbrace{\lim_{X \to \infty} \frac{1}{X} \sum_{i=1}^{N} \left[J_{ii} \right]_{-X/2}^{X/2}}_{\displaystyle \sum_{i=1}^{N} \frac{A_i^2}{2\Omega_i} \Omega_i} + \underbrace{\lim_{X \to \infty} \frac{1}{X} \sum_{i,j=1}^{N} \left[J_{ij} \right]_{-X/2}^{X/2}}_{0} = \frac{1}{2} \sum_{i=1}^{N} A_i^2
$$

(2.23)

On the other hand, the variance of a sinusoidal approximation to a random road profile is given by Equation (2.16). So, a road profile $z_R = z_R(s)$ described by Equation (2.18) will have a given psd $\Phi(\Omega)$ if the amplitudes are generated according to

$$
A_i = \sqrt{2\,\Phi(\Omega_i)\,\Delta\Omega}, \quad i = 1(1)N
$$

(2.24)

and the wavenumbers Ω_i are chosen to lie at N equal intervals $\Delta\Omega$.

A realization of the country road with a psd of $\Phi_0 = 10 * 10^{-6}$ m^2/(rad/m) and a waviness of $w = 2$ is shown in Figure 2.7. The pseudo-random road profile $z = z(s)$

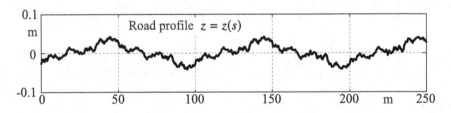

FIGURE 2.7
Realization of a country road.

was generated according to Equations (2.18) and (2.24) by $N = 500$ sine waves in the frequency range from $\Omega_1 = 2\pi/100 = 0.0628$ rad/m to $\Omega_N = 2\pi/0.1 = 62.83$ rad/m which covers wave lengths in the range of 0.1 m $\leq L \leq 100$ m. Unlike real random signals, a pseudo-random approximation is always periodic.

2.3.4   Example

A slight modification of the function dyn_01_qcm_step.m provided in Listing 1.4 makes the pseudo-random road profile available to the quarter car model. The corresponding changes are restricted to the first five lines and provided in Listing 2.2.

Listing 2.2

Function fun_02_qcm_road: quarter car model dynamics with random road input

```
1  function [xdot,out] = fun_02_qcm_road(t,x,p)
2  % quarter car model with trailing arm suspension on random road
3
4  % pseudo random road input
5  s = p.v*t; u = sum( p.Amp.*sin(p.Om*s+p.Psi) ); out.zr = u;
6
7  % state variables
8  z  = x(1);   be  = x(2);   phi  = x(3);
9  zd = x(4);   bed = x(5);   phid = x(6);
10
11 % torque in revolute joint
12 Ts = - ( p.Ts0 + p.cs*be + p.ds*bed );
13
14 % tire deflection (static tire radius)
15 rS = p.h + z - p.b + p.a*sin(be) - u ;
16
17 % longitudinal tire force (adhesion assumed)
18 Fx = - p.cx *( p.a*(1-cos(be))    - rS*phi  ) ...
19      - p.dx *( p.a*sin(be)*bed - rS*phid ) ;
20
21 % vertical tire force (contact assumed)
22 Fz =  p.cz *( p.r0 - rS );
23
24 % negative wheel loads not possible --> liftoff
25 if Fz < 0, Fz=0; end;  out.Fz=Fz;
26
27 % limit longitudinal force to maximum friction force
28 Fx_max = p.mu*Fz;
29 if Fx < -Fx_max, Fx = -Fx_max; end
30 if Fx > +Fx_max, Fx = +Fx_max; end
31 out.Fx=Fx;
32
33 % mass matrix
34 Massma=[ p.mC+p.mK+p.mW              (p.s*p.mK+p.a*p.mW)*cos(be)    0 ; ...
35   (p.s*p.mK+p.a*p.mW)*cos(be) p.ThetaK+p.s^2*p.mK+p.a^2*p.mW  0 ; ...
36            0                          0                  p.ThetaW ];
37
38 % vector of generalized forces and torques
39 q=[Fz-(p.mC+p.mK+p.mW)*p.g+(p.s*p.mK+p.a*p.mW)*sin(be)*bed^2 ; ...
40    Ts-(p.s*p.mK+p.a*p.mW)*cos(be)*p.g+p.a*(Fx*sin(be)+Fz*cos(be)); ...
41    -rS*Fx ];
42
43 % state derivatives
44 xdot = [ zd;  bed;  phid;  Massma\q ];
45
46 end
```

The actuator displacements u now represent the pseudo-random road profile $z_R = z_R(s)$ defined by Equation (2.18). By assuming a constant driving velocity $v = const.$, the relationship $s = v\,t$ will compute the spatial coordinate as a function of time. The additional structure element p.v provides the velocity of the vehicle. The structure vector elements p.Amp, p.Om, and p.Psi define the amplitudes, the spatial frequencies, and the randomly distributed phase angles of the random road approximation. These vectors are computed in the MATLAB-Script given by Listing 2.3. The function dyn_02_qcm_road.m now takes the wheel liftoff into account. At first, the wheel load is restricted to positive values, $F_z \geq 0$. Then, the longitudinal force is restricted to the maximum friction force $F_x^{max} = \mu F_z$ and assigned to an additional element of the output structure out. The additional element p.mu defined in Listing 2.3 provides the coefficient of friction between tire and road or tire and actuator respectively. More complex tire models, like the one presented in Chapter 3, automatically take a potential wheel liftoff into account and apply a sophisticated sliding model to the longitudinal tire force.

Listing 2.3
Script app_02_qcm_road.m: quarter car model exposed to pseudo-random excitation

```
1 p    = uty_par_qcmta(); % assign vehicle parameter to structure
2 p.mu = 1.0;             % add tire/road friction coefficient
3 p.v  = 80/3.6;          % add vehicle velocity, km/h --> m/s
4
5 % specify spectral density and waviness of random road
6 Phi0 = 10e-6; w = 2;
7 % range of frequencies and number of samples
8 Omin = 2*pi/100;  Omax = 2*pi/0.1;  n = 500;
9
10 % calculate amplitudes and random phases for pseudo random road
11 dOm = (Omax-Omin)/(n-1);  Om  = Omin:dOm:Omax;
12 Om0 = 1; Phi = Phi0.*(Om./Om0).^(-w);  Amp = sqrt(2*Phi*dOm);
13 rng('default'); Psi = 2*pi*rand(size(Om));
14
15 % add road data to parameter structure
16 p.Om = Om; p.Amp = Amp; p.Psi = Psi;
17
18 % default initial states: x0 = [ zC; be; phi; zCd; bed; phid ]
19 x0 = [ 0; 0; 0; 0; 0; 0];
20
21 % simulation interval
22 t0=0; tst=2; tE=tst+6;
23 % adjust initial state to road
24 s_V = p.v*t0; u = sum( Amp.*sin(Om*s_V+Psi) ); x0(1)=u;
25 % time simulation (up to steady state + stationary drive)
26 [ts,xs] = ode45(@(t,x) fun_02_qcm_road(t,x,p),[t0,tst],x0);
27 [to,xo] = ode45(@(t,x) fun_02_qcm_road(t,x,p),[tst,tE],xs(end,:).');
28
29 % get road height, wheel load and chassis acceleration
30 zr=zeros(size(to)); fz=zeros(size(to)); zdd=zeros(size(to));
31 for i=1:length(to)
32   [xdot,out] = fun_02_qcm_road(to(i),xo(i,:).',p);
33   zr(i)= out.zr;  fz(i)=out.Fz;  zdd(i)=xdot(4);
34 end
35
36 % compute mean value and standard deviation of wheel load
37 fz_m = mean(fz);  fz_std = std(fz);
38 disp(['wheel load: mean value =',num2str(fz_m)])
39 disp(['wheel load: standard deviation =',num2str(fz_std)])
40
41 % generate histogram of wheel load and reduce edges to center points
42 nh=20; [nfz,fze]=histcounts(fz,nh,'Normalization','pdf');
43 fzc=0.5*(fze(2:end)+fze(1:end-1));
44
45 % generate gaussian density function
46 fzi=linspace(min(fz),max(fz),201);
```

```
47 pfz=exp(-(fzi-fz_m).^2./(2*fz_std.^2))./(fz_std*sqrt(2*pi));
48
49 % plot results
50 subplot(3,2,1)
51 plot(to,zr), grid on,xlabel('t / s'),legend('road: z_R / m')
52 subplot(3,2,2)
53 plot(to,xo(:,2)*180/pi),grid on,xlabel('t / s'),legend('\beta / deg')
54 subplot(3,2,3)
55 plot(to,xo(:,1)),grid on,xlabel('t / s'),legend('z_C / m')
56 subplot(3,2,4)
57 plot(to,xo(:,3)*180/pi),grid on,xlabel('t / s'),legend('\phi / deg')
58 subplot(3,2,5)
59 plot(to,fz/1000),grid on,hold on,xlabel('t / s')
60 plot([tst,tE],[1,1]*fz_m/1000,'--k'), ylim([min(fze),max(fze)]/1000)
61 legend('wheel load F_z / kN','mean value')
62 subplot(3,2,6)
63 barh(fzc/1000,nfz,'FaceColor','[0.95,0.95,0.95]'), hold on
64 plot(pfz,fzi/1000,':k','Linewidth',1);ylim([min(fze),max(fze)]/1000)
65 legend('Histogram F_z','Gauss'), h=gca; h.XAxis.Visible ='off';
```

As done in the corresponding script provided by Listing 1.3, the vehicle data are defined by the function uty_par_qcmta given in Listing 1.2. The MATLAB-Function rand[1] is used to generate uniformly distributed pseudo-random numbers. The command rng('default') initializes the random number generator and produces the same random sequence at each run.

To reduce too large initial disturbances, the chassis displacement z represented by the first state variable $x(1)$ is adjusted to the initial actuator displacement $u = u(t_0)$ at first. The simulation from $t = t_0 = 0$ s to $t = t_{st} = 2$ s eliminates the transient response. Then, the simulation from $t = t_0 = t_{st} = 2$ s to $t = t_E = t_{st} + 6$ s generates the stationary response of a vehicle exposed to road undulations that correspond with a drive at constant velocity on a rough road.

MATLAB simulations performed by any ode-solver just provide the state variables at different time steps stored here in the vector to and in the matrix xo. The loop in lines 33 to 36 calls the MATLAB-Function app_02_qcm_road.m which provides the state derivatives xdot and the output structure out at each time step and makes it possible to assign the road height, the wheel load, and the chassis acceleration to the vectors zr, fz, zd, which are pre-allocated in line 32 to speed up the loop.

The MATLAB-Functions mean and std applied to the vector fz provide the mean value of fz_m = 2437.7 N and the standard deviation fz_std = 743.1 N of the wheel load. As expected, the mean values correspond quite well to the steady-state value $F_z^{st} = 2452.5$ N, which results from the first part of Equation 1.54 and corresponds to the weight of the quarter car model. The MATLAB-Function histcounts generates a histogram of the wheel load, which is normalized to a probability density function (pdf). The edges fze generated by histcounts are then converted to the center values fzc and plotted by the MATLAB-Function barh as a horizontal bar plot next to the time history of the wheel load, Figure 2.8. The horizontal bar plot corresponds quite well with the Gaussian probability density function (pdf) which was computed according to (2.8) and plotted as a broken line in addition.

---

[1]Note: Different MATLAB-Version may generate different random numbers.

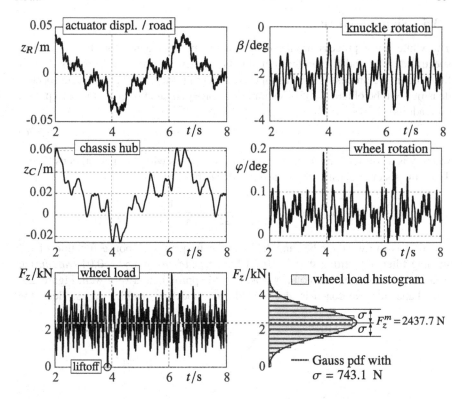

FIGURE 2.8
Simulation results with a pseudo-random excitation.

The time histories of the chassis and the actuator displacements clearly indicate the filtering effects of the wheel suspension. The quarter car model is here driven with a velocity of $v_V = 80 \; km/h$ over a rather rough road that produces large wheel load variations. According to Equation (2.10), the wheel loads will be with a probability of 99.7% in the range of

$$2437.7 - 3*743.1 < F_z < 2437.7 + 3*743.1 \quad \text{or} \quad 208.4 \, N < F_z < 4667 \, N \quad (2.25)$$

which takes a Gaussian normal distribution for granted. The lower limit, which is close to zero, indicates the danger of wheel liftoff. On a closer look the time history of the wheel load $F_z = F_z(t)$ actually shows at $t = 3.8$ s a singular liftoff event.

To improve the statistical significance of the simulation results, a more sophisticated random road approximation may be applied where the range of the wavenumbers is divided not in equally spaced intervals but in intervals where the bandwidth $b_i = (\Omega_i + \Delta\Omega_i)/\Omega_i$ is kept constant.

2.3.5   Shaping Filter

The white noise process produced by random number generators has a uniform spectral density and is therefore not suitable to describe real road profiles. But, if the white noise process is used as input to a shaping filter, more appropriate spectral densities will be obtained [35]. A simple first-order shaping filter for the road profile z_R reads as

$$\frac{d}{ds} z_R(s) = -\gamma \, z_R(s) + w(s) \qquad (2.26)$$

where γ is a constant and $w(s)$ is a white noise process with the spectral density Φ_w. Then, the spectral density of the road profile is obtained from

$$\Phi_R = H(\Omega) \, \Phi_W \, H^T(-\Omega) = \frac{1}{\gamma + i\,\Omega} \, \Phi_W \, \frac{1}{\gamma - i\,\Omega} = \frac{\Phi_W}{\gamma^2 + \Omega^2} \qquad (2.27)$$

where Ω is the wavenumber and $H(\Omega)$ is the frequency response function of the shaping filter. By setting $\Phi_W = 10 * 10^{-6} \ m^2/(rad/m)$ and $\gamma = 0.01 \ rad/m$, the measured psd of a typical country road can be approximated very well, Figure 2.9. The shape filter approach is also suitable for modeling parallel tracks. Here, the

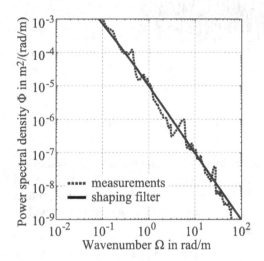

FIGURE 2.9
Shaping filter as approximation to measured psd.

crosscorrelation between the irregularities of the left and right track have to be taken into account also [43].

The white noise process $w(s)$ is discontinuous. Hence, solving the shape filter differential equations like Equation (2.26) with standard ode-solvers is extremely time consuming or will cause severe problems. That is why the shape filter approach usually is applied within analytical calculations only.

2.3.6 Two-Dimensional Model

The generation of fully two-dimensional road profiles $z_R = z_R(x, y)$ via a sinusoidal approximation is very laborious. Because a shaping filter is a dynamic system, the resulting road profile realizations are not reproducible. By adding band-limited white noise processes and taking the momentary position x, y as seed for the random number generator, a reproducible road profile can be generated [40]. By assuming the same

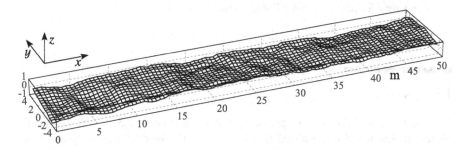

FIGURE 2.10
Two-dimensional road profile.

statistical properties in longitudinal and lateral directions, two-dimensional profiles, like the one in Figure 2.10, can be obtained.

On a two-dimensional road profile, the road normal and hence the wheel load F_z change their directions according to the local inclination of the profile, Figure 2.11.

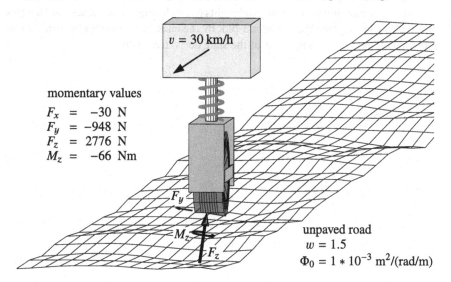

$v = 30$ km/h

momentary values

$$F_x = -30 \text{ N}$$
$$F_y = -948 \text{ N}$$
$$F_z = 2776 \text{ N}$$
$$M_z = -66 \text{ Nm}$$

unpaved road
$w = 1.5$
$\Phi_0 = 1 * 10^{-3}$ m^2/(rad/m)

FIGURE 2.11
Quarter car model driving on unpaved road modeled by a two-dimensional profile.

That is why a small longitudinal tire force F_x which keeps the wheel rolling is in general accompanied by a lateral tire force F_y and a turn torque M_z. As explained in detail in Chapter 3 the lateral tire force and the turn torque are induced by the camber angle which corresponds here with the lateral inclination of the road because the wheel is kept perfectly upright in this application.

Hence driving on rough roads will not only produce vertical vehicle motions but impact the vehicle in all directions. That is why, three-dimensional vehicle models are mandatory for accurate and reliable virtual tests.

Exercises

2.1 Use the MATLAB-Script `app_02_qcm_road.m` (Listing 2.3) together with the corresponding functions provided by Listing 1.2 and Listing 2.2 to study the influence of the vehicle velocity on the wheel load. Replace the histogram-plot by the longitudinal tire force F_x which is available in `out.fx` as an element of the output structure which is provided by the function `fun_02_qcm_road`.

Perform simulations with different values of the road power spectral density.

2.2 Modify the MATLAB-Script in Listing 2.3 and the corresponding function in Listing 2.2 so that the displacements of the actuator correspond with a cosine-shaped bump. Use the function in Listing 2.1 or at least the relevant statements in it. The basic vehicle data are still defined by the function `uty_par_qcmta` given in Listing 1.2.

Perform simulations with different heights (including negative values) and lengths of the obstacle. Replace the histogram-plot by the chassis acceleration which will be provided by the fourth component of the state derivatives `xdot`.

3

Tire

CONTENTS

3.1   Introduction

3.1.1   Tire Development

Some important milestones in the development of pneumatic tires are shown in Table 3.1.

TABLE 3.1
Milestones in tire development

1839	Charles Goodyear: vulcanization
1845	Robert William Thompson: first pneumatic tire (several thin inflated tubes inside a leather cover)
1888	John Boyd Dunlop: patent for bicycle (pneumatic) tires
1893	The Dunlop Pneumatic and Tyre Co. GmbH, Hanau, Germany
1895	André and Edouard Michelin: pneumatic tires for Peugeot Paris-Bordeaux-Paris (720 mi) finished with 50 tire deflations and 22 complete inner tube changes
1899	Continental: "long-lived" tires (approx. 500 km)
1904	Carbon added: black tires.
1908	Frank Seiberling: grooved tires with improved road traction
1922	Dunlop: steel cord thread in the tire bead
1942	Synthetic rubber: extremely important during World War II
1943	Continental: patent for tubeless tires
1946	Radial tire
1952	High-quality nylon tire
$\vdots$	

Of course tire development did not stop in 1952, but modern tires are still based on these achievements. Today, run-flat tires are under investigation. A run-flat tire enables the vehicle to continue to be driven at reduced speeds (i.e., 80 km/h or 50 mp/h) and for limited distances (80 km or 50 mi). The introduction of run-flat tires makes it mandatory for car manufacturers to fit a system where the drivers are made aware that the run-flat has been damaged.

3.1.2   Tire Composites

Tires are very complex. They combine dozens of components that must be formed, assembled, and cured together. And their ultimate success depends on their ability to blend all of the separate components into a cohesive product that satisfies the driver's needs.

A modern radial tire for passenger cars usually consists of the tread, the headbands and the carcass consisting of textile cord layers and steel belts, a butyl cap, as well as the sidewalls with fillers, an alloy band, and the bead wires, Figure 3.1.

The weight of passenger car tires in the range from 13-inch tires for small city cars to 20-inch models for SUVs and sports cars is in between 6.5 kg to 15 kg whereas

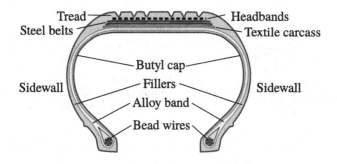

FIGURE 3.1
Tire cross section of a typical passenger car tire.

truck tires range in weights from 30 kg to 80 kg [32]. The largest share is represented by the tread, which accounts for approximately 35% to 40% of the total tire weight. The top layer of the tread wears out over time, thus reducing its weight.

3.1.3 Tire Forces and Torques

In any point of contact between the tire and the road surface, normal and friction forces are transmitted. According to the tire's profile design, the contact patch forms a not necessarily coherent area, Figure 3.2.

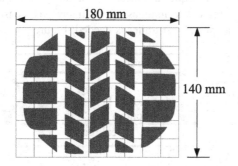

FIGURE 3.2
Footprint of a test tire of size 205/55 R 16 at $F_z = 4700$ N and $p = 2.5$ bar.

The effect of the contact forces can be fully described by a resulting force vector applied at a specific point of the contact patch and a torque vector. The vectors are described in a track-fixed coordinate system. The z-axis is normal to the track, and the x-axis is perpendicular to the z-axis and perpendicular to the wheel rotation axis e_{yR}. Then, the demand for a right-handed coordinate system also fixes the y-axis.

The components of the contact force vector are named according to the direction of the axes, Figure 3.3. A nonsymmetric distribution of the forces in the contact patch causes torques around the x- and y-axes. A cambered tire generates a tilting torque

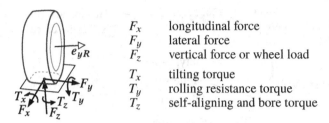

F_x	longitudinal force
F_y	lateral force
F_z	vertical force or wheel load
T_x	tilting torque
T_y	rolling resistance torque
T_z	self-aligning and bore torque

FIGURE 3.3
Forces and torques acting in the tire contact patch.

T_x. The torque T_y includes the rolling resistance of the tire. In particular, the torque around the z-axis is important in vehicle dynamics. It consists of two parts,

$$T_z = T_B + T_S \tag{3.1}$$

The rotation of the tire around the z-axis causes the bore torque T_B. The self-aligning torque T_S takes into account that, in general, the resulting lateral force is not acting in the center of the contact patch.

3.1.4   Measuring Tire Forces and Torques

In general, tire forces and torques are measured under quasi-static operating conditions. Different measurement techniques are available. To measure tire forces and torques on the road, a special test trailer is needed, Figure 3.4. Here, the measure-

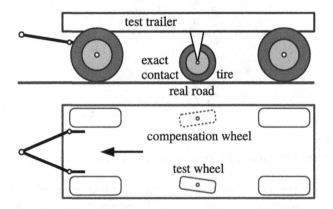

FIGURE 3.4
Layout of a typical tire test trailer.

ments are performed under real operating conditions. Arbitrary surfaces like asphalt or concrete and different environmental conditions like dry, wet, or icy are possible.

Measurements with test trailers are quite costly and in general, restricted to passenger car tires.

Indoor measurements of tire forces and torques can be performed on drums or on a flatbed, Figure 3.5. On drum test rigs, the tire is placed either inside or outside

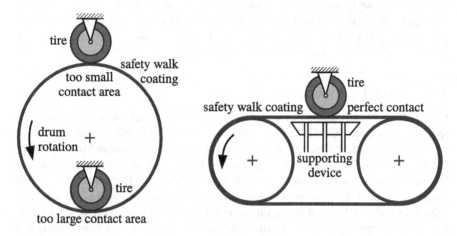

FIGURE 3.5
Tire test rigs: Inner and outer drum as well as flatbed.

the drum. In both cases the shape of the contact area between tire and drum is not correct. That is why one cannot rely on the measured self-aligning torque. Due to its simple and robust design, wide applications including measurements of truck tires are possible. The flatbed tire test rig is more sophisticated. Here, the contact patch is as flat as on the road. But, the safety walk coating that is attached to the steel bed does not generate the same friction conditions as on a real road surface.

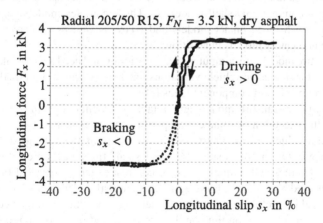

FIGURE 3.6
Typical results of tire measurements.

The slip, a dimensionless quantity that is discussed in Section 3.3.4, is used to quantify the sliding condition in the area where the tire is in contact with the road or the testing device, respectively. Usually, the velocity or angular velocity of the testing device (trailer, drum, or flatbed) is kept constant. Then, different sliding conditions are generated by slowly increasing or decreasing the angular velocity or slowly varying the alignment of the test wheel. In consequence, the measurements for increasing and decreasing sliding conditions will result in different graphs, Figure 3.6. In general, the mean values are then taken as steady-state results.

3.1.5   Modeling Aspects

For the dynamic simulation of on-road vehicles, the model-element "tire/road" is of special importance, according to its influence on the achievable results. It can be said that the sufficient description of the interactions between tire and road is one of the most important tasks of vehicle modeling, because all the other components of the chassis influence the vehicle dynamic properties via the tire contact forces and torques. Therefore, in the interest of balanced modeling, the precision of the complete vehicle model should stand in reasonable relation to the performance of the applied tire model. At present, two groups of models can be identified: handling models and structural or high-frequency models [25]. A general overview of standard tire models and their application to vehicle dynamics is provided in [14].

Structural tire models are very complex. Within RMOD-K, the tire is modeled by four circular rings with mass points that are also coupled in the lateral direction [31]. Multi-track contact and the pressure distribution across the belt width are taken into account. The tire model FTire consists of an extensible and flexible ring that is mounted to the rim by distributed stiffnesses in the radial, the tangential, and the lateral direction [13]. The ring is approximated by a finite number of belt elements to which a number of mass-less tread blocks are assigned, Figure 3.7.

Instead of this lumped mass approach, the CDTire model approximates belt and sidewalls by a complete 3D shell model [7]. These complex models apply sophisticated local brush type contact models and FTire as well as CDTire incorporate temperature and cavity models in addition.

Complex tire models are computer time consuming and need a lot of data. Usually, they are used for stochastic vehicle vibrations occurring during rough road rides and causing strength-relevant component loads [38].

Comparatively lean tire models are suitable for vehicle dynamics simulations, while, with the exception of some elastic partial structures such as twist-beam axles in cars or the vehicle frame in trucks, the elements of the vehicle structure can be seen as rigid. On the tire's side, "semi-physical" tire models prevail, where the description of forces and torques relies, in contrast to purely physical tire models, also on measured and observed force-slip characteristics. This class of tire models is characterized by useful compromise between user friendliness, model complexity, and efficiency in computation time on the one hand, and precision in representation on the other hand.

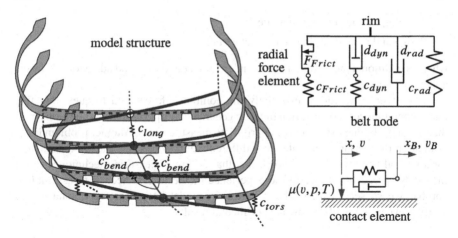

FIGURE 3.7
Structure of the FTire model with radial force and brush contact element.

In vehicle dynamics practice, often there exists the problem of data provision for a special type of tire for the examined vehicle. Considerable amounts of experimental data for car tires has been published or can be obtained from the tire manufacturers. If one cannot find data for a special tire, its characteristics can be guessed at least by an engineer's interpolation of similar tire types, Figure 3.8. In the field of truck tires, there is still a considerable backlog in data provision. These circumstances must be respected in conceiving a user-friendly tire model.

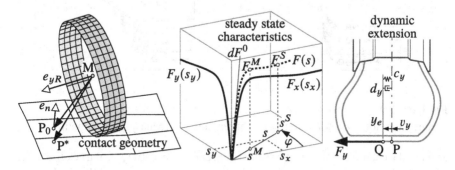

FIGURE 3.8
Basic model concept of the TMeasy handling tire model.

For a special type of tire, usually the following sets of experimental data are provided:

- Longitudinal force versus longitudinal slip (mostly just brake-force)
- Lateral force versus slip angle

- Aligning torque versus slip angle
- Radial and axial compliance characteristics

whereas additional measurement data under camber and low road adhesion are favorable special cases.

Any other correlations, especially the combined forces and torques, effective under operating conditions, often must be generated by appropriate assumptions with the model itself, due to the lack of appropriate measurements. Another problem is the evaluation of measurement data from different sources (i.e., measuring techniques) for a special tire [16]. It is a known fact that different measuring techniques result in widely spread results. Here the experience of the user is needed to assemble a "probably best" set of data as a basis for the tire model from these sets of data, and to verify it eventually with his own experimental results.

3.1.6   Typical Tire Characteristics

Standard measurements usually performed for different wheel loads F_z provide the longitudinal force F_x as a function of the longitudinal slip s_x, Figure 3.9, and the lateral force F_y as well as the self-aligning torque T_S as a function of the slip angle α, which is related to the lateral slip by $\tan \alpha = s_y$, Figure 3.10. Although similar in

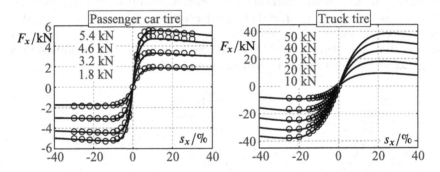

FIGURE 3.9
Longitudinal force $F_x = F_x(s_x)$: ∘ Measurements, − TMeasy approximation.

general, the characteristics of a typical passenger car tire and a typical truck tire differ quite a lot in some details. Usually, truck tires are designed for durability and not for generating large lateral forces. The characteristic curves $F_x = F_x(s_x)$, $F_y = F_y(\alpha)$, and $T_S = T_S(\alpha)$ for the passenger car and truck tire can be approximated quite well by the tire handling model TMeasy [17].

Within the TMeasy model approach, one-dimensional characteristics are automatically converted to two-dimensional combined-slip characteristics, Figure 3.11.

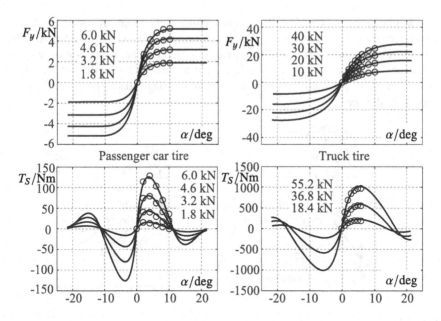

FIGURE 3.10
Lateral force $F_y = F_y(\alpha)$ and self-aligning torque $T_S = T_S(\alpha)$:
o Measurements, − TMeasy approximation.

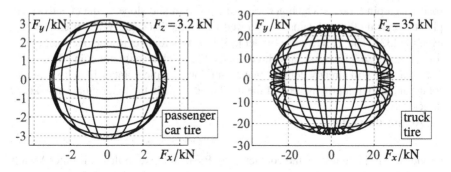

FIGURE 3.11
Combined forces $F_y = F_y(F_x)$ at different longitudinal slips and slip angles
$|s_x| = 1, 2, 4, 6, 10, 15$ % and $|\alpha| = 1, 2, 4, 6, 10, 14°$.

3.2 Contact Geometry

3.2.1 Geometric Contact Point

Within handling tire models, the area where a tire is in contact with the road is
approximated by an effective local road plane which is defined by the surface normal

e_n and one specific point P called contact point, right graph in Figure 3.12. The geometric contact point is defined by the point on the intersection line between the rim center plane and the effective local road plane that has the shortest distance to the rim center. Technically its calculation is straightforward. But, the effective local road plane changes its orientation from point to point on rough roads and is therefore not known in advance.

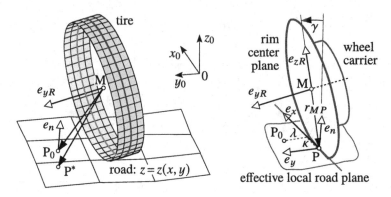

FIGURE 3.12
Tire in contact with an uneven road.

An appropriate sampling technique can overcome this problem. A rather complex road enveloping model which generates the position and slope of the effective road plan by several elliptical cams is employed within the Pacejka/SWIFT tire model for example [33]. The TMeasy tire model [55] just uses four representative road points for that purpose. Of course, any three points that by chance do not coincide or form a straight line will already define a plane. But, four sample points will deliver a more robust and better approximation of the local road slope in the longitudinal and lateral directions.

At first, the vector

$$r_{MP^*} = -r_0\, e_{n0} \tag{3.2}$$

provides a simple guess of the contact point $P \approx P^*$ relative to the rim center M, left graph in Figure 3.12. The approximation makes use of the unloaded tire radius r_0 and the unit vector e_{n0} which points somehow upwards. For standard road profiles $e_{n0,0} = [\,0\ 0\ 1\,]^T$ represents the global track normal which can be used for that purpose. The vector

$$r_{0P^*,0} = r_{0M,0} + r_{MP^*,0} = \begin{bmatrix} x^* \\ y^* \\ z^* \end{bmatrix} \tag{3.3}$$

determines the position of the estimated contact point P^* with respect to the origin 0 of the earth-fixed x_0-y_0-z_0-reference frame, where the vector r_{0M} describes the absolute position of the rim center M and the comma separated index 0 indicates that the vector components x^*, y^*, z^* are expressed in the earth-fixed reference frame.

The irregularities of the road profile can be described by an arbitrary function of two spatial coordinates,

$$z = z(x, y) \tag{3.4}$$

Usually, the estimated contact point P* does not lie on the road because Equation (3.2) generates a vector where neither the deflection of the tire nor its potential liftoff is taken into account. The corresponding road point P_0 is computed by

$$r_{0P_0,0} = \begin{bmatrix} x^* \\ y^* \\ z(x^*, y^*) \end{bmatrix} \tag{3.5}$$

where Equation (3.4) was used to adjust the z-coordinate to the conforming height of the road profile.

The road normal generated at this particular point P_0 is not of much use, because it does not represent the overall slope of the effective road plane that spans over the whole contact area. That is why, the road profile is sampled by four points R_1 to R_4 in order to determine an representative local road normal, Figure 3.13. At first, the

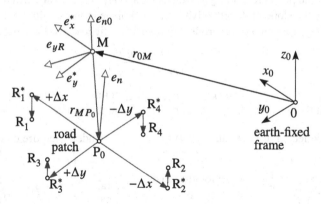

FIGURE 3.13
Road sample points close to estimated contact point.

longitudinal or circumferential and the lateral tire directions are approximated by

$$e_x^* = \frac{e_{yR} \times e_{n0}}{|e_{yR} \times e_{n0}|} \quad \text{and} \quad e_y^* = e_{n0} \times e_x^* \tag{3.6}$$

where the unit vectors e_{yR} and e_{n0} define the wheel rotation axis and the upwards direction. Calculating the unit vector e_x^* demands a normalization because e_{yR} is not perpendicular to the upwards direction e_{n0} in general. Then, the sample points R_1^* to R_4^* defined by

$$r_{MR_{1,2}^*} = r_{0P_0,0} \pm \Delta x\, e_x^* \quad \text{and} \quad r_{MR_{3,4}^*} = r_{0P_0,0} \pm \Delta y\, e_y^* \tag{3.7}$$

will be located close to the road surface in the front, in the rear, to the left, and to the right of the first contact point guess P_0 provided by (3.3). The distances Δx and Δy in the longitudinal and the lateral directions are adjusted to the unloaded radius r_0 and the width b_0 of the tire in order to sample the contact patch appropriately. By setting $\Delta x = 0.1\, r_0$ and $\Delta y = 0.3\, b_0$, a quite realistic behavior even on track grooves could be achieved [66].

Similar to Equation (3.5), the corresponding sample points located on the road R_1 to R_4 are given by

$$r_{0R_i,0} = \begin{bmatrix} x_i^* \\ y_i^* \\ z\left(x_i^*,\, y_i^*\right) \end{bmatrix}, \quad i = 1(1)4 \tag{3.8}$$

where x_i^* and y_i^* are the x- and y-components of the vectors

$$r_{0R_i^*,0} = r_{0M,0} + r_{MR_i^*,0} = \begin{bmatrix} x_i^* \\ y_i^* \\ z_i^* \end{bmatrix}, \quad i = 1(1)4 \tag{3.9}$$

which specify the absolute position of the sample points R_1^* to R_4^*. The lines fixed by the road sample points R_1 and R_2 as well as R_3 and R_4, respectively, will now be used to determine the slope of the effective local road plane in the longitudinal and lateral directions. A normalized cross product delivers the effective road normal as

$$e_n = \frac{r_{R_2R_1} \times r_{R_4R_3}}{\left| r_{R_2R_1} \times r_{R_4R_3} \right|} \tag{3.10}$$

where $r_{R_2R_1} = r_{0R_1} - r_{0R_2}$ and $r_{R_4R_3} = r_{0R_3} - r_{0R_4}$ define vectors pointing from R_2 to R_1 and from R_4 to R_3 respectively, Figure 3.14. Then, the estimated unit vectors in the

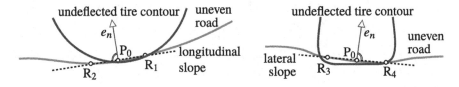

FIGURE 3.14
Inclination of local track plane in longitudinal and lateral directions.

tire's longitudinal or circumferential and lateral directions can be updated. According to (3.6) one gets

$$e_x = \frac{e_{yR} \times e_n}{\left| e_{yR} \times e_n \right|} \quad \text{and} \quad e_y = e_n \times e_x \tag{3.11}$$

where the computed effective road normal vector e_n defined in (3.10) replaces the approximation e_{n0}. The unit vector e_x into the longitudinal or circumferential direction coincides with the direction of the intersection line between the rim center plane and the local road plane, right graph in Figure 3.12. Now, the vector from the rim center M to the contact point P_0 is split into three parts,

$$r_{MP_0} = -r_S\, e_{zR} + \kappa\, e_x + \lambda\, e_y \tag{3.12}$$

where κ and λ are distances measured in the longitudinal or circumferential and the lateral directions, right graph in Figure 3.12. The first part

$$r_{MP} = -r_S \, e_{zR} \tag{3.13}$$

defines the geometric contact point P that lies on the intersection line between the rim center plane and the effective local road plane and has the shortest distance to the rim center M. The unit vector

$$e_{zR} = e_x \times e_{yR} \tag{3.14}$$

which is mutually perpendicular to the intersection line e_x and the wheel rotation axis e_{yR} defines the radial direction of the tire and r_S denotes the loaded or static tire radius.

The mean value of the sample track points

$$r_{0P_0,0} = \frac{1}{4} \left(r_{0R_1,0} + r_{0R_2,0} + r_{0R_3,0} + r_{0R_4,0} \right) \tag{3.15}$$

will define the contact point P as the average position of the effective local road plane. Then, the vector

$$r_{MP_0} = r_{0P_0,0} - r_{0M_0,0} \tag{3.16}$$

delivers its position relative to the rim center. A scalar multiplication of Equation (3.12) with the unit vector e_n results in

$$e_n^T \, r_{MP_0} = -r_S \, e_n^T \, e_{zR} + \kappa \, e_n^T \, e_x + \lambda \, e_n^T \, e_y \tag{3.17}$$

As the unit vectors e_x and e_y are perpendicular to e_n, Equation (3.17) simplifies to

$$e_n^T \, r_{MP_0} = -r_S \, e_n^T \, e_{zR} \tag{3.18}$$

Hence, the loaded or static tire radius is just given by

$$r_S = -\frac{e_n^T \, r_{MP_0}}{e_n^T \, e_{zR}} \tag{3.19}$$

Finally, inspecting the right graph in Figure 3.12, the tire camber angle is provided by

$$\cos(90° - \gamma) = \sin \gamma = e_{yR}^T \, e_n \quad \text{or} \quad \gamma = \arcsin\left(e_{yR}^T \, e_n\right) \tag{3.20}$$

It describes the inclined position of the rim center plane defined by the unit vector e_{yR} against the track normal e_n and has a significant impact in the lateral tire force and the turn or bore torque.

Like a real tire, the effective local road plane defined by the effective normal e_n and the geometric contact point P performs a road filtering process. Sharp bends or discontinuities, which will occur at very rough roads or step- and ramp-sized obstacles, will be smoothed by this approach.

3.2.2   Static Contact Point and Tire Deflection

Assuming that the shape of the deflected tire area corresponds with the pressure distribution in the lateral direction of the contact patch, the acting point of the resulting vertical tire force F_z will be shifted from the geometric contact point P to the static contact point Q, Figure 3.15.

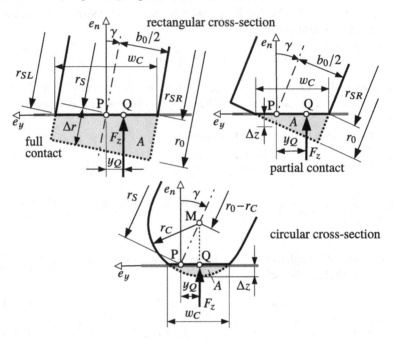

FIGURE 3.15
Static contact point in different contact situations.

If y_Q defines the center of the pressure distribution, the location of the static contact point Q is defined by the vector

$$r_{0Q} = r_{0P} + y_Q\, e_y \tag{3.21}$$

where r_{0P} defines the position of the geometric contact point P and e_y is the unit vector pointing into the lateral direction. The lateral shift y_Q and the equivalent vertical tire deflection Δz, which will be used to calculate the vertical tire force F_z later on, strongly depend on the shape of the cross section of the unloaded tire and on the camber angle γ in addition.

Whereas passenger car or truck tires usually have a nearly rectangular cross section, motor-cycle tires often do have a circular or a more rounded cross section. That is why, the contact scenarios of a cambered tire with perfectly rectangular and perfectly circular cross sections are illustrated in, Figure 3.15. As demonstrated in [55] the lateral shift y_Q, the equivalent vertical tire deflection Δz, and the actual width of the contact area w_C can be computed as functions of the unloaded radius r_0, the tire width b_0, and the camber angle γ for these particular cases.

Real tire cross sections are somewhere in between a perfect rectangle and a circle. By introducing a simple roundness parameter in the range of $0 \le R_N \le 1$, a real tire cross section can be approximated by a weighted superposition of a perfectly rectangular and a circular cross section. The images in Figure 3.16 illustrate how

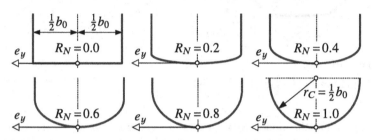

FIGURE 3.16
Different shapes of tire cross sections controlled by a roundness parameter.

increasing values of the roundness parameter R_N will morph a rectangular cross section ($R_N = 0$) continuously into a circular one ($R_N = 1$). As a consequence, the shift y_Q from P to Q, the generalized tire deflection Δz, and the actual width w_C of the contact patch are calculated as weighted mean values of the special cases $R_N = 0$ and $R_N = 1$ that define perfectly rectangular and perfectly circular tire cross sections.

Appropriate values for the roundness parameter R_N may be set quite simply by matching the shapes generated by different values of R_N with the real cross section of a tire or just by estimating a proper value.

3.2.3 Length of Contact Patch

In normal contact situations the difference between the unloaded and the static tire radius delivers the tire deformation $\Delta_z = r_0 - r_S$. However, this overall deformation consists of two parts

$$\Delta z = \Delta z_F + \Delta z_B \tag{3.22}$$

where Δz_F and Δz_B denote the average flank and the belt deformation of the tire, Figure 3.17. Approximating the belt deflection Δz_B by truncating a circle with the radius of the unloaded tire r_0 results in

$$\left(\frac{L}{2}\right)^2 + (r_0 - \Delta z_B)^2 = r_0^2 \tag{3.23}$$

Under normal operating conditions, the belt deflections will be comparatively small. Taking $\Delta z_B \ll r_0$ for granted, Equation (3.23) simplifies to

$$\frac{L^2}{4} = 2 r_0 \Delta z_B \quad \text{or} \quad L = \sqrt{8 r_0 \Delta z_B} \tag{3.24}$$

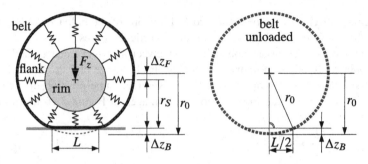

FIGURE 3.17
Tire deformation and length of contact patch.

Complex structural tire models compute these deflections explicitly. Within a handling tire model, a practical approximation is required. By simply assuming that the belt deformation roughly equals the flank deformation

$$\Delta z_F \approx \Delta z_B \approx \frac{1}{2} \Delta z \tag{3.25}$$

will hold according to (3.22). Then

$$L = \sqrt{8\, r_0\, \Delta z_B} = \sqrt{8\, r_0 \frac{1}{2} \Delta z} = 2\sqrt{r_0\, \Delta z} \tag{3.26}$$

provides the contact length as a function of the unloaded tire radius r_0 and the tire deflection Δz.

Inspecting the passenger car tire footprint in Figure 3.2 leads to a contact patch length of $L \approx 140$ mm. For this tire, the load, the vertical stiffness, and the unloaded radius are specified with $F_z = 4700$ N, $c_z = 265\,000$ N/m, and $r_0 = 316.9$ mm. The overall tire deflection can be estimated by $\Delta z = F_z/c_z = 4700$ N / $265\,000$ N/m $= 0.0177$ m in this particular case. Then, Equation (3.26) produces with $L = 2\sqrt{0.3169 \text{ m} * 0.0177 \text{ m}} = 0.1498$ m ≈ 150 mm a contact patch length that corresponds quite well with the given length of the tire footprint.

3.2.4   Contact Point Velocity

The velocity of the contact point will be needed to calculate the tire forces and torques that are generated by friction. The momentary position of the static contact point Q defined by Equation (3.21) is given by

$$r_{0Q} = r_{0M} + r_{MQ} \tag{3.27}$$

where the vector r_{0M} describes the position of the rim center M with respect to the origin 0 of an earth-fixed coordinate system, and r_{MQ} denotes the vector from the rim center M to the static contact point Q. The absolute velocity of the contact point will be obtained from

$$v_{0Q,0} = \dot{r}_{0Q,0} = \dot{r}_{0M,0} + \dot{r}_{MQ,0} \tag{3.28}$$

where $\dot{r}_{0M,0} = v_{0M,0}$ names the absolute velocity of the rim center. The vector r_{MQ} contains the tire deflection Δz normal to the road. It changes its length and orientation in all those motions of the wheel carrier that do not contain elements of the wheel rotation. Hence, the time derivative of $r_{MQ,0}$ may be expressed as

$$\dot{r}_{MQ,0} = \omega^*_{0R,0} \times r_{MQ,0} + \Delta\dot{z}\, e_{n,0} \tag{3.29}$$

where e_n describes the effective road normal, $\Delta\dot{z}$ denotes the change of the tire deflection, and

$$\omega^*_{0R} = \omega_{0R} - e^T_{yR}\, \omega_{0R}\, e_{yR} \tag{3.30}$$

is the angular velocity of the wheel rim without any component in the direction of the wheel rotation axis e_{yR}. Now Equation (3.28) reads as

$$v_{0Q,0} = v_{0M,0} + \omega^*_{0R,0} \times r_{MQ,0} + \Delta\dot{z}\, e_{n,0} \tag{3.31}$$

As the point Q lies on the track, $v_{0Q,0}$ must not contain any component normal to the track,

$$e^T_{n,0}\, v_{0Q,0} = e^T_{n,0}\, v_{0M,0} + e^T_{n,0}\, \omega^*_{0R,0} \times r_{MQ,0} + \Delta\dot{z}\, e^T_{n,0}\, e_{n,0} = 0 \tag{3.32}$$

In addition, $e^T_{n,0}\, e_{n,0} = 1$ will hold because $e_{n,0}$ is a unit vector. Then, the time derivative of the vertical tire deformation is simply given by

$$\Delta\dot{z} = -e^T_{n,0} \left(v_{0M,0} + \omega^*_{0R,0} \times r_{MQ,0} \right) \tag{3.33}$$

Finally, the components of the contact point velocity in the longitudinal and the lateral directions are obtained from

$$v_x = e^T_{x,0}\, v_{0Q,0} = e^T_{x,0} \left(v_{0M,0} + \omega^*_{0R,0} \times r_{MQ,0} \right) \tag{3.34}$$

and

$$v_y = e^T_{y,0}\, v_{0Q,0} = e^T_{y,0} \left(v_{0M,0} + \omega^*_{0R,0} \times r_{MQ,0} \right) \tag{3.35}$$

where the relationships $e^T_{x,0} e_{n,0} = 0$ and $e^T_{y,0} e_{n,0} = 0$ were used to simplify the expressions. The contact point velocities v_x and v_y will characterize the sliding situation of a non-rotating tire.

3.2.5 Dynamic Rolling Radius

If a rigid disc of radius r_D performs a roll motion on a flat surface, then the constraint equation

$$v = r_D\, \Omega, \tag{3.36}$$

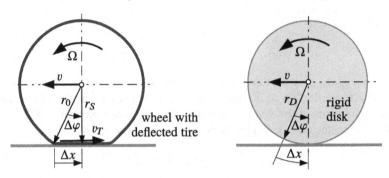

FIGURE 3.18
Dynamic rolling radius.

also known as rolling condition, will couple the velocity v of the disc center to the angular velocity Ω of the disc. On a rolling wheel, the tire deflection must be taken into account somehow. To do so, the roll motion of a wheel with r_0 and r_S as unloaded and loaded or static radius will be compared to a rolling disc of the fictitious or dynamic rolling radius r_D, Figure 3.18. With $v = dx/dt$ and $\Omega = d\varphi/dt$, the rolling condition (3.36) for the rigid disc results in

$$\frac{dx}{dt} = r_D \frac{d\varphi}{dt} \quad \text{or} \quad dx = r_D \, d\varphi \quad \text{or} \quad \Delta x = r_D \, \Delta\varphi \tag{3.37}$$

where Δx denotes the traveling distance if the rolling disc of radius r_D rotates with a specific angle $\Delta\varphi$. Supposing that the rotation angle $\Delta\varphi$ causes the wheel of the unloaded radius r_0 to perform a rolling motion where a tread particle is transported from the beginning of the contact patch to the center of the contact patch, then the traveling distance of the wheel will be given by

$$\Delta x = r_0 \sin \Delta\varphi \tag{3.38}$$

The comparison with a rolling disc on the basis of the same traveling distance delivers

$$r_D \, \Delta\varphi = r_0 \sin \Delta\varphi \tag{3.39}$$

Hence, the dynamic tire radius is determined by

$$r_D = r_0 \frac{\sin \Delta\varphi}{\Delta\varphi} \tag{3.40}$$

At small, yet finite angular rotations, the sine function can be approximated by the first terms of its Taylor expansion. Then, Equation (3.40) reads as

$$r_D = r_0 \frac{\Delta\varphi - \frac{1}{6}\Delta\varphi^3}{\Delta\varphi} = r_0 \left(1 - \frac{1}{6}\Delta\varphi^2\right) \tag{3.41}$$

Using the relationship

$$\cos \Delta\varphi = \frac{r_S}{r_0} \tag{3.42}$$

which follows by inspecting Figure 3.18 and expanding the cosine function in a Taylor series too, one gets

$$\cos \Delta\varphi = 1 - \frac{1}{2}\Delta\varphi^2 \quad \text{or} \quad \Delta\varphi^2 = 2\,(1 - \cos \Delta\varphi) = 2\left(1 - \frac{r_S}{r_0}\right) \qquad (3.43)$$

Then, Equation (3.41) finally reads as

$$r_D = r_0 \left(1 - \frac{1}{3}\left(1 - \frac{r_S}{r_0}\right)\right) = \frac{2}{3}r_0 + \frac{1}{3}r_S \qquad (3.44)$$

which will represent the dynamic rolling radius as a weighted mean value of the unloaded and the loaded or static radius of a tire.

Note: For a tire close to liftoff $r_S = r_0$ will hold. Then, Equation (3.44) provides a dynamic rolling radius of $r_D = r_0$ that corresponds to the trivial solution of Equation (3.40) when the sine function is simply approximated by its argument.

By means of the dynamic rolling radius, the roll motion of a wheel can be transferred to the rolling of a rigid disc. But, as a deflected tire is in contact with the road over the whole length of the contact patch, the rolling condition (3.36) must be modified to

$$v_T = r_D \, \Omega \qquad (3.45)$$

where v_T denotes the average velocity at which the tread particles are transported through the contact patch. As the loaded or static tire radius r_S depends on the wheel load F_z, the dynamic rolling radius r_D will automatically be a function of the wheel load too: $r_D = r_D(F_z)$.

3.3 Steady-State Forces and Torques

3.3.1 Wheel Load

The vertical tire force F_z can be calculated as a function of the tire deflection Δz and its time derivative $\Delta\dot{z}$

$$F_z = F_z(\Delta z, \Delta\dot{z}) \qquad (3.46)$$

Because the tire can only apply pressure forces to the road, the normal force will be restricted to $F_z \geq 0$. In a first approximation, F_z is separated into a static and a dynamic part,

$$F_z = F_z^{st} + F_z^{D} \qquad (3.47)$$

The static part is described as a nonlinear function of the tire deflection,

$$F_z^{st} = a_1 \, \Delta z + a_2 \, (\Delta z)^2 \qquad (3.48)$$

The constants a_1 and a_2 may be calculated from the radial stiffness at nominal and double payload,

$$c_{z1} = \left.\frac{d\,F_z^{st}}{d\,\Delta z}\right|_{F_z^{st}=F_z^N} \quad \text{and} \quad c_{z2} = \left.\frac{d\,F_z^{st}}{d\,\Delta z}\right|_{F_z^{st}=2F_z^N} \tag{3.49}$$

The derivative of Equation (3.48) results in

$$\frac{d\,F_z^{st}}{d\,\Delta z} = a_1 + 2\,a_2\Delta z \tag{3.50}$$

From Equation (3.48) one gets

$$\Delta z = \frac{-a_1 \pm \sqrt{a_1^2 + 4a_2 F_z^{st}}}{2a_2} \tag{3.51}$$

Because the tire deflection is always positive, the minus sign in front of the square root has no physical meaning and thus can be omitted. Hence, Equation (3.50) can be written as

$$\frac{d\,F_z^{st}}{d\,\Delta z} = a_1 + 2\,a_2\left(\frac{-a_1 + \sqrt{a_1^2 + 4a_2 F_z^{st}}}{2a_2}\right) = \sqrt{a_1^2 + 4a_2 F_z^{st}} \tag{3.52}$$

Combining Equations (3.49) and (3.52) results in

$$\begin{aligned} c_{z1} &= \sqrt{a_1^2 + 4a_2 F_z^N} \quad &\text{or} \quad a_1^2 + 4a_2 F_z^N &= c_{z1}^2, \\ c_{z2} &= \sqrt{a_1^2 + 4a_2 2F_z^N} \quad &\text{or} \quad a_1^2 + 8a_2 F_z^N &= c_{z2}^2 \end{aligned} \tag{3.53}$$

and finally leads to

$$a_1 = \sqrt{2\,c_{z1}^2 - c_{z2}^2} \quad \text{and} \quad a_2 = \frac{c_{z2}^2 - c_{z1}^2}{4\,F_z^N} \tag{3.54}$$

The parabolic approximation in Equation (3.48) of the static wheel load F_z^{st} as a function of the tire deflection Δz fits very well with measurements, Figure 3.19. The characteristic for the passenger car tire is nearly linear. The radial tire stiffness of $c_{z1} = 190$ N/mm at the payload of $F_z = 3\,200$ N slightly increases to the value of $c_{z2} = 206$ N/mm at double the payload.

The MATLAB-Script in Listing 3.1 performs a least squares approximation to a measured wheel load characteristic of a truck tire, calculates the stiffness at the given payload and its double, and plots the approximated characteristic as well as the given pairs of measured values.

Listing 3.1

Script app_03_fz_approx.m: nonlinear wheel load characteristic

```
1 % measured wheel load versus tire deflection
2 Fz = [ 0.0;  10.9;   21.5;   32.2;   46.1 ]; % wheel load [kN]
3 dz = [ 0.0;  15.5;   27.2;   36.4;   47.2 ]; % tire deflection [mm]
4 % least square approximation by parabola: Fz = a(1)*dz + a(2)*dz^2
5 A = [dz, dz.^2]; a = A\Fz;
6 % get deflections at payload FzN and double the payload 2*FzN
7 FzN = 35;  % set tire payload (truck tire)
8 dz_FzN   = ( -a(1) + sqrt(a(1)^2+4*a(2)*FzN)   ) / (2*a(2));
9 dz_2FzN = ( -a(1) + sqrt(a(1)^2+4*a(2)*2*FzN) ) / (2*a(2));
10 % calculate stiffness at FzN and 2*FzN
11 cz1 = ( a(1) + 2*a(2)*dz_FzN  ); disp(['cz1=',num2str(cz1),' kN/mm'])
12 cz2 = ( a(1) + 2*a(2)*dz_2FzN ); disp(['cz2=',num2str(cz2),' kN/mm'])
13 % plot approximated characteristic and compare to measurements
14 d_z = linspace(0,2*dz_FzN,201);  F_z = a(1)*d_z + a(2)*d_z.^2;
15 plot(d_z,F_z,'r','Linewidth',1), hold on, plot(dz,Fz,'ok'), grid on
16 xlabel('\Delta z / mm'), ylabel('F_z / kN')
```

The characteristic of the truck tire shows a significant progressive nonlinearity. Here, the radial stiffness at the payload of $F_z = 35$ kN is given by $c_{z1} = 1.25$ kN/mm and it rises to $c_{z2} = 1.68$ kN/mm at double the payload.

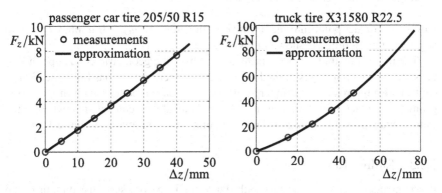

FIGURE 3.19

Static wheel load as a function of the tire deflection.

If no measurements are available at all, an appropriate wheel load characteristic can intuitively be modeled by estimating the stiffness c_{z1} at the payload $F_z = F_z^N$ and by setting the value for the stiffness c_{z2} at double the payload to the same or an increased value depending on the desired or estimated linear or progressive behavior.

Finally, the dynamic part of the wheel load is roughly approximated by

$$F_z^D = d_R \Delta\dot{z} \tag{3.55}$$

where d_R is a constant describing the radial tire damping, and the derivative of the tire deformation $\Delta\dot{z}$ is given by Equation (3.33). In reality the dynamic part is much more complex and will depend on the driving velocity v or the wheel rotation Ω respectively. However, the simple linear approach (3.55) is in general of sufficient accuracy in particular for passenger cars and trucks which are equipped with a suspension system that dissipates most of the vibration energy.

3.3.2 Tipping Torque

The lateral shift y_Q of the vertical tire force F_z from the geometric contact point P to the static contact point Q is equivalent to a force applied in P and the tipping torque

$$T_x = F_z \, y_Q \tag{3.56}$$

acting around a longitudinal axis in P, Figure 3.20. A positive camber angle $\gamma > 0$

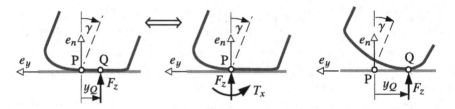

FIGURE 3.20
Tipping torque at full and partial contact.

moves the contact point into the negative y-direction. As $y_Q < 0$ implies $T_x < 0$, a negative tipping torque is shown in Figure 3.20.

The use of the tipping torque instead of shifting the geometric contact point P to the static contact point Q is limited to those cases where the tire has full or nearly full contact with the road. If a cambered tire has only partial contact with the road, the geometric contact point P may even be located outside the contact area, whereas the static contact point Q is still a real contact point, see right image in Figure 3.20.

3.3.3 Rolling Resistance

If a non-rotating tire has contact with flat ground, the pressure distribution in the contact patch will be symmetric from the front to the rear, Figure 3.21. The resulting

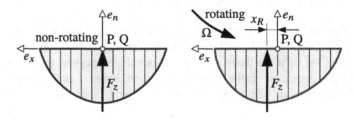

FIGURE 3.21
Pressure distribution at a non-rotating and rotating tire.

vertical force F_z is applied in the center of the contact patch that coincides in the xz-plane with the geometric and static contact points P and Q. Hence, it will generate no torque around the y-axis.

If the tire rotates, tread particles will be stuffed into the front of the contact patch, which causes a slight pressure increase, Figure 3.21. Now the resulting vertical force F_z is applied in front of the contact points P and Q and generates the rolling resistance torque

$$T_y = -\frac{r_D \, \Omega}{|r_D \, \Omega|} \, F_z \, x_R \qquad (3.57)$$

where the term $r_D \, \Omega / |r_D \, \Omega|$ ensures that T_y always acts against the sign of the transport velocity $v_T = r_D \, \Omega$ or the wheel angular velocity Ω respectively. A small modification

$$\frac{r_D \, \Omega}{r_D \, |\Omega|} \quad \longrightarrow \quad \frac{r_D \, \Omega}{|r_D \, \Omega| + v_N} \qquad (3.58)$$

avoids discontinuities at $\Omega = 0$, but requires a small fictitious velocity $v_N > 0$ to be chosen appropriately. The distance x_R from the contact points P and Q to the working point of F_z usually is related to the unloaded tire radius r_0,

$$f_R = \frac{x_R}{r_0} \qquad (3.59)$$

According to [26], the dimensionless rolling resistance coefficient slightly increases with the traveling velocity v of the vehicle

$$f_R = f_R(v) \qquad (3.60)$$

Under normal operating conditions, 20 km/h $< v <$ 200 km/h, the rolling resistance coefficient for typical passenger car tires is in the range of $0.01 < f_R < 0.02$. The rolling resistance hardly influences the handling properties of a vehicle, but it plays a major part in fuel consumption.

3.3.4 Longitudinal Force and Longitudinal Slip

To get a certain insight into the mechanism generating tire forces in the longitudinal direction, we consider a tire on a flatbed test rig. The rim rotates with angular velocity Ω and the flatbed runs with velocity v_x. The distance between the rim center and the flatbed is controlled to the loaded tire radius corresponding to the wheel load F_z, Figure 3.22.

A tread particle enters at the time $t = 0$ the contact patch. If we assume adhesion between the particle and the track, then the top of the particle will run with the bed velocity v_x and the bottom with the average transport velocity $v_T = r_D \, \Omega$. Depending on the velocity difference $\Delta v = r_D \, \Omega - v_x$ the tread particle is deflected in the longitudinal direction,

$$u = (r_D \, \Omega - v_x) \, t \qquad (3.61)$$

The time a particle spends in the contact patch is given by

$$T = \frac{L}{r_D \, |\Omega|} \qquad (3.62)$$

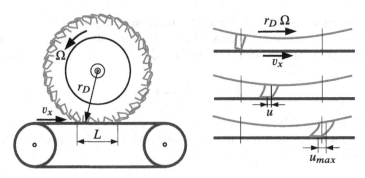

FIGURE 3.22
Tire on flatbed test rig.

where L denotes the length of the contact patch and $T > 0$ is assured by $|\Omega|$. Maximum deflection occurs when the tread particle leaves the contact patch at time $t = T$,

$$u_{max} = u(t = T) = (r_D \Omega - v_x)T = (r_D \Omega - v_x)\frac{L}{r_D |\Omega|} \tag{3.63}$$

The deflected tread particle applies a force to the tire. In a first approximation, we get

$$F_x^t = c_x^t u \tag{3.64}$$

where c_x^t represents the stiffness of one tread particle in the longitudinal direction. On normal wheel loads, more than one tread particle will be in contact with the track, Figure 3.23a. The number p of tread particles can be estimated by

$$p = \frac{L}{s + a} \tag{3.65}$$

where s is the length of one particle and a denotes the distance between the particles. Particles entering the contact patch are undeformed, whereas the ones leaving

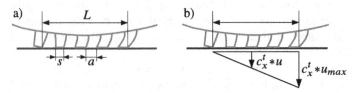

FIGURE 3.23
a) Tread particles, b) force distribution.

have the maximum deflection. According to Equation (3.64), this results in a linear force distribution versus the contact length, Figure 3.23b. The resulting force in the longitudinal direction for p particles is then given by

$$F_x = \frac{1}{2} p c_x^t u_{max} \tag{3.66}$$

Inserting Equations (3.65) and (3.63) results in

$$F_x = \frac{1}{2} \frac{L}{s+a} c_x^t (r_D \Omega - v_x) \frac{L}{r_D |\Omega|} \qquad (3.67)$$

The contact length L was previously calculated in Equation (3.24). Approximating the belt deformation by $\Delta z_B \approx \frac{1}{2} F_z / c_z$ results in

$$L^2 \approx 4 r_0 \frac{F_z}{c_z} \qquad (3.68)$$

where c_z denotes the vertical tire stiffness, and nonlinearities and dynamic parts in the tire deformation were neglected. Now Equation (3.66) can be written as

$$F_x = 2 \frac{r_0}{s+a} \frac{c_x^t}{c_z} F_z \frac{r_D \Omega - v_x}{r_D |\Omega|} \qquad (3.69)$$

The nondimensional relation between the sliding velocity of the tread particles in the longitudinal direction $v_x^S = v_x - r_D \Omega$ and the average transport velocity $r_D |\Omega|$ form the longitudinal slip,

$$s_x = \frac{-(v_x - r_D \Omega)}{r_D |\Omega|} \qquad (3.70)$$

This slip is not simply defined in order to achieve an appropriate dimensionless quantity, it is, moreover, the plain results from a simple physical approach.

If the tire properties r_0, s, a, c_x^t, and c_z are summarized in the constant k, Equation (3.69) will simplify to

$$F_x = k F_z s_x \qquad (3.71)$$

Hence, the longitudinal force F_x will be proportional to the wheel load F_z and to the longitudinal slip s_x in this first approximation.

But, Equation (3.71) will hold only as long as all tread particles stick to the track. At moderate slip values, the particles at the end of the contact patch start sliding; and at high slip values, only the parts at the beginning of the contact patch will still stick to the road, Figure 3.24. The resulting longitudinal force characteristics as typical for

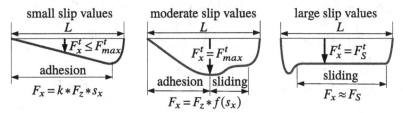

FIGURE 3.24
Longitudinal force distribution for different slip values.

standard tires is plotted in Figure 3.25. The nonlinear function of the longitudinal force F_x versus the longitudinal slip s_x can be defined by only five parameters: the initial inclination (driving stiffness) dF_x^0, the location s_x^M, and the magnitude of the maximum F_x^M, the start of full sliding s_x^S, and the magnitude of the sliding force F_x^S.

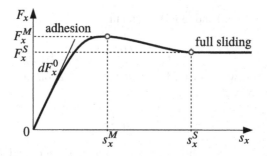

FIGURE 3.25
Typical longitudinal force characteristic.

3.3.5   Lateral Slip, Lateral Force, and Self-Aligning Torque

The wheel rotation with the angular velocity Ω affects the longitudinal slip only. Hence, similar to Equation (3.70), the lateral slip s_y and simultaneously the slip angle α_S will be given by

$$\tan \alpha_S \;=\; s_y \;=\; \frac{-v_y}{r_D\,|\Omega|} \tag{3.72}$$

where the lateral component of the contact point velocity v_y is defined in Equation (3.35). As long as the tread particles stick to the road (small amounts of slip), an almost linear distribution of the forces along the length L of the contact patch will appear. At moderate slip values, the particles at the end of the contact patch start sliding; and at high slip values, only the parts at the beginning of the contact patch stick to the road, Figure 3.26. The nonlinear characteristic of the lateral force F_y versus the

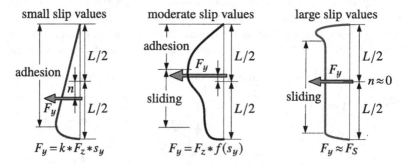

FIGURE 3.26
Lateral force distribution over contact patch.

lateral slip s_y can be described by the initial inclination (cornering stiffness) dF_y^0, the location s_y^M and the magnitude F_y^M of the maximum, the beginning of full sliding s_y^S, and the magnitude F_y^S of the sliding force. The distribution of the lateral forces over the contact patch length also defines the point of application of the resulting

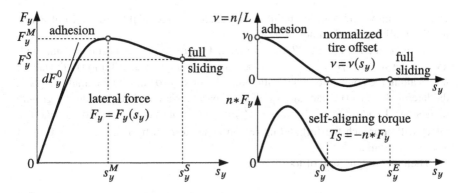

FIGURE 3.27
Typical lateral force characteristic, normalized tire offset, and self-aligning torque.

lateral force. At small slip values, this point lies behind the center of the contact patch (contact points P and Q). With increasing slip values, it moves forward, sometimes even before the center of the contact patch. At extreme slip values, when practically all particles are sliding, the resulting force is applied at the center of the contact patch. The resulting lateral force F_y with the dynamic tire offset or pneumatic trail n as a lever generates the self-aligning torque,

$$T_S = -n\,F_y \qquad (3.73)$$

The lateral force as well as the dynamic tire offset are functions of the lateral slip. Typical plots of the lateral force $F_y = F_y(s_y)$, the normalized dynamic tire offset $v = v(s_y) = n(s_y)/L$, and the self-aligning torque $T_S = T_S(s_y)$ are shown in Figure 3.27.

The dynamic tire offset n has been normalized by the length of the contact patch L for practical reasons. The normalized dynamic tire offset $v = n/L$ starts at $s_y = 0$ with a positive value $v_0 > 0$ and it tends to zero ($v \to 0$) at large slip values ($s_y \geq s_y^E$). Sometimes the normalized dynamic tire offset overshoots to negative values before it reaches zero again. This behavior can be modeled by introducing the slip value s_y^0 as an additional model parameter, where the normalized dynamic tire offset v passes the s_y-axis and approaches zero again at $s = s_y^E$.

At small values of the lateral slip $s_y \approx 0$, one gets, in a first approximation, a triangular distribution of the lateral forces over the contact patch length (cf. Figure 3.26). The working point of the resulting force is then given by

$$n_0 = n(s_y \approx 0) = \frac{1}{6}L \qquad (3.74)$$

Because the triangular force distribution will take for granted a constant pressure in the contact patch, the value $v_0 = n_0/L = 1/6 = 0.167$ can serve as a first approximation only. In reality, the pressure will drop to zero in the front and in the rear of the contact patch, Figure 3.21. Because low pressure means low friction forces, the triangular force distribution will be rounded to zero in particular in the rear of the contact patch, which will move the working point of the resulting force slightly to the front which implies realistic initial values of $v_0 < \frac{1}{6}$.

As indicated in Figure 3.27, the normalized tire offset v will most likely start with a vanishing slope $dv/ds_y(s_y = 0) = 0$ because the triangular distribution of the lateral forces along the contact length is valid not only for $s_y = 0$ but also for small values of the lateral slip $s_y \approx 0$ where almost all tread particles are still sticking to the ground (adhesion). In addition, $dv/ds_y(s_y = s_y^E) = 0$ will hold because the normalized tire offset will approach the full sliding region characterized by $v(s_y > s_y^E) = 0$ in an asymptotic manner.

Then, the polynomial of fourth order

$$v(s_y) = v_0 \left(1 + a\, s_y^2 + b\, s_y^3 + c\, s_y^4 \right) \tag{3.75}$$

can be used to model the normalized tire offset as a function of the lateral slip. It provides the initial values $v(s_y = 0) = v_0$ and $dv/ds_y(s_y = 0) = 0$ automatically. The demands

$$v(s_y^0) = 0; \qquad v(s_y^E) = 0; \qquad \left.\frac{dv}{ds_y}\right|_{s_y = s_y^E} = 0 \tag{3.76}$$

make it possible to adjust the coefficients a, b, and c to the parameter s_y^0 and s_y^E. The function provided in Listing 3.2 computes the normalized tire offset as a function of the lateral slip according to Equations (3.75) and (3.76).

Listing 3.2
Function `tmy_tireoff`: normalized tire offset

```
1  function n2L=tmy_tireoff ... % normalized tire offset
2  ( sy ...          % lateral slip
3  , n2L0 ...        % normalized caster offset n/L @ sy=0
4  , sy0 ...         % lateral slip where n/L passes sy-axis
5  , syE )           % lateral slip where full sliding starts
6
7    sya  = abs(sy); % absolute value because n(sy) = n(-sy)
8
9    if sya < syE                    % 4th-order polynomial 0<=sy<syE
10       a  = -( 2*sy0^2 + (sy0+syE)^2 ); % coefficients
11       b  =  ( 2*(sy0+syE)^2 ) / syE;   % normalized to
12       c  = -( 2*sy0 + syE ) / syE;     % (sy0*syE)^2
13       n2L = n2L0*( 1 + (sya/(sy0*syE))^2 * ( a + sya*( b + sya*c ) ) );
14    else
15       n2L = 0;                     % full sliding
16    end
17
18 end
```

At least the value of `n2L0` representing the initial value of the normalized tire offset $v_0 = n_0/L$ can be estimated quite well as explained above. If no measurements are available, the slip values s_y^0 and s_y^E where the tire offset passes and finally approaches the x-axis again must be estimated. Usually the value for s_y^0 is somewhat higher than the slip value s_y^M where the lateral force reaches its maximum.

3.4 Combined Forces

3.4.1 Combined Slip and Combined Force Characteristic

The longitudinal force as a function of the longitudinal slip $F_x = F_x(s_x)$ and the lateral force depending on the lateral slip $F_y = F_y(s_y)$ can be defined by their characteristic parameter: initial inclination dF_x^0, dF_y^0, location s_x^M, s_y^M and magnitude of the maximum F_x^M, F_y^M as well as sliding limit s_x^S, s_y^S, and sliding force F_x^S, F_y^S, Figure 3.28. During general driving situations, e.g., acceleration or deceleration in

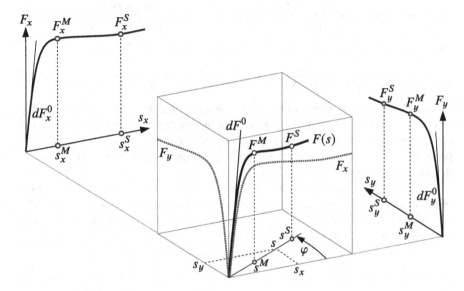

FIGURE 3.28
Combined tire forces.

curves, longitudinal s_x and lateral slip s_y appear simultaneously. The combination of the more or less differing longitudinal and lateral tire forces requires a normalization process, cf. [24], [34].

The longitudinal slip s_x and the lateral slip s_y can vectorially be added to the combined slip

$$s = \sqrt{\left(\frac{s_x}{\hat{s}_x}\right)^2 + \left(\frac{s_y}{\hat{s}_y}\right)^2} = \sqrt{\left(s_x^N\right)^2 + \left(s_y^N\right)^2} \tag{3.77}$$

where the slips were normalized, $s_x \to s_x^N$ and $s_y \to s_y^N$, in order to achieve a nearly equally weighted contribution to the combined slip. The normalizing factors

$$\hat{s}_x = \frac{s_x^M}{s_x^M + s_y^M} + \frac{F_x^M/dF_x^0}{F_x^M/dF_x^0 + F_y^M/dF_y^0} \tag{3.78}$$

and

$$\hat{s}_y \;=\; \frac{s_y^M}{s_x^M + s_y^M} \;+\; \frac{F_y^M / dF_y^0}{F_x^M / dF_x^0 + F_y^M / dF_y^0} \tag{3.79}$$

take characteristic properties of the longitudinal and lateral tire force characteristics into account. If the longitudinal and lateral tire characteristics do not differ too much, the normalizing factors will be approximately equal to one.

If the wheel locks, the average transport velocity will vanish, $r_D |\Omega| = 0$. Hence, longitudinal, lateral, and combined slip will tend to infinity, $s \to \infty$. To avoid this problem, the normalized slips s_x^N and s_y^N are modified to

$$s_x^N \;=\; \frac{s_x}{\hat{s}_x} \;=\; \frac{-(v_x - r_D\,\Omega)}{r_D\,|\Omega|\,\hat{s}_x} \quad\Rightarrow\quad s_x^N \;=\; \frac{-(v_x - r_D\,\Omega)}{r_D\,|\Omega|\,\hat{s}_x + v_N} \tag{3.80}$$

and

$$s_y^N \;=\; \frac{s_y}{\hat{s}_y} \;=\; \frac{-v_y}{r_D\,|\Omega|\,\hat{s}_y} \quad\Rightarrow\quad s_y^N \;=\; \frac{-v_y}{r_D\,|\Omega|\,\hat{s}_y + v_N} \tag{3.81}$$

In normal driving situations, where $r_D |\Omega| \gg v_N$ holds, the difference between the original slips and the modified slips are hardly noticeable. However, a small and positive fictitious velocity $v_N > 0$ avoids the singularities at $r_D |\Omega| = 0$ and will produce in this particular case a combined slip that points exactly in the direction of the sliding velocity of a locked wheel.

Similar to the graphs of the longitudinal and lateral forces, the graph $F = F(s)$ of the combined tire force can be defined by the characteristic parameter dF^0, s^M, F^M, s^S, and F^S. These parameters are calculated from the corresponding values of the longitudinal and lateral force characteristics

$$dF^0 \;=\; \sqrt{\left(dF_x^0\,\hat{s}_x \cos\varphi\right)^2 + \left(dF_y^0\,\hat{s}_y \sin\varphi\right)^2} \tag{3.82}$$

$$s^M = \sqrt{\left(\frac{s_x^M}{\hat{s}_x}\cos\varphi\right)^2 + \left(\frac{s_y^M}{\hat{s}_y}\sin\varphi\right)^2}, \quad F^M = \sqrt{\left(F_x^M \cos\varphi\right)^2 + \left(F_y^M \sin\varphi\right)^2} \tag{3.83}$$

$$s^S = \sqrt{\left(\frac{s_x^S}{\hat{s}_x}\cos\varphi\right)^2 + \left(\frac{s_y^S}{\hat{s}_y}\sin\varphi\right)^2}, \quad F^S = \sqrt{\left(F_x^S \cos\varphi\right)^2 + \left(F_y^S \sin\varphi\right)^2} \tag{3.84}$$

where the slip normalization must also be considered at the initial inclination. The angular functions

$$\cos\varphi \;=\; \frac{s_x^N}{s} \quad \text{and} \quad \sin\varphi \;=\; \frac{s_y^N}{s} \tag{3.85}$$

grant a smooth transition from the characteristic curve of longitudinal to the curve of lateral forces in the range of $\varphi = 0$ to $\varphi = 90°$. The longitudinal and the lateral forces follow then from the corresponding projections in longitudinal

$$F_x \;=\; F \cos\varphi \;=\; F\,\frac{s_x^N}{s} \;=\; \frac{F}{s}\,s_x^N \;=\; f\,s_x^N \tag{3.86}$$

and lateral directions

$$F_y = F \sin \varphi = F \frac{s_y^N}{s} = \frac{F}{s} s_y^N = f s_y^N \qquad (3.87)$$

where $f = F/s$ describes the global derivative of the combined tire force characteristic. In addition, the normalized dynamic tire offset $n = n(s_y)$ described by Equation 3.75 and based on a pure lateral slip situation will be extended by

$$n(s) = n(s_y) \sin \varphi = n(s_y) \frac{s_y^N}{s} \qquad (3.88)$$

to a modified normalized tire offset $n = n(s)$, which, similar to the lateral force, is also affected by the combined slip.

3.4.2 Suitable Approximation and Results

The combined tire force characteristic $F = F(s)$ is now approximated in intervals by appropriate functions, Figure 3.29. In the first interval the rational fraction

$$F(s) = \frac{s}{1 + \frac{s}{s^M} \left(\frac{s}{s^M} + \frac{dF^0 s^M}{F^M} - 2 \right)} dF^0, \quad 0 \le s \le s^M \qquad (3.89)$$

is used, which is defined by the initial inclination dF^0 and the location s^M and the magnitude F^M of the maximum tire force. When fixing the parameter values, one just

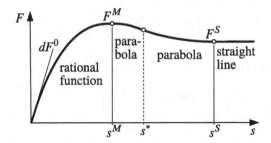

FIGURE 3.29
Approximation of the combined tire force characteristic.

has to make sure that the condition $dF^0 \ge 2 F^M / s^M$ is fulfilled, because otherwise the function will have a turning point in the interval of interest. It can be seen that the global derivative of the combined tire force $f = F/s$ is well defined at a vanishing slip and coincides in this particular case with the initial inclination of the combined tire force characteristic $f(s = 0) = dF^0$. In the interval $s^M \le s \le s^S$, the combined tire force characteristic is smoothly continued by two parabolas

$$F(s) = \begin{cases} F^M - a \left(s - s^M \right)^2, & s^M \le s \le s^* \\ F^S + b \left(s^S - s \right)^2, & s^* \le s \le s^S \end{cases} \qquad (3.90)$$

until it finally reaches the sliding area $s \geq s^S$, where the combined tire force is approximated by a straight line

$$F(s) = F^S \tag{3.91}$$

The curve parameters a, b, and s^* defining the two parabolas are determined by the demands

$$\frac{d^2 F}{d s^2}\bigg|_{s \to s^M} = \frac{d^2 F}{d s^2}\bigg|_{s^M \leftarrow s} \tag{3.92}$$

$$F(s \to s^*) = F(s^* \leftarrow s) \quad \text{and} \quad \frac{d F}{d s}\bigg|_{s \to s^*} = \frac{d F}{d s}\bigg|_{s^* \leftarrow s} \tag{3.93}$$

To calculate the second derivative of the rational function at $s = s^M$, the first derivative is needed first. Abbreviating the denominator by

$$D = 1 + \frac{s}{s^M}\left(\frac{s}{s^M} + \frac{dF^0\, s^M}{F^M} - 2\right) \tag{3.94}$$

one gets

$$\frac{d F}{d s} = \frac{D - s\left(\dfrac{1}{s^M}\left(\dfrac{s}{s^M} + \dfrac{dF^0\, s^M}{F^M} - 2\right) + \dfrac{s}{s^M}\dfrac{1}{s^M}\right)}{D^2}\, dF^0 \tag{3.95}$$

which can be simplified to

$$\frac{d F}{d s} = \frac{1 - \left(s/s^M\right)^2}{D^2}\, dF^0 \tag{3.96}$$

A further derivative yields

$$\frac{d^2 F}{d s^2} = \frac{D^2\left(-2 s/s^M\, 1/s^M\right) - \left(1 - \left(s/s^M\right)^2\right) 2D \dfrac{d D}{d s}}{D^4}\, dF^0 \tag{3.97}$$

At $s = s^M$, the abbreviation D simplifies to

$$D\left(s = s^M\right) = D_M = 1 + \frac{s^M}{s^M}\left(\frac{s^M}{s^M} + \frac{dF^0\, s^M}{F^M} - 2\right) = \frac{dF^0\, s^M}{F^M} \tag{3.98}$$

and Equation (3.97) finally results in

$$\frac{d^2 F}{d s^2}\bigg|_{s \to s^M} = dF^0\, \frac{-2/s^M}{D_M^2} = -2\, \frac{dF^0}{s^M}\left(\frac{F^M}{dF^0\, s^M}\right)^2 \tag{3.99}$$

The second derivative of the first parabola defined in Equation (3.90) just yields the value $-2a$. Hence, the parameter

$$a = \frac{dF^0}{s^M}\left(\frac{F^M}{dF^0\, s^M}\right)^2 \tag{3.100}$$

will grant a smooth transition from the rational function to the first parabola. Now the parameter s^* and b can be calculated. The first demand in Equation (3.93) on the parabolas defined by Equation (3.90) results in

$$F^M - a\left(s^* - s^M\right)^2 = F^S + b\left(s^S - s^*\right)^2 \tag{3.101}$$

and the second one yields

$$-2a\left(s^* - s^M\right) = 2b\left(s^S - s^*\right)(-1) \quad \text{or} \quad a\left(s^* - s^M\right) = b\left(s^S - s^*\right) \tag{3.102}$$

After multiplication with the term $s^S - s^*$, it can be inserted in Equation (3.101), which then will read

$$F^M - a\left(s^* - s^M\right)^2 = F^S + a\left(s^* - s^M\right)\left(s^S - s^*\right) \tag{3.103}$$

Rearranging some terms, we get

$$F^M - F^S = a\left(s^* - s^M\right)\left[\left(s^S - s^*\right) + \left(s^* - s^M\right)\right] \tag{3.104}$$

which finally provides the location of the point where the parabolas are connected to each other

$$s^* = s^M + \frac{F^M - F^S}{a\left(s^S - s^M\right)} \tag{3.105}$$

and in addition the parameter

$$b = \frac{s^* - s^M}{s^S - s^*}\, a \tag{3.106}$$

as a result from Equation (3.102). Note that for realistic tire characteristics, the connecting point s^* will be located in the transition interval $s^M \le s^* \le s^S$ in general. In the very unrealistic case of $s^* > s^S$, a third-order polynomial instead of the two connected parabolas will realize an "emergency"-transition from the maximum to the sliding force, Listing 3.3.

Listing 3.3
Function `tmy_fcombined`: combined tire force

```
1 function [ f, fos ] = tmy_fcombined ... % comb. force and global deriv.
2 ( s ...    % generalized slip
3 , df0 ...  % initial inclination of gen. force char.
4 , fm ...   % maximum force value
5 , sm ...   % slip where f(sm) = fm
6 , fs ...   % sliding force value
7 , ss )     % slip where f(ss) = fs
8
9 % defaults (df0=0, s=0)
10   f = 0.0; fos = df0;
11
12 % adjust smloc and ssloc if df0 is too small
13   if df0 > 0.0
14     smloc = max([2.0*fm/df0, sm]);
15     ssloc = ss + (smloc-sm);
```

```
16   else
17     return;
18   end
19
20 % normal operating conditions
21   if s > 0.0 && smloc > 0.0
22     if  s > ssloc
23 %      full sliding
24       f = fs;
25       fos = f/s;
26     else
27       if s < smloc && fm > 0.0 % adhesion
28         p = df0*smloc/fm - 2.0;
29         sn = s/smloc;
30         fos = df0 / ( 1.0 + ( sn + p ) * sn );
31         f = fos * s ;
32       else % adhesion --> sliding (two parabolas or cubic function)
33         a = (fm/smloc)*(fm/smloc) / (df0*smloc);
34         sstar = smloc + (fm-fs)/(a*(ssloc-smloc));
35         if sstar <= ss
36           if s <= sstar
37             f = fm - a*(s-smloc)*(s-smloc); % sm < s < sstar
38           else
39             b = a*(sstar-smloc)/(ssloc-sstar);
40             f = fs + b*(ssloc-s)*(ssloc-s); % sstar < s < ss
41           end
42         else % one cubic function (just in case)
43           sn = (s-smloc)/(ssloc-smloc);
44           f = fm - (fm-fs) * sn*sn *(3.0-2.0*sn);
45         end
46         fos = f/s; % global derivative
47       end
48     end
49   end
50
51 end
```

The function `tmy_fcombined` introduces local values for the location of the maximum and the slip where full sliding starts. These local values `smloc` and `ssloc` are adjusted appropriately if the parameter happens to violate the demand $dF^0 \geq 2\, F^M / s^M$ that ovoids a turning point in the first section.

Within the TMeasy model approach, the one-dimensional tire characteristics $F_x = F_x(s_x)$ and $F_y = F_y(s_y)$ are automatically converted to two-dimensional characteristics.

The MATLAB-Script provided in Listing 3.4 defines for a typical passenger car tire the parameter for the longitudinal and lateral tire force as well as the dynamic tire offset. Then it computes the steady-state tire force characteristics as well as the self-aligning torque for different values of the longitudinal and lateral slips using the functions `tmy_tireoff` and `tmy_fcombined` provided in Listing 3.2 and Listing 3.3. Finally, the results are plotted for selected slip values.

Listing 3.4

MATLAB-Script `app_03_steady_state_tire_chars.m`: steady-state tire forces

```
1 % tire properties: unloaded radius, payload, vertical stiffness
2 r0 = 0.293; fz0 = 3500; cz=190000;
3
4 % tire characteristics in long. and lat. directions
5 dfx0 = 100000;   dfy0 =   80000;   % init slopes in N/-
```

```
 6 fxm  =    3900;   fym  =    3650;   % maximum forces in N
 7 sxm  =   0.110;   sym  =   0.160;   % sm where f(sm) = fm
 8 fxs  =    3600;   fys  =    3600;   % sliding forces in N
 9 sxs  =   0.400;   sys  =   0.500;   % ss where f(ss) = fs
10
11 % dynamic tire offset parameter
12 n0  = 0.180;  % normalized tire offset n/L @ sy = 0
13 sy0 = 0.190;  % sy where n/L passes sy-axis
14 syE = 0.350;  % sy where n/L approaches sy-axis again
15
16 % compute slip normalizing factors
17 hsxn = sxm/(sxm+sym) + (fxm/dfx0)/(fxm/dfx0+fym/dfy0);
18 hsyn = sym/(sxm+sym) + (fym/dfy0)/(fxm/dfx0+fym/dfy0);
19
20 nvar = 101;  % generate various slip values
21 sx = linspace(-1,1,nvar)*1.1*max(sxs);
22 sy = linspace(-1,1,nvar)*1.1*max(sys);
23 % pre-allocate fx, fy and ts to speed up loop
24 fx = zeros(length(sx),length(sy)); fy=fx; ts=fx;
25
26 % compute tire forces, tire offset, and self-align. torque
27 L = 2*sqrt(r0*fz0/cz); % contact length
28 for j = 1:nvar
29    to  = L*tmy_tireoff( sy(j), n0,sy0,syE );
30    syn = sy(j)/hsyn;
31    for i = 1:nvar
32 %    combined slip
33       sxn = sx(i)/hsxn;
34       sc = sqrt ( sxn^2 + syn^2 );
35       if sc > 0
36          cphi=sxn/sc; sphi=syn/sc;
37       else
38          cphi=sqrt(2)/2; sphi=sqrt(2)/2;
39       end
40 %    combined characteristic for normalized slip values
41       df0 = sqrt( (dfx0*hsxn*cphi)^2 + (dfy0*hsyn*sphi)^2 );
42       fm  = sqrt( (fxm*cphi)^2       + (fym*sphi)^2       );
43       sm  = sqrt( (sxm/hsxn*cphi)^2  + (sym/hsyn*sphi)^2  );
44       fs  = sqrt( (fxs*cphi)^2       + (fys*sphi)^2       );
45       ss  = sqrt( (sxs/hsxn*sphi)^2  + (sys/hsyn*sphi)^2  );
46 %    combined tire force
47       [ f, ~ ] = tmy_fcombined( sc, df0,fm,sm,fs,ss );
48 %    split into longitudinal and lateral forces
49       fx(i,j) = f*cphi; fy(i,j) = f*sphi;
50 %    compute self-aligning torque
51       ts(i,j) = -to*fy(i,j);
52    end
53 end
54
55 fx=fx/1000; fy=fy/1000;   % N --> kN
56 hsp1=subplot(2,2,1); is=1:2:nvar; % reduced samples
57 plot(fx(:,is),fy(:,is),'k','LineWidth',1), hold on
58 plot(fx(is,:).',fy(is,:).','k','LineWidth',1), grid on
59 xlabel('f_x/kN'), ylabel('f_y/kN'),   axis equal
60 hsp2=subplot(2,2,2); is=round((nvar+1)/2):10:nvar; % selection
61 plot(sy,ts(is,:).'), grid on, title('t_s=t_s(s_y) @ selected s_x')
62 xlabel('s_y'), ylabel('Nm'), legend(num2str(sx(is).'))
63 hsp3=subplot(2,2,3);
64 plot(sx,fx(:,is)), grid on, xlabel('s_x'), ylabel('kN')
65 title('f_x=f_x(s_x) @ selected s_y'), legend(num2str(sy(is).'))
66 hsp4=subplot(2,2,4);
67 plot(sy,fy(is,:).'), grid on, xlabel('s_y'), ylabel('kn')
68 title('f_y=f_y(s_y) @ selected s_x'), legend(num2str(sx(is).'))
```

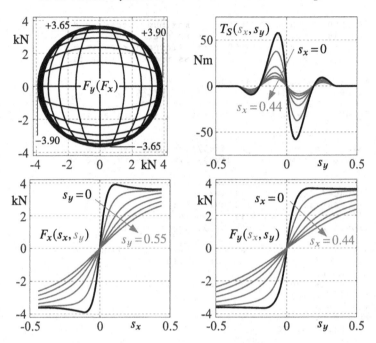

FIGURE 3.30
Combined forces, self-aligning torque, and tire force characteristics.

The combined force characteristic $F_y = F_y(F_x)$ in Figure 3.30 demonstrates the friction limits of the tire. In the early days of vehicle dynamics, the limit curve was assumed to form a circle. The engineer Wunibald Irmin Erich Kamm was one of the first who recognized the consequences of these combined force limits on the dynamics of vehicles. That is why, the limit curve is often addressed as Kamm's circle. However, typical for most passenger car and truck tires, the maximum longitudinal force is larger than the maximum lateral force. In this particular case $F_x^M = 3.90$ kN and $F_y^M = 3.65$ kN holds. Hence, the limit curve of the combined forces is more elliptical in shape. The reason is founded in the friction law between tire and road which will react in a degressive manner to the increase of pressure. Longitudinal forces $F_x \neq 0$ cause a circumferential deflection of the tire, but they will have nearly no influence on the pressure distribution. However, the lateral tire deflection caused by lateral forces $F_y \neq 0$ will result in a pressure distribution that is downscaled on one side and magnified accordingly on the other, Figure 3.31. The change from the nearly constant pressure distribution to a trapezoidal shaped one will then reduce the maximum transmittable friction force in the lateral direction due to the degressive friction law. As there is no similar effect in longitudinal direction, $F_y^M < F_x^M$ will be the consequence.

The last row in Figure 3.30 shows the influence of the lateral slip s_y on the longitudinal force F_x and the influence of the longitudinal slip s_x on the lateral force

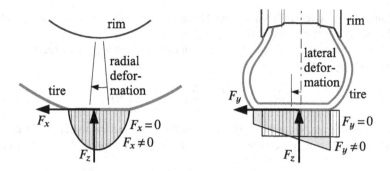

FIGURE 3.31
Effects of F_x and F_y on pressure distribution in the contact patch.

F_y. These mutual influences have a strong effect when a vehicle is accelerated or braked in a turn.

According to Equation (3.75) the normalized tire offset was modeled as a function of the lateral slip only. However, the self-aligning torque is generated by multiplying the dynamic tire offset with the lateral force. The latter depends on the longitudinal slip too. That is why the self-aligning torque $T_S = -n(s_y) F_y(s_x, s_y)$ depends in general and in particular in this model approach also on the longitudinal slip, upper right plot in Figure 3.30.

3.5 Bore Torque

The wheel performs an angular rotation about the axis defined by the unit vector e_{yR} and it is forced to participate on all rotations of the wheel carrier or knuckle that will not coincide with the wheel rotation axis. Hence, the angular velocity of the wheel may be split into two parts,

$$\omega_{0W} = \omega_{0R}^* + \Omega \, e_{yR} \tag{3.107}$$

where the wheel rotation itself is represented by the angular velocity Ω and the vector ω_{0R}^* describes the angular velocities of the knuckle without any parts in the direction of the wheel rotation axis. In particular during steering motions, the angular velocity of the wheel ω_{0W} has a component in the direction of the track normal e_n

$$\omega_n = e_n^T \, \omega_{0W} \neq 0 \tag{3.108}$$

which will cause a bore motion of the tire contact patch. If the wheel moves in the longitudinal and lateral direction too, then a very complicated deflection profile of the tread particles in the contact patch will occur. However, by a simple approach, the resulting bore torque can be approximated quite reasonably by the parameter of the combined tire force characteristic.

At first, the complex shape of a tire's contact patch is approximated by a circle, Figure 3.32. By setting

$$R_P = \frac{1}{2}\left(\frac{L}{2} + \frac{B}{2}\right) = \frac{1}{4}(L + B) \qquad (3.109)$$

the radius of the circle can be adjusted to the length L and the width B of the actual contact patch. During pure bore motions, circumferential forces F are generated at

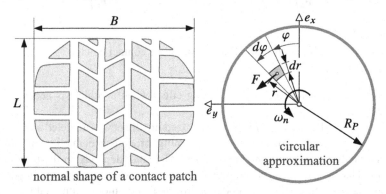

normal shape of a contact patch

FIGURE 3.32
Bore torque approximation.

each patch element dA at the radius r. The integration over the contact patch A,

$$T_B = \frac{1}{A}\int_A F\, r\, dA \qquad (3.110)$$

will then produce the resulting bore torque. At large bore motions, all particles in the contact patch are sliding. Then, the circumferential force F will be equal to the sliding force F^S of the combined tire characteristic and will be constant. Then, Equation (3.110) simplifies to

$$T_B^{max} = \frac{1}{A}F^S\int_A r\, dA \qquad (3.111)$$

The areas of the circle and the infinitesimal element are given by $A = R_P^2\,\pi$ and $dA = r\, d\varphi\, dr$. Then, Equation (3.111) results in

$$T_B^{max} = \frac{1}{R_P^2\,\pi}F^S\int_0^{R_P}\int_0^{2\pi} r\, r\, d\varphi\, dr = \frac{2}{R_P^2}F^S\int_0^{R_P} r^2\, dr = \frac{2}{3}R_P\,F^S = R_B\,F^S \quad (3.112)$$

where

$$R_B = \frac{2}{3}R_P \qquad (3.113)$$

can be considered the equivalent bore radius of the contact patch.

Regarding the maximum bore torque, the circular patch can be substituted by a thin ring with the radius r_B. On pure bore motions, the sliding velocity in the ring is given by $R_B \, \omega_n$ and, similar to Equations (3.70) and (3.72), the corresponding bore slip will then be determined by

$$s_B = \frac{-R_B \, \omega_n}{r_D \, |\Omega| + v_N} \tag{3.114}$$

where a small fictitious velocity v_N added to the denominator avoids numerical problems, when $\Omega = 0$ holds at a locked wheel. Replacing the sliding force F^S by the slip depending force $F(s_B)$, Equation (3.112) will provide the bore torque as

$$T_B = R_B \, F(s_B) \tag{3.115}$$

where $F(s_B)$ is determined by the combined force characteristic. On large bore slips, $F(s_B) \rightarrow F^S$ holds and will limit the bore torque automatically to the maximum torque defined in Equation (3.112).

3.6 Generalized or Three-Dimensional Slip

Even on steering maneuvers at standstill, a longitudinal s_x, a lateral s_y, and a bore slip s_B will occur simultaneously, Figure 3.33. By extending the combined slip s defined

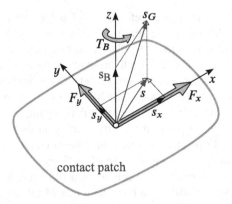

FIGURE 3.33
Generalized or three-dimensional slip.

in Equation (3.77) with the bore slip s_B to the generalized slip

$$s_G = \sqrt{s^2 + s_B^2} = \sqrt{s_x^2 + s_y^2 + s_B^2} \tag{3.116}$$

the effects of the bore motion on the combined tire forces and vice versa can be taken into account. Similar to the procedure described in Section 3.4, where the combined force was decomposed in the longitudinal and the lateral force, the generalized force characteristic $F_G = F(s_G)$ will now provide, by

$$F = F_G \frac{s}{s_G} = f_G s \quad \text{and} \quad T_B = R_B F_G \frac{s_B}{s_G} = f_G R_B s_B \quad (3.117)$$

the combined force F and the bore torque T_B in the corresponding parts of F_G or its global derivative $f_G = F_G/s_G$ respectively. According to (3.86) and (3.87)

$$F_x = F \frac{s_x^N}{s} = f_G s \frac{s_x^N}{s} = f_G s_x^N \quad \text{and} \quad F_y = F \frac{s_y^N}{s} = f_G s \frac{s_y^N}{s} = f_G s_y^N \quad (3.118)$$

provide the forces in the longitudinal and lateral directions.

This simple steady-state bore torque model will serve as a rough approximation only. In particular, it is less accurate at slow bore motions ($s_B \approx 0$) that will occur at parking maneuvers. However, a straightforward extension to a dynamic bore torque model will generate more realistic parking torques later on.

3.7   Different Influences on Tire Forces and Torques

3.7.1   Wheel Load

The resistance of a real tire against deformations has the effect that with increasing wheel load, the distribution of pressure over the contact patch becomes more and more uneven. The tread particles are deflected just as they are transported through the contact patch. The pressure peak in the front of the contact patch cannot be used, for these tread particles are far away from the adhesion limit because of their small deflection. In the rear part of the contact patch, the pressure drop leads to a reduction of the maximally transmittable friction force. With rising imperfection of the pressure distribution over the contact patch, the ability to transmit forces of friction between tire and road lessens. In practice, the tire characteristics are not just scaled by the wheel load as indicated in the simple approach in Sections 3.3.5 and 3.3.4 but they will also depend on it in a much more complicated way. In order to respect this fact in a tire model, the characteristic data for at least two wheel loads must be specified. Within TMeasy the payload F_z^N and its double $2 F_z^N$ are used for this purpose.

The influence of the wheel load F_z on the tire force characteristics $F_x(s_x)$ and $F_y(s_y)$ is then described by the data set given in Table 3.2. Assume that at vanishing wheel loads, no tire forces can be transmitted,

$$dF_x^0(F_z=0) = 0, \quad F_x^M(F_z=0) = 0, \quad F_x^S(F_z=0) = 0 \quad (3.119)$$

and

$$dF_y^0(F_z=0) = 0, \quad F_y^M(F_z=0) = 0, \quad F_y^S(F_z=0) = 0 \quad (3.120)$$

TABLE 3.2

Characteristic tire parameter with degressive wheel load influence

Longitudinal Force F_x		Lateral Force F_y	
$F_z = 4.0\,kN$	$F_z = 8.0\,kN$	$F_z = 4.0\,kN$	$F_z = 8.0\,kN$
$dF_x^0 = 120\,kN$	$dF_x^0 = 200\,kN$	$dF_y^0 = 55\,kN$	$dF_y^0 = 80\,kN$
$s_x^M = 0.110$	$s_x^M = 0.100$	$s_y^M = 0.200$	$s_y^M = 0.220$
$F_x^M = 4.40\,kN$	$F_x^M = 8.70\,kN$	$F_y^M = 4.20\,kN$	$F_y^M = 7.50\,kN$
$s_x^S = 0.500$	$s_x^S = 0.800$	$s_y^S = 0.800$	$s_y^S = 1.000$
$F_x^S = 4.25\,kN$	$F_x^S = 7.60\,kN$	$F_y^S = 4.15\,kN$	$F_y^S = 7.40\,kN$

will hold. As the corresponding values at $F_z = F_z^N$ and $F_z = 2F_z^N$ are provided in Table 3.2, the initial inclinations, the maximal forces, and the sliding forces for arbitrary wheel loads F_z may be inter- or extrapolated by quadratic functions

$$Y(F_z) = \frac{F_z}{F_z^N}\left[2Y(F_z^N) - \tfrac{1}{2}Y(2F_z^N) - \left(Y(F_z^N) - \tfrac{1}{2}Y(2F_z^N) \right)\frac{F_z}{F_z^N} \right] \quad (3.121)$$

where Y stands for dF_x^0, dF_y^0, F_x^M, F_y^M, F_x^S and F_y^S. However, the location of the maxima s_x^M, s_y^M and the slip values s_x^S, s_y^S at which full sliding will appear cannot be specified at vanishing wheel loads automatically. The corresponding data in Table 3.2 make only a linear inter- or extrapolation possible. Using X as a placeholder for s_x^M, s_y^M, s_x^S, and s_y^S, we obtain then

$$X(F_z) = X(F_z^N) + \left(X(2F_z^N) - X(F_z^N)\right)\left(\frac{F_z}{F_z^N} - 1\right) \quad (3.122)$$

The resulting tire characteristics at different wheel loads are plotted in Figure 3.34, where the relationship $\tan\alpha = s_y$, previously provided in Equation (3.72), was used to convert the lateral slip s_y into the slip angle α.

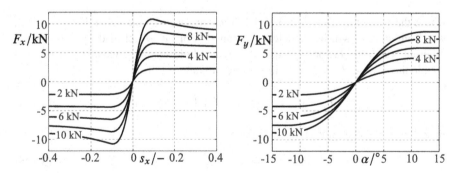

FIGURE 3.34

Tire characteristics at the wheel loads $F_z = [\,2,\,4,\,6,\,8,\,10\,]$ kN.

According to Equation (3.73), the self-aligning torque is modeled via the lateral force and the dynamic tire offset. The lateral force characteristics are specified in

Table 3.2. In addition, the characteristic curve parameter describing the dynamic tire offset will be provided for the single and double payload too. The resulting self-aligning torque as well as typical data are shown in Figure 3.35. Similar to

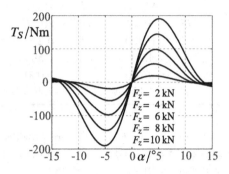

Tire offset parameter	
$F_z = 4.0\,kN$	$F_z = 8.0\,kN$
$(n/L)_0 = 0.178$	$(n/L)_0 = 0.190$
$s_y^0 = 0.200$	$s_y^0 = 0.225$
$s_y^E = 0.350$	$s_y^E = 0.375$

FIGURE 3.35
Self-aligning torque at different wheel loads F_z and tire offset parameter.

Equation (3.122), the parameters for arbitrary wheel loads were calculated by linear inter- or extrapolation. The degressive influence of the wheel load on the self-aligning torque can be seen here as well. With the parameter for the description of the tire offset, it has been assumed that at the payload $F_z = F_z^N$, the related tire offset reaches the value of $(n/L)_0 = 1/6 = 0.167$ at vanishing slip values $s_y = 0$. The slip value s_y^0, at which the tire offset passes the x-axis, has been estimated. Usually the value is somewhat higher than the position of the lateral force maximum. With increasing wheel load, it will move to higher values. The values for s_y^S are estimated too.

The dynamic rolling radius r_D is also affected by the wheel load. In extension to Equation (3.44), it will be approximated in the tire model TMeasy by

$$r_D = \lambda r_0 + (1 - \lambda) r_S \tag{3.123}$$

where the weighting factors $2/3$ and $1/3$ are replaced by the generalized factor λ, which in addition will be modeled as a function of the wheel load F_z. Introducing different weighting factors λ^N and λ^{2N}, which hold for the payload $F_z = F_z^N$ and double the payload $F_z = 2F_z^N$, a linear interpolation results in

$$\lambda = \lambda^N + \left(\lambda^{2N} - \lambda^N\right)\left(F_z/F_z^N - 1\right) \tag{3.124}$$

In addition, the static tire radius is replaced by

$$r_S = r_0 - \Delta r \approx r_0 - F_z^S/c_z \tag{3.125}$$

where the radial tire deformation Δr is approximated by the quotient of the static wheel load F_z^{st} and the vertical tire stiffness c_z. Then, the dynamic rolling radius can be modeled as a pure function of the static wheel load

$$r_D = \lambda(F_z^{st}) r_0 + \left(1 - \lambda(F_z^{st})\right)\left(r_0 - F_z^{st}/c_z\right) \tag{3.126}$$

where for the sake of simplicity the weighting factor λ determined in Equation (3.124) as a function of wheel load F_z is evaluated for the static wheel load F_z^{st} too. Depending on the values for λ^N and λ^{2N}, which according to Equation (3.124) will provide the load-dependent weighting factor λ, Equation (3.126) may produce a dynamic tire radius r_D that starts to decrease while the static wheel load F_z^{st} is further increasing. This nonphysical behavior can be avoided by keeping r_D constant when F_z^{st} will start to produce decreasing values for r_D.

This simple but effective model approach fits very well to measurements. As done in Figure 3.36, the deviation of the dynamic rolling radius r_D from the unloaded tire radius r_0 is plotted versus the wheel load F_z in general. For this typical passenger

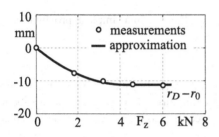

weighting factors
for dynamic rolling radius
at payload and its double

| $F_z = F_z^N$ | $\lambda^N = 0.375$ |
| $F_z = 2F_z^N$ | $\lambda^{2N} = 0.750$ |

FIGURE 3.36
Wheel load influence on the dynamic rolling radius.

car tire $F_z^N = 3.2$ kN, $r_0 = 0.315$ m, and $c_z = 190$ kN/m define the payload, the unloaded radius, and the vertical tire stiffness. In this particular case, the dynamic tire radius is kept constant at the value of $r_D(F_z^{st} = 4.2667$ kN$) = 0.304$ m even on larger static wheel loads. Note that the average of the weighting factors λ^N and λ^{2N} at the payload and its double $\overline{\lambda} = (0.375 + 0.750)/2 = 0.5625$ is rather close to the value of $\lambda = 2/3 = 0.6667$, which was the result of the simple model approach in Section 3.2.5.

3.7.2 Friction

The tire characteristics are valid for one specific tire-road combination only. Hence, different tire-road combinations will demand different sets of model parameters. A reduced or changed friction coefficient mainly influences the maximum force and the sliding force, whereas the initial inclination will remain unchanged. So, by setting

$$s^M \rightarrow \frac{\mu_L}{\mu_0} s^M, \quad F^M \rightarrow \frac{\mu_L}{\mu_0} F^M, \quad s^S \rightarrow \frac{\mu_L}{\mu_0} s^S, \quad F^S \rightarrow \frac{\mu_L}{\mu_0} F^S, \quad (3.127)$$

the essential tire model parameters, primarily dependent on the friction coefficient μ_0, are adjusted to the new or a local friction coefficient μ_L, where $\mu = \mu_L/\mu_0$ defines the scaling parameter. The result of this simple approach is shown in Figure 3.37.

If the road model will not only provide the road height z as a function of the coordinates x and y but also the local friction coefficient μ_L, then braking on μ-split maneuvers can easily be simulated [48].

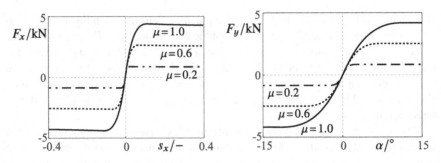

FIGURE 3.37
Force characteristics at different friction coefficients in a simple approximation.

3.7.3 Camber

At a cambered tire the angular velocity of the wheel Ω about the wheel rotation axis e_{yR} generates independently from any steer motion a bore angular velocity about the axis e_n normal to the local road plane. According to Figure 3.38 it is defined by

$$\Omega_n = \Omega \sin \gamma \qquad (3.128)$$

where γ denotes the camber angle.

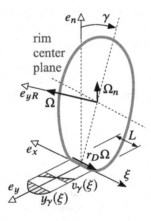

FIGURE 3.38
Velocity state of tread particles at cambered tire.

Now the tread particles in the contact patch have a lateral velocity that depends on their longitudinal position ξ within the contact patch and is provided by

$$v_\gamma(\xi) = -\Omega_n \frac{L}{2} \frac{\xi}{L/2} = -\Omega \sin \gamma\, \xi, \qquad -L/2 \leq \xi \leq L/2 \qquad (3.129)$$

At the contact point ($\xi = 0$) it vanishes, whereas at the end of the contact patch ($\xi = \frac{L}{2}$) it takes on the same value as at the beginning ($\xi = -\frac{L}{2}$), but pointing in the opposite direction.

Assuming that the tread particles stick to the track, the deflection profile will then be defined by

$$\dot{y}_\gamma(\xi) = v_\gamma(\xi) \tag{3.130}$$

The time derivative of the lateral deflection y can be transformed to a space derivative

$$\dot{y}_\gamma(\xi) = \frac{d\, y_\gamma(\xi)}{d\xi} \frac{d\xi}{dt} = \frac{d\, y_\gamma(\xi)}{d\xi} |r_D\,\Omega| \tag{3.131}$$

where the absolute value of the average transport velocity $r_D\,\Omega$ was used to make the result independent of the sign of the wheel rotation. Now Equation (3.130) can be written as

$$\frac{d\, y_\gamma(\xi)}{d\xi} r_D\,|\Omega| = -\Omega \sin\gamma\,\xi \quad \text{or} \quad \frac{d\, y_\gamma(\xi)}{d\xi} = -\frac{\Omega \sin\gamma}{r_D\,|\Omega|} \frac{L}{2} \frac{\xi}{L/2} \tag{3.132}$$

where the term $L/2$ was used to achieve dimensionless terms. Similar to the lateral slip s_y, which is defined by Equation (3.72), we can introduce a camber slip now

$$s_\gamma = \frac{-\Omega \sin\gamma}{r_D\,|\Omega|} \frac{L}{2} \tag{3.133}$$

Then, Equation (3.132) reads as

$$\frac{d\, y_\gamma(\xi)}{d\xi} = s_\gamma \frac{\xi}{L/2} \tag{3.134}$$

The shape of the lateral displacement profile is obtained by integration over the contact length

$$y_\gamma = s_\gamma \frac{1}{2} \frac{L}{2} \left(\frac{\xi}{L/2} \right)^2 + C \tag{3.135}$$

The boundary condition $y\left(\xi = \frac{1}{2}L\right) = 0$ can be used to determine the integration constant C. One gets

$$C = -s_\gamma \frac{1}{2} \frac{L}{2} \tag{3.136}$$

Then, Equation (3.135) finally results in

$$y_\gamma(\xi) = -s_\gamma \frac{1}{2} \frac{L}{2} \left[1 - \left(\frac{\xi}{L/2} \right)^2 \right] \tag{3.137}$$

The lateral displacements of the tread particles caused by a camber slip are compared now with the ones caused by pure lateral slip, Figure 3.39. At a tire with pure lateral slip, each tread particle in the contact patch possesses the same lateral velocity, which results in

$$dy_y/d\xi\, r_D\,|\Omega| = v_y, \tag{3.138}$$

a) camber slip

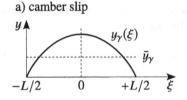

b) lateral slip

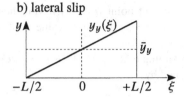

FIGURE 3.39
Displacement profiles of tread particles.

where according to Equation (3.131) the time derivative $\dot{y}_y$ was transformed to the space derivative $dy_y/d\xi$. Hence, the deflection profile is linear and reads as

$$y_y = v_y/(r_D |\Omega|)\xi = -s_y \xi, \qquad (3.139)$$

where the definition in Equation (3.72) was used to introduce the lateral slip s_y. Then, the average deflection of the tread particles under pure lateral slip will be given by

$$\bar{y}_y = -s_y \frac{L}{2}. \qquad (3.140)$$

The average deflection of the tread particles under pure camber slip is obtained from

$$\bar{y}_\gamma = -s_\gamma \frac{1}{2} \frac{L}{2} \frac{1}{L} \int_{-L/2}^{L/2} \left[1 - \left(\frac{x}{L/2}\right)^2\right] d\xi = -\frac{1}{3} s_\gamma \frac{L}{2}. \qquad (3.141)$$

Assuming that the hereby-generated lateral forces are proportional to the average deflections of the tread particles,

$$-s_y \frac{L}{2} \equiv -\frac{1}{3} s_\gamma \frac{L}{2} \quad \text{or} \quad s_y \equiv \frac{1}{3} s_\gamma \qquad (3.142)$$

will be the consequence. In normal driving conditions, the camber angle and thus the lateral camber slip are limited to small values, $s_y^\gamma \ll 1$. So, the lateral camber force may be modeled by

$$F_y^\gamma = dF_y^0 \frac{1}{3} s_\gamma. \qquad (3.143)$$

By replacing the initial inclination of the lateral tire force characteristic with the global derivative of the combined tire force characteristic,

$$dF_y^0 \longrightarrow \frac{F}{s} = f(s) \qquad (3.144)$$

the camber force F_y^γ will then be automatically reduced when the combined slip s is approaching the sliding area.

The angular velocity Ω_n defined in Equation (3.128) generates a bore slip and as a consequence a bore torque T_B. The resulting tire torque around an axis normal

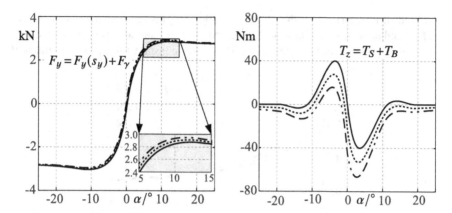

FIGURE 3.40
Camber influence on lateral force and torque around vertical axis: $\gamma = 0, 2, 4°$.

to the local road plane is then generated by the self-aligning and the bore torques, $T_z = T_S + T_B$. The resulting torque as well as the lateral force characteristic are plotted for a typical passenger car tire in Figure 3.40.

As the camber angle affects the pressure distribution in the contact patch and it changes the shape of the contact patch from rectangular to trapezoidal, it is extremely difficult, if not impossible, to quantify the camber influence with the aid of such a simple model approach. But, it turns out that the results are very realistic. By introducing a load-dependent weighting factor in Equation (3.143), the camber force can be adjusted to measurements.

3.8 First-Order Tire Dynamics

3.8.1 Simple Dynamic Extension

Measurements show that the dynamic reaction of tire forces and torques to disturbances can be approximated quite well by first-order systems [21]. Then, the dynamic tire forces F_x^D, F_y^D and the dynamic tire torque T_z^D are given by first-order differential equations

$$\tau_x \, \dot{F}_x^D + F_x^D = F_x^{st}, \quad \tau_y \, \dot{F}_y^D + F_y^D = F_y^{st}, \quad \tau_\psi \, \dot{T}_z^D + T_z^D = T_z^{st} \quad (3.145)$$

which are driven by the steady values F_x^{st}, F_y^{st} and T_z^{st}. The time constants τ_x, τ_y, τ_ψ can be derived from corresponding relaxation lengths r_x, r_y, r_ψ. Because the tread particles of a rolling tire move with transport velocity $r_D|\Omega|$ through the contact patch,

$$\tau_i = \frac{r_i}{r_D|\Omega|} \quad i = x, y, \psi \quad (3.146)$$

will hold. But, it turned out that these relaxation lengths are functions of the wheel load F_z and depend in addition on the longitudinal and lateral slip s_x, s_y or the side slip angle α respectively, Figure 3.41. Therefore, constant relaxation lengths will

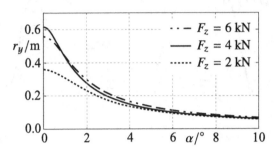

FIGURE 3.41
Measured lateral force relaxation length for a typical passenger car tire [21].

approximate the real tire behavior in zero-order approximation only. An appropriate model for the dynamic tire performance would be of great advantage because then the cumbersome task of deriving the relaxation lengths from measurements can be avoided.

Hence, the tire dynamics is significantly faster at larger than at smaller slip values because the relaxation decreases with increasing slip or side slip values. The tire test procedure proposed in [63] makes use of this specific tire feature by applying different sweep rates at smaller and larger slip angles when measuring steady-state tire forces in a highly efficient quasi-static approximation.

3.8.2   Enhanced Dynamics

The tire forces F_x and F_y as well as the torque T_z about the axis normal to the local road plane acting in the contact patch deflect the tire in the longitudinal, lateral and torsional directions, Figure 3.42. Due to these deflections, the sliding velocities as well as the bore velocity change to

$$v_x \longrightarrow v_x + \dot{x}_e, \quad v_y \longrightarrow v_y + \dot{y}_e, \quad \omega_n \longrightarrow \omega_n + \dot{\psi}_e \qquad (3.147)$$

where x_e, y_e, and ψ_e describe the tire deflections in the longitudinal, lateral, and torsional directions. As a consequence the normalized slips defined in (3.80) and (3.81) as well as the bore slip given by (3.114) will change to

$$s_x^N = \frac{-(v_x + \dot{x}_e - r_D\,\Omega)}{r_D\,|\Omega|\,\hat{s}_x + v_N} \quad \text{and} \quad s_y^N = \frac{-(v_y + \dot{y}_e)}{r_D\,|\Omega|\,\hat{s}_y + v_N} \qquad (3.148)$$

as well as

$$s_B = \frac{-R_B\left(\omega_n + \dot{\psi}_e\right)}{r_D\,|\Omega| + v_N} \qquad (3.149)$$

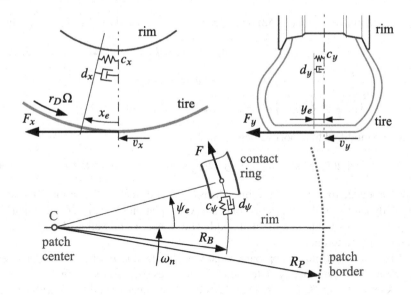

FIGURE 3.42
Tire deflections in the longitudinal, lateral, and torsional directions.

Then, in a first approximation (3.118) will deliver the dynamic tire forces in the longitudinal and lateral directions

$$F_x^D = f_G \frac{-(v_x + \dot{x}_e - r_D\,\Omega)}{r_D\,|\Omega|\,\hat{s}_x + v_N} \quad \text{and} \quad F_y^D = f_G \frac{-\left(v_y + \dot{y}_e\right)}{r_D\,|\Omega|\,\hat{s}_y + v_N} \qquad (3.150)$$

Finally, the second part of Equation 3.117 yields the dynamic bore torque as

$$T_B^D = f_G\,R_B\, \frac{-R_B\left(\omega_n + \dot{\psi}_e\right)}{r_D\,|\Omega| + v_N} \qquad (3.151)$$

In this first-order approach, the global derivative f_G of the generalized force characteristic $F_G = F_G(s_G)$ is computed via the steady-state slip values which keeps the model as simple as possible.

On the other hand, the tire compliance provides the dynamic forces in the longitudinal and lateral directions as well as the dynamic bore torque as

$$F_x^D = c_x\,x_e + d_x\,\dot{x}_e, \quad F_y^D = c_y\,y_e + d_y\,\dot{y}_e, \quad T_B^D = c_\psi\,\psi_e + d_\psi\,\dot{\psi}_e \qquad (3.152)$$

where c_x, c_y, c_ψ and d_x, d_y, d_ψ represent the corresponding stiffness and damping properties of the tire.

Combining Equation (3.152) with Equations (3.150) and (3.151) finally results in

$$\left(v_{Tx}^* d_x + f_G\right) \dot{x}_e = -v_{Tx}^* c_x x_e - f_G \left(v_x - r_D \Omega\right) \tag{3.153}$$

$$\left(v_{Ty}^* d_y + f_G\right) \dot{y}_e = -v_{Ty}^* c_y y_e - f_G v_y \tag{3.154}$$

$$\left(v_T^* d_\psi + R_B^2 f_G\right) \dot{\psi}_e = -v_T^* c_\psi \psi_e - R_B^2 f_G \omega_n \tag{3.155}$$

where the modified transport velocities

$$v_{Tx}^* = r_D |\Omega| \hat{s}_x + v_N, \quad v_{Ty}^* = r_D |\Omega| \hat{s}_y + v_N, \quad v_T^* = r_D |\Omega| + v_N \tag{3.156}$$

were introduced to shorten the terms. The first-order differential equations character-ize the dynamics of the tire deflections and via Equation (3.152) also the dynamics of the tire forces in the longitudinal and lateral directions as well as the dynamic of the bore or turn torque.

This first-order model approach is completely defined by the global derivative f_G of the generalized steady-state tire characteristics F_G, and the stiffness c_x, c_y, c_ψ and damping d_x, d_y, d_ψ properties of the tire. Via the steady-state tire characteristics, the dynamics of the tire deflections and hence the dynamics of the tire forces will automatically depend on the wheel load F_z and the longitudinal and lateral slip.

As described in Section 3.3.5, the self-aligning torque is computed as the product of the tire offset and the lateral force. By neglecting possible dynamics of the tire offset, the dynamic self-aligning torque can be approximated by

$$T_S^D = -n F_y^D \tag{3.157}$$

where n denotes the steady-state tire offset and F_y^D names the dynamic tire force. In this approach, the dynamics of the self-aligning torque is controlled by the dynamics of the lateral tire force only.

According to Equation (3.146), the relaxation length for the tire deflections and hence for the corresponding tire forces and the bore torque are now given by

$$r_{x,y,\psi} = r_D |\Omega| \tau_{x,y,\psi} \tag{3.158}$$

where the corresponding time constants

$$\tau_{x,y} = \frac{v_{Tx,y}^* d_{x,y} + f_G}{v_{Tx,y}^* c_{x,y}} = \frac{d_{x,y}}{c_{x,y}} + \frac{f_G}{v_{Tx,y}^* c_{x,y}} \tag{3.159}$$

and

$$\tau_\psi = \frac{v_T^* d_\psi + R_B^2 f_G}{v_T^* c_\psi} = \frac{d_\psi}{c_\psi} + \frac{R_B^2 f_G}{v_T^* c_\psi} \tag{3.160}$$

can easily be derived from Equations (3.153) to (3.155).

This simple model approach leads to relaxation lengths that are automatically adapted to the corresponding tire parameter. Figure 3.43 shows the results in lateral direction for a typical passenger car tire. A comparison with Figure 3.41 shows that

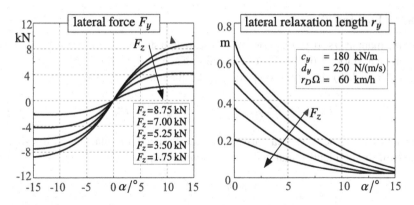

FIGURE 3.43
Lateral force characteristic and relaxation length at different wheel loads.

magnitude and the overall behavior of the lateral relaxation length are reproduced quite well. But of course a perfect matching cannot be expected. However, by introducing nonlinear characteristics for the corresponding tire stiffness properties or by adding appropriate weighting functions, a better fitting to measured relaxation lengths would be possible. By introducing a dynamic lateral stiffness modeled by a Maxwell element a further improvement is possible [15].

As shown in [21], the bore motion is characterized by a relaxation length r_ψ that depends less distinctly on the wheel load. Again, this model approach generates a quite realistic dynamic performance, Figure 3.44. As no further data are given in

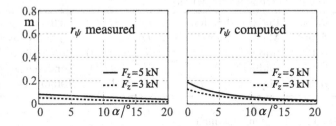

FIGURE 3.44
Measured [21] and computed relaxation length of the dynamic bore torque.

[21], estimated model parameters that represent typical passenger car tires were used to compute the relaxation length. The corresponding tire parameters that apply for a payload of $F_z^N = 4000$ N are given in Table 3.3. In addition, a transport velocity of $r_D|\Omega| = 40$ km/h was assumed.

3.8.3 Parking Torque

Parking maneuvers are often performed close to or in standstill situations. At standstill, the angular velocity of the wheel Ω is zero. Then, the modified transport velocity defined in (3.156) simplifies to $v_T^* = v_N$ and Equation (3.155) reads as

$$\left(v_N \, d_\psi + R_B^2 \, f_G \right) \dot{\psi}_e = -v_T \, c_\psi \, \psi_e - R_B^2 \, f_G \, \omega_n \qquad (3.161)$$

The fictitious velocity v_N was just introduced to avoid singularities in the slips at standstill. A sufficiently small value will certainly imply,

$$v_N \, d_\psi \ll R_B^2 \, f_G \qquad (3.162)$$

If the bore motion performed with a certain amount of the angular rotation of the wheel carrier $\omega_n \ne 0$ starts at a vanishing twist angle $\psi = 0$, then the differential equation (3.161) will merge into

$$R_B^2 \, f_G \, \dot{\psi} = -R_B^2 \, f_G \, \omega_n \quad \text{or} \quad \dot{\psi} = -\omega_n \qquad (3.163)$$

Hence, at the beginning of the bore motion, the torsional tire deflection represented by the twist angle ψ is just increased or decreased, depending on the magnitude and sign of the steering motion ω_n, which corresponds to a simple spring model.

    The simple dynamic torque model operates with parameters that are mostly derived from the steady-state tire properties. In particular, the parameters given in Table 3.3 are needed. To measure the parking effort, the wheel is rotated at standstill

TABLE 3.3
Dynamic bore torque model parameters

dF^0	= 90 000	Initial inclination	v_N	= 0.010	Fictitious velocity
F^M	= 4250	Maximum force	r_D	= 0.310	Dynamic tire radius
s^M	= 0.18	Slip where $F=F^M$	R_B	= 0.060	Bore radius
F^S	= 4100	Sliding force	c_ψ	= 2700	Torsional stiffness
s^S	= 0.50	Slip where $F=F^S$	d_ψ	= 1.8	Torsional damping

($\Omega = 0$) with a low frequent sine input

$$\psi_W = \psi_W^0 \sin 2\pi f_E t \quad \text{or} \quad \dot{\psi}_W = \omega_n = 2\pi \psi_W^0 \cos 2\pi f_E t \qquad (3.164)$$

around an axis perpendicular to the contact patch. The left plot in Figure 3.45 indicates an amplitude of $\psi_W^0 = 20°$ and an excitation frequency of $f_E = 0.1$ Hz. The torsional stiffness of the tire $c_\psi = 2700$ Nm/rad and the maximum achievable bore torque, which according to Equation (3.112) is given by $T_B^{max} = R_B F^S = 0.06 * 4100 = 246$ Nm limits the twist angle to a value of $|\psi| \le 246/2700 = 0.091$ rad $= 5.2°$. The simple dynamic torque model is able to generate periodic cycles that are quite similar to measurements. Only in the beginning the curves will differ because the simple bore torque model approximates the contact patch by one rigid ring that is coupled to the rim by a spring damper combination and while rotating is exposed to a generalized

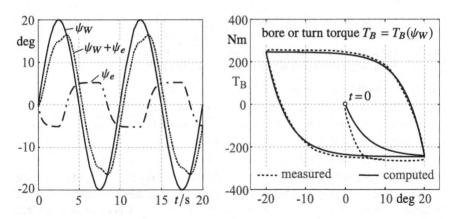

FIGURE 3.45
Measured [21] and computed parking torque at standstill.

friction force. In reality, all tread particles will stick to the road at first and then pass over to a sliding motion from the outer parts of the contact patch to the inner.

The tire model TMeasy can even handle real parking maneuvers where the vehicle is close to or in a standstill position and the steered wheels rotate not just about an axis normal to the local road plain but about the inclined kingpin axis. Despite such a complex sliding situation, where a longitudinal slip, a lateral slip, and the bore slip are present simultaneously, the performance of the TMeasy tire model is extremely smooth and the results are very practical [55].

Exercises

3.1 The position of the wheel rotation axis with respect to the earth-fixed system is defined by the unit vector $e_{yR,0} = [0.097; 0.995; -0.024]^T$ in a particular driving situation. Calculate the unit vectors $e_{x,0}$ and $e_{y,0}$ pointing in the direction of the longitudinal and the lateral tire force as well as the tire camber angle γ when $e_{n,0} = [0; 0; 1]^T$ defines the track normal.

3.2 A tire with an unloaded radius of $r_0 = 0.546\,m$ is exposed to a vertical force of $F_z = 35\,kN$. Calculate the vertical tire stiffness c_z and the contact length L if a loaded or static tire radius of $r_S = 0.510\,m$ is measured.

3.3 The following table lists the vertical tire deflection of a passenger car at different loads.

Δz mm	0	5	10	15	20	25	30	35	40
F_z kN	0	0.85	1.75	2.60	3.60	4.60	5.60	6.55	7.55

Plot the tire deflection versus the wheel load. Calculate the tire stiffness at the payload $F_z^N = 3.2\ kN$ and its double. Estimate the length of the contact patch at $F_z = F_z^N$ and the equivalent bore radius when the unloaded tire radius and the width of the tire are given by $r_0 = 293\ mm$ and $b = 205\ mm$, respectively.

3.4 Measurements at a payload of $F_z^N = 3.2\ kN$ result in

s_x [%]	0	2	4	6	8	10	14	18	22	26	30
F_x [kN]	0	2.00	2.95	3.25	3.30	3.35	3.35	3.32	3.30	3.28	3.26
α [deg]	0	2	4	6	8	10					
F_y [kN]	0	1.75	2.55	2.92	3.05	3.14					

Plot the tire characteristics $F_x(s_x)$ and $F_y(s_y)$ by converting the slip angle α into the corresponding lateral slip s_y. Deduce by simple inspection the characteristic data tire model data dF_x^0, F_x^M, s_x^M, F_x^S, s_x^S and dF_y^0, F_y^M, s_y^M, F_y^S, s_y^S. Generate the combined force characteristics using the MATLAB-Script given in Listing 3.4.

3.5 Estimate the tire characteristics in the longitudinal and lateral directions of a scooter tire defined by the code 140/60 – 13 at a payload of $F_z^N = 1500\ N$ that will characterize the tire on dry road. Assume that a scooter tire will perform somehow softer than a passenger car tire.

Generate the combined force characteristic by changing Listing 3.4 accordingly.

4

Drive Train

CONTENTS

4.1   Components and Concepts

4.1.1   Conventional Drive Train

The drive train serves two functions: it transmits power from the engine to the drive wheels, and it varies the amount of torque. The main parts of a drive train for conventional ground vehicles are the engine, clutch, transmission, differentials, shafts, brakes, and wheels, Figure 4.1.

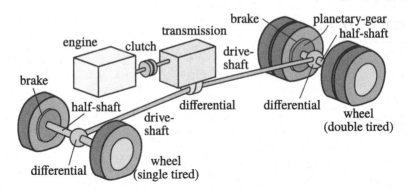

FIGURE 4.1
Components of a conventional drive train.

On heavy trucks, planetary gears are embedded into the wheels in order to reduce the amount of torque transmitted by the drive- and half-shafts. Most passenger cars have rear- or front-wheel drive. All-wheel drive is often used on upper-class cars and sport utility vehicles. Front-wheel drive is very common on light trucks. Advanced drive trains make use of electronically controlled differentials in order to transfer the driving torque properly to the axles and wheels.

Different kinds of driving concepts can be found on heavy trucks. Here, the notation $w \times d$, where d names the number of wheels in total and d the number of driven wheels, is usually used to specify the driving concept. Hence, 4×4 stands for all-wheel drive on a truck with two axles, and 8×4 names a truck with four axles (= eight wheels) in total, where two axles (= four wheels) are driven. Note that the number of tires and the number of wheels may be different because on solid axles usually on one wheel, two tires (double tires) are mounted.

4.1.2   Hybrid Drive

Nowadays hybrid driven cars have become very popular. The hybrid is a compromise. It attempts to significantly increase the mileage and reduce the emissions of a gasoline-powered car while overcoming the shortcomings of an electric car. At present, many hybrid models are already available. A hybrid drive train combines two or more power sources in general. Usually a conventional combustion engine is combined with an electric motor where the drive sources may operate in series or parallel, Figure 4.2.

In the serial arrangement, the vehicle is driven by the electric motor, which provides an adequate torque over a wide speed range and does not necessarily require a bulky transmission. As the combustion engine will drive the electric generator only, it can be run at a constant and efficient rate even as the vehicle changes speed. Series-hybrids can also be fitted with a capacitor or a flywheel to store regenerative braking energy.

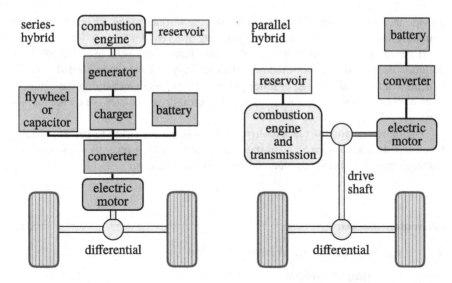

FIGURE 4.2
Typical layout of serial and parallel hybrid drives.

In the parallel layout, both drive sources can be used jointly or separately to accelerate the vehicle. The possibility for power addition permits relatively small dimensioning of the machines, without having to accept disadvantages in the driving performance. As only one electric motor is integrated, it can only be either generative or motive.

The powersplit hydro drive combines the serial and parallel hybrid in a strong hybrid drive. A transmission (planetary gear) splits the power of the internal combustion engine in one portion that is transmitted directly (mechanically) to the drive shaft, while the other portion is converted into electrical energy and optionally stored in the battery or consumed by the electric motor. Both the combustion engine and the electric motor can be used to drive the vehicle. Even a purely electric drive will be possible.

On "extended range" electric vehicles the batteries are charged from the grid and by a small internal combustion engine that powers a generator which powers the batteries.

Today plug-in hybrid electric vehicles have become popular. In this case the battery can be recharged not only by its on-board combustion engine and the generator, but also by plugging it into an external source of electric power.

4.1.3   Electric Drive

In an electric-drive vehicle, the torque is supplied to the wheels by one or more electric motors. Attaching electric motors to each driven wheel will require no drive line at all but will increase the unsprung mass of the wheel significantly. The possibility to provide individual and easily controllable driving torques at the wheels offers a lot options to improve the traction and the handling of a vehicle.

If the electric motor is powered solely by a battery, the vehicle will create less pollution than a gasoline-powered one. A large array of batteries is needed in order to achieve a reasonable range of miles per charge. Although recharging the batteries takes a relatively long time, more and more electric-drive vehicles are on the market now.

The onward march of electric-drive systems is not restricted to passenger cars. Electric motors at some or all semi-trailer axles are used only recently to increase the traction capacity of heavy commercial vehicle combinations.

4.2   Eigendynamics of Wheel and Tire

4.2.1   Equation of Motion

Besides the longitudinal tire force F_x that generates a torque around the wheel rotation axis e_{yR} via the static tire radius r_S and the rolling resistance torque T_y, the rotation of a wheel is influenced by the torques T_D and T_B, Figure 4.3. The driving torque T_D

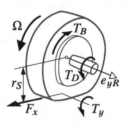

FIGURE 4.3
Wheel and tire.

is transmitted to the wheel by the half-shaft. It may accelerate or decelerate the wheel according to the torque generated in the combustion engine or an electric motor. The torque T_B acts between the wheel body and the wheel. It is generated by a wheel hub motor or by a brake and is just able to decelerate the wheel in the latter case.

The dynamics of the wheel rotation is governed by the angular momentum around the wheel rotation axis

$$\Theta \dot{\Omega} = T_D - T_B - r_S F_x + T_y \tag{4.1}$$

where Θ and Ω denote the inertia and the angular velocity of the wheel.

4.2.2 Steady-State Tire Forces

The dynamics of a wheel that is neither driven nor braked ($T_D = 0$ and $T_B = 0$) may be simplified to

$$\Theta \dot{\Omega} = -r_S F_x \qquad (4.2)$$

Here, the rolling resistance torque T_y was neglected, because in comparison to the term $r_S F_x$ it is small in general. Within handling tire models like TMeasy, the longitudinal tire force F_x is described as a function of the longitudinal slip s_x. For vanishing lateral slips, the normalization factor $\hat{s}_x$ in Equation (3.80) can be set to one. Then, the longitudinal slip is given by

$$s_x = \frac{-(v_x - r_D \,\Omega)}{r_D \,|\Omega| + v_N} \qquad (4.3)$$

where v_x denotes the longitudinal component of the contact point velocity and the small but positive fictitious velocity $v_N > 0$ avoids numerical problems when the wheel will be locked, $\Omega = 0$. Now, the angular velocity of the wheel is approximated by

$$\Omega = \frac{v_x}{r_D} + \Delta\Omega \quad \text{or} \quad r_D \,\Omega = v_x + r_D \,\Delta\Omega \qquad (4.4)$$

where in addition

$$|r_D \Delta\Omega| \ll |v_x| \qquad (4.5)$$

is assumed, which means that the angular velocity $\Delta\Omega$ describes small deviations from the rolling condition $r_D \Omega = v_x$. Then, the longitudinal slip defined by Equation (4.3) simplifies to

$$s_x = \frac{-(v_x - (v_x + r_D \,\Delta\Omega))}{|v_x + r_D \,\Delta\Omega| + v_N} \approx \frac{-r_D \,\Delta\Omega}{|v_x| + v_N} \qquad (4.6)$$

According to Equation (4.5), the longitudinal slip will be small too, $s_x \ll 1$, and the steady-state longitudinal tire force can be approximated by

$$F_x = F_x^{st} \approx dF_x^0 \, s_x = dF_x^0 \, \frac{r_D \,\Delta\Omega}{|v_x| + v_N} \qquad (4.7)$$

where dF_x^0 describes the initial inclination of the longitudinal tire characteristic $F_x = F_x(s_x)$. Now the angular momentum of the wheel defined in Equation (4.2), simplifies to a linear first-order differential equation,

$$\Theta \,\Delta\dot{\Omega} = -r_S \, dF_x^0 \, \frac{r_D \,\Delta\Omega}{|v_x| + v_N} \qquad (4.8)$$

The dynamics of this simple wheel tire model is then characterized by the eigenvalue

$$\lambda = -\frac{dF_x^0}{|v_x| + v_N} \, \frac{r_S^2}{\Theta} \qquad (4.9)$$

where $r_S \approx r_D$ was assumed in addition. In drive away or braking to standstill maneuvers where $v_x = 0$ will hold, the eigenvalue is proportional to $1/v_N$. This strong dependency on the fictitious velocity causes problems, because small values for v_N will result in a very large eigenvalue, which indicates a stiff differential equation for the wheel rotation. On the other hand, too large values for v_N will produce results with poor accuracy [46]. Hence, the approach where the longitudinal force is modeled as a function of the longitudinal slip is not practical at all when the vehicle comes close to standstill.

4.2.3 Dynamic Tire Forces

However, a simple but effective extension to first-order dynamic tire forces gets rid of the strong influence of the fictitious velocity v_N and produces quite good and practical results in any driving situation [41]. As shown in Section 3.8, the dynamic longitudinal tire force can be modeled by

$$F_x = F_x^D = c_x\, x_e + d_x\, \dot{x}_e \tag{4.10}$$

where x_e describes the longitudinal tire deflection, and c_x and d_x denote the corresponding stiffness and damping properties. Then, the angular momentum of the wheel provided in Equation (4.2) reads as

$$\Theta\, \Delta\dot{\Omega} = -r_S\, (c_x\, x_e + d_x\, \dot{x}_e) \tag{4.11}$$

The tire deflection x_e is defined by the first-order differential equation (3.153), which in the case of pure longitudinal slip will read as

$$\left(v_{Tx}^*\, d_x + dF_x^0\right) \dot{x}_e = -v_{Tx}^*\, c_x\, x_e - dF_x^0\, (v_x - r_D\, \Omega) \tag{4.12}$$

where the global derivative f_G of the generalized tire force characteristic was approximated by the initial inclination of the longitudinal tire force characteristic dF_x^0. As no lateral slip is considered here, a slip normalization is obsolete ($\hat{s}_x = 1$) and Equation (3.156) will deliver by

$$v_{Tx}^* = r_D\, |\Omega| + v_N \tag{4.13}$$

the modified transport velocity slightly simplified. Then, Equation (4.12) can be written as

$$\left(\left(|v_x + r_D\, \Delta\Omega| + v_N\right) d_x + dF_x^0\right) \dot{x}_e = $$
$$- \left(|v_x + r_D\, \Delta\Omega| + v_N\right) c_x\, x_e - dF_x^0\, \left(v_x - (v_x + r_D\, \Delta\Omega)\right) \tag{4.14}$$

where Equation (4.4) was used in addition to introduce the small angular velocity $\Delta\Omega$ describing the deviation from the rolling condition. In normal driving situations, the longitudinal tire deflection x_e and its time derivative $\dot{x}_e$ will remain small too. Then, Equation (4.14) simplifies further to

$$\left(|v_x + v_N|\, d_x + dF_x^0\right) \dot{x}_e = r_D\, dF_x^0\, \Delta\Omega - |v_x + v_N|\, c_x\, x_e \tag{4.15}$$

Now the time derivative of Equation (4.15) can be combined with Equation (4.11) to one differential equation,

$$\left(|v|\, d_x + dF_x^0 \right) \ddot{x}_e \;=\; \frac{r_D\, dF_x^0}{\Theta}\left(-r_S\, c_x\, x_e - r_S\, d_x\, \dot{x}_e\right) \;-\; |v|\, c_x\, \dot{x}_e \tag{4.16}$$

where the abbreviation $v = v_x + v_N$ was introduced in addition. Collecting the terms with $\dot{x}_e$ and isolating the steady-state longitudinal tire force $c_x x_e$, the second-order differential equation defined in Equation (4.16) can be written as

$$\underbrace{\left(\frac{|v|\, d_x}{dF_x^0} + 1\right)\frac{\Theta}{r_S r_D}}_{m}\ \underbrace{\ddot{x}_e}_{\ddot{x}} \;+\; \underbrace{\left(d_x + \frac{|v|\, c_x\, \Theta}{dF_x^0\, r_S r_D}\right)}_{d}\ \underbrace{\dot{x}_e}_{\dot{x}} \;+\; \underbrace{c_x}_{c}\ \underbrace{x_e}_{x} \;=\; 0 \tag{4.17}$$

which corresponds with a single mass oscillator. As a consequence, the eigenvalues given by

$$\lambda_{1,2} \;=\; -\frac{d}{2m} \pm i\sqrt{\frac{c}{m} - \left(\frac{d}{2m}\right)^2} \tag{4.18}$$

will characterize the dynamics of the wheel and the longitudinal tire force. The MATLAB-Script in Listing 4.1 calculates the eigenvalues, the eigenfrequencies, and the damping ratios[1] for different driving velocities. The results are plotted in Figure 4.4.

Listing 4.1
Script app_04_eigenvalues_wheel_dynamics.m: wheel tire dynamics

```
1  Theta = 1.2;          % [kgm^2] inertia of wheel (passenger car)
2  r = 0.3;              % [m] wheel radius (dynamic approx static)
3  dfx0 = 100000;        % [N/-] initial incl. force char.
4  cx = 160000;          % [N/m] longitudinal tire stiffness
5  dx = 500;             % [N/(m/s)] longitudinal tire damping
6  vN = 0.01;            % [m/s] fictitious velocity
7
8  % set velocity range and allocate space to speed up loop
9  vi=1:1:60;la1=zeros(size(vi));la2=la1;freq=la1;dratio=la1;
10
11 for i=1:length(vi)
12 % corresponding mass oscillator
13    v = vi(i) + vN;
14    m = Theta/r^2*(1+v*dx/dfx0); d = dx + v*cx*Theta/(r^2*dfx0); c = cx;
15 % eigenvalues
16    la1(i) = -d/(2*m) + sqrt( (d/(2*m))^2 - c/m );
17    la2(i) = -d/(2*m) - sqrt( (d/(2*m))^2 - c/m );
18 % frequency [Hz] and damping ratio [-]
19    freq(i)   = imag(la1(i))/(2*pi);
20    dratio(i) = d / ( 2*sqrt(c*m) );
21 end
22
23 % plots
```

---

[1]The damping parameter $d = d_C = 2\sqrt{c\, m}$, which happens to satisfy the condition $(d/(2m))^2 = c/m$, is called critical damping because Equation (4.18) will deliver only one real double eigenvalue $\lambda_1 = \lambda_2$.

```
24 subplot(2,2,[1,3])
25 plot(real(la1),imag(la1),'k'), hold on, grid on
26 plot(real(la2),imag(la2),'k'), title('Eigenvalues'), axis equal
27 subplot(2,2,2)
28 plot(vi,freq,'k'), grid on, title('Eigenfrequencies [Hz]')
29 subplot(2,2,4)
30 plot(vi,dratio,'k'), grid on, title('Damping ratio [-]')
```

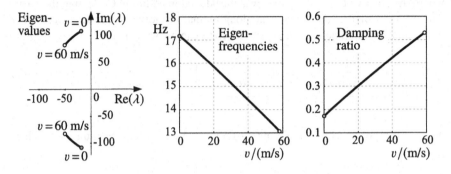

FIGURE 4.4
Wheel tire dynamics for different driving velocities.

The wheel tire dynamics is not sensitive to the fictitious velocity v_N as long as

$$\frac{v_N d_x}{dF_0} \ll 1 \quad \text{or} \quad v_N \ll \frac{dF_0}{d_x} \tag{4.19}$$

and

$$\frac{v_N c_x \Theta}{dF_0 r_S r_D} \ll d_x \quad \text{or} \quad v_N \ll \frac{d_x dF_0 r_S r_D}{c_x \Theta} \tag{4.20}$$

will be granted. In this case the demands will deliver

$$v_N \ll 100\,000 \text{ (N/–)} / 500 \text{ N/(m/s)} = 200 \text{ m/s}$$

and

$$v_N \ll \frac{500 \text{ Ns/m} * 100 \text{ (kN/–)} * 0.3 \text{ m} * 0.3 \text{ m}}{160 \text{ kN/m} * 1.2 \text{ kgm}^2} = 23.4 \text{ m/s}$$

in particular. Note that for standard wheel tire data, any value of $v_N < 1$ m/s will be appropriate; $v_N = 0.01$ m/s was chosen here.

Indirect tire-pressure monitoring systems (iTPMS) of second generation analyze a wheel tire vibration of approximately 120 Hz, which is sensitive to the tire inflation pressure. To investigate this kind of system, at least the first two eigenmodes of the tire belt have to be taken into account [56].

4.3 Simple Vehicle Wheel Tire Model

4.3.1 Equations of Motion

The wheel tire model in Section 4.2.3 is now supplemented by a chassis mass and will be put on a grade, Figure 4.5. The mass m includes the mass of the wheel suspension

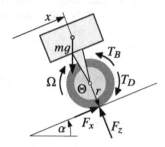

FIGURE 4.5
Simple vehicle model.

system and represents the part of the overall vehicle mass that is related to one wheel. The inertia of the wheel is denoted by Θ, the grade angle is named by α, and r serves as a simple approximation for the static ($r_S \approx r$) as well as the dynamic tire radius ($r_D \approx r$). The rolling resistance of the tire is neglected in this simple model. The vehicle is supposed to move along the grade only. Its actual position is determined by the coordinate x, and

$$F_z = mg \cos \alpha \tag{4.21}$$

will provide a constant wheel load here. Then, the equations of motion consisting of the linear momentum for the vehicle, the angular momentum for the wheel, and the dynamics of the longitudinal tire deflection read as

$$m\ddot{x} = \overbrace{c_x x_e + d_x \dot{x}_e}^{F_x^D} - mg \sin \alpha \tag{4.22}$$

$$\Theta \dot{\Omega} = T_D - T_B - r \underbrace{(c_x x_e + d_x \dot{x}_e)}_{F_x^D} \tag{4.23}$$

$$\left(v_{Tx}^* d_x + f_G \right) \dot{x}_e = -v_{Tx}^* c_x x_e - f_G \left(v_x - r_D \Omega \right) \tag{4.24}$$

where c_x and d_x describe the stiffness and damping properties of the tire in the longitudinal direction. In addition, the special case of pure longitudinal slip was assumed, which according to Equation (4.13) will simplify the modified transport velocity to $v_{Tx}^* = r_D |\Omega| + v_N$ and reduce the global derivative of the generalized force characteristic to the corresponding derivative of the longitudinal force characteristic. For given driving and braking torques, the equation of motion can be solved numerically.

4.3.2 Driving Torque

Usually, the driving torque T_D is transmitted by the half-shaft. By modeling the torsional flexibility of the drive-shaft by a linear spring damper model, one gets

$$T_D = T_S = -c_S \, \Delta\varphi_S - d_S \, (\Omega - \omega_S) \tag{4.25}$$

where c_S and d_S describe the torsional stiffness and damping properties of the half-shaft, Ω denotes the angular velocity of the wheel, and ω_S is the angular velocity of the half-shaft. Finally, the twist angle $\Delta\varphi_S$ of the half-shaft is defined by the differential equation

$$\frac{d}{dt}(\Delta\varphi_S) = \Omega - \omega_S \tag{4.26}$$

If, however, a wheel hub motor is used as the driving source instead, the engine torque T_E will be applied directly to the wheel. Then

$$T_D = T_E \tag{4.27}$$

simply holds.

4.3.3 Braking Torque

The braking torque applied to the wheel usually is generated by friction, Figure 4.6. However, a simple dry friction model will cause severe numerical problems because

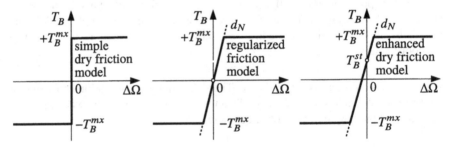

FIGURE 4.6
Coulomb dry friction model and enhanced brake torque model.

it is not defined in a locking situation ($\Delta\Omega = 0$). The regularized model that is mostly used in commercial software packages avoids this problem but becomes less accurate when approaching the locking situation. The enhanced dry friction model avoids the jump at $\Delta\Omega = 0$ and provides an appropriate locking torque [39]. So, the braking torque will here be modeled by

$$T_B = T_B^{st} + d_N \, \Delta\Omega \quad \text{and} \quad \left| T_B \right| \leq T_B^{mx} \tag{4.28}$$

where T_B^{st} names the static or locking torque, $d_N > 0$ is a constant with the dimension of Nm/(rad/s), T_B^{mx} denotes the maximum braking torque, and

$$\Delta\Omega = \Omega - \omega_K \tag{4.29}$$

describes the relative angular velocity between the wheel and the body where the brake caliper is mounted. Usually this will be the knuckle.

The static part provides a steady-state locking torque when the relative angular velocity is vanishing, $T_B^{st}(\Delta\Omega = 0) = T_B^{st}$. In the steady state when $\dot{\Omega} = 0$ holds in addition, Equation (4.23) delivers

$$0 = T_D - T_B^{st} - r F_x^D \tag{4.30}$$

Hence, the static braking torque

$$T_B^{st} = T_D - r F_x^D \tag{4.31}$$

will counteract appropriately the resulting torque applied to the wheel, namely consisting of the driving torque T_D and the torque $r F_x^D$ generated by the longitudinal tire force. Just like the overall braking torque T_B, the steady-state part is bound to the maximum braking torque

$$\left| T_B^{st} \right| \le T_B^{mx} \tag{4.32}$$

The numeric constant d_N is chosen such that the dynamics of a fully braked and freely rolling wheel will be similar. According to Equation (4.23), the dynamics of a fully braked wheel will be approximately described by

$$\Theta \dot{\Omega} = - \left(T_B^{st} + d_N \Omega \right) - r F_x^S \tag{4.33}$$

where in a fully braking maneuver the dynamic longitudinal force F_x^D can be replaced by the steady-state sliding force F_x^S and a vanishing driving torque ($T_D = 0$) was stated in addition. Within this simple vehicle model, the knuckle rotation is not taken into account. That is why the relative angular velocity $\Delta\Omega$ required in the braking torque model in Equation (4.28) may be substituted by the absolute angular wheel velocity Ω here. The eigenvalue of the first-order differential equation provided by Equation (4.33), is simply given by

$$\lambda = d_N / \Theta \tag{4.34}$$

The eigenvalues of a freely rolling wheel are provided in Equation (4.18). Their absolute value is given by

$$|\lambda_{1,2}| = \sqrt{(Re(\lambda))^2 + (Im(\lambda))^2} = \sqrt{\left(-\frac{d}{2m}\right)^2 + \frac{c}{m} - \left(\frac{d}{2m}\right)^2} = \sqrt{\frac{c}{m}} \tag{4.35}$$

where the generalized mass m and the stiffness $c = c_x$ are related to the wheel tire data by Equation (4.17). Hence, by setting

$$d_N = \Theta \sqrt{c/m} \tag{4.36}$$

the dynamics of the braking torque model can be adjusted to the freely rolling wheel that features the highest eigenfrequencies at vanishing driving velocities. At $v = 0$, the generalized mass simplifies to $m = \Theta/(r_s r_D)$ or to $m = \Theta/r^2$ within this simple model approach. Then

$$d_N = \Theta \sqrt{\frac{c_x}{\Theta/r^2}} = r\sqrt{c_x \Theta} \qquad (4.37)$$

will provide an appropriate value for the "damping" constant required in the enhanced braking torque model.

4.3.4 Simulation Results

The equations of motion for the simple vehicle wheel tire model including the enhanced braking torque model are provided by the function in Listing 4.2 as a set of first-order differential equations.

Listing 4.2
Function `fun_04_chassis_wheel_tire`: vehicle wheel tire model dynamics

```
1 function [xp,out] = fun_04_chassis_wheel_tire(t,x,p)
2 % simple vehicle model including wheel and tire dynamics
3
4 % get states
5 v = x(1);   o = x(2);   % vehicle velocity and angular velocity of wheel
6 xe= x(3);   xv= x(4);   % longitudinal tire deflection and vehicle pos.
7
8 % actual driving and maximum braking torque via linear interpolation
9 out.tq_D = interp1(p.t_Drive,p.Drive_Torque,t);
10 tq_Bmx  = interp1(p.t_Brake,p.Brake_Torque,t);
11
12 % modified transport velocity and long. slip without normalization
13 vt = abs(p.r*o) + p.vN;   vs = v-p.r*o;   out.sx = -vs/vt;
14
15 % generalized tire characteristics (sx only)
16 [~,fos] = tmy_fcombined( abs(out.sx), p.dfx0,p.fxm,p.sxm,p.fxs,p.sxs);
17
18 % time deriv. of long. tire deflection and long. dynamic tire force
19 xedot=-(p.cx*xe*vt+fos*vs)/(p.dx*vt+fos); out.fxd=p.cx*xe+p.dx*xedot;
20
21 % applied braking torque (enhanced dry friction model)
22 out.tq_B = out.tq_D - p.r*out.fxd+p.dN*o;
23 out.tq_B = sign(out.tq_B)*min(abs(out.tq_B),tq_Bmx);
24
25 % derivatives
26 odot = ( out.tq_D - out.tq_B - p.r*out.fxd ) / p.Theta;   % wheel
27 vdot = ( out.fxd - p.mass*p.gravity*sin(p.al) ) / p.mass; % chassis
28 xp = [ vdot; odot; xedot; v ]; % state derivatives
29
30 end
```

The function `tmy_fcombined` that generates the combined force characteristic is given in Listing 3.3. The time history of the driving and the maximum braking torques is stored as lookup tables in the vectors `p.t_Drive` and `p.Drive_Torque` as well as `p.t_Brake` and `p.Brake_Torque`. The MATLAB-Function `interp1` calculates the actual values via a linear interpolation. The MATLAB-Script in Listing 4.3 performs a simulation with the simple vehicle wheel tire model. The results are plotted in Figure 4.7.

Listing 4.3

Script app_04_chassis_wheel_tire.m: vehicle wheel tire model

```
 1 p.gravity = 9.81;        % [m/s^2] constant of gravity
 2 p.al = 20/180*pi;        % [degree-->rad] grade angle
 3 p.mass  = 400;           % [kg] chassis mass
 4 p.Theta = 1.2;           % [kgm^2] inertia of wheel
 5 p.r = 0.3;               % [m] wheel radius
 6
 7 % set paylaod and compute actual wheel load
 8 fzN = 3100;   fz = p.mass*p.gravity*cos(p.al);
 9
10 % long. tire characteristic with linear wheel load influence
11 p.dfx0 = 100000*fz/fzN;  % [N/-] initial incl. long. force char.
12 p.fxm  = 3200*fz/fzN;    % [N] maximum long. force
13 p.sxm  = 0.1;            % [-] sx where fx=fxm
14 p.fxs  = 3000*fz/fzN;    % [N] long. sliding force
15 p.sxs  = 0.8;            % [-] sx where fx=fxs
16 p.cx   = 160000;         % [N/m] longitudinal tire stiffness
17 p.dx   = 500;            % [N/(m/s)] longitudinal tire damping
18 p.vN   = 0.01;           % [m/s] fictitious velocity
19
20 % appropriate "damping" constant for enhanced braking torque model
21 p.dN = p.r * sqrt(p.cx*p.Theta);
22 % set driving torque (time [s], torque [Nm])
23 p.t_Drive      = [ 0.0   0.9   1.1   3.9   4.1   20.0 ];
24 p.Drive_Torque = [ 0.0   0.0   1.0   1.0   0.0   0.0  ]*p.fxm*p.r;
25 % set braking torque (time [s], torque [Nm])
26 p.t_Brake      = [ 0.0   5.95  6.05  17.95  18.05   20.0 ];
27 p.Brake_Torque = [ 0.0   0.00  1.50  1.50   0.00    0.0 ]*p.fxm*p.r;
28 % perform simulation
29 tE = min(max(p.t_Drive),max(p.t_Brake)); % duration
30 x0=[ 0; 0; 0; 0]; % simple initial states
31 [tout,xout]=ode23(@(t,x) fun_04_chassis_wheel_tire(t,x,p),[0,tE],x0);
32 % get additional output quantities
33 sxi=tout; fxi=tout; tqb=tout; tqd=tout;
34 for i=1:length(tout)
35   [xp,out] = fun_04_chassis_wheel_tire(tout(i),xout(i,:).',p);
36   sxi(i)=out.sx; fxi(i)=out.fxd; tqb(i)=out.tq_B; tqd(i)=out.tq_D;
37 end
38 % plot results
39 subplot(2,2,1), plot(tout,xout(:,1),'--r','Linewidth',1), hold on
40                 plot(tout,p.r*xout(:,2),'b','Linewidth',1), grid on
41                 xlabel('s'),ylabel('m/s'), legend('v','r*\Omega')
42 subplot(2,2,2), plot(tout,xout(:,4),'b','Linewidth',1), grid on
43                 xlabel('s'),ylabel('m'), legend('x_V')
44 subplot(2,2,3), plot(tout,tqd,'--r','Linewidth',1), hold on
45                 plot(tout,tqb,'b','Linewidth',1), grid on
46                 xlabel('s'),ylabel('Nm'), legend('T_D','T_B')
47 subplot(2,2,4), plot(tout,fxi/1000,'b','Linewidth',1), grid on
48                 xlabel('s'),ylabel('kN'), legend('F_x^D')
```

The vehicle starts from a standstill on a grade. At first it rolls backward, $v = r\Omega < 0$; then the vehicle is accelerated by the driving torque T_D that causes the wheel to spin, $r\Omega > v$. After a short period, a braking torque T_B is applied that forces the wheel to lock in an instant, $r\Omega = 0$. When the vehicle comes to a standstill $r\Omega = 0$, $v = 0$ at $t \approx 6$ s, the enhanced braking torque model automatically changes the sign in order to prevent the vehicle from moving downhill again. As the brake is not released yet, the vehicle oscillates for some time in the longitudinal direction. During this period where the wheel is locked, the system vehicle and tire represent a damped oscillator. The stiffness and damping properties of the tire in the longitudinal direction $c_x = 160\,000$ N/m and $d_x = 500$ N/(m/s), together with the corresponding vehicle

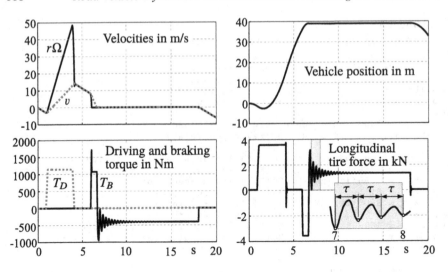

FIGURE 4.7
Driving and braking on a grade.

mass of $m = 400$ kg, result in a frequency $f = \sqrt{c/m - (d/(2m))^2}/(2pi) = 3.2$ Hz. Inspecting the time history of the longitudinal force $F_x(t)$ in Figure 4.7, one counts approximately three cycles in the interval $7 \text{ s} \leq t \leq 8$ s, which is enlarged in the lower right corner of the graph. This results in a vibration period of $\tau = 1/3$ s or a frequency of $f = 1/\tau = 3$ Hz, which corresponds quite well with the predicted result. Finally, the vehicle comes to a complete standstill which implies $v = 0$ and $\Omega = 0$. The tire force of $F_x \approx 1340$ N that is needed to compensate the downhill force $mg \sin \alpha = 1342$ N is maintained as long as the brake is applied. At $t = 18$ s, the brake is released and the vehicle starts to roll downhill again.

The MATLAB-Solver ode23, which is an implementation of an explicit low-order (2,3) Runge-Kutta algorithm, was used for the simulation because the brake torque model features sharp bends when reaching the maximum. The MATLAB-Script in Listing 4.3 calls ode23 at line 31 by specifying the starting time $t_0 = 0$ and the final time t_E only. Then, the ode solver will adjust the integration step size appropriately in order to match the default relative error tolerance 1e-3 and the default absolute tolerance of 1e-6 for each component of the state vector. Supplementing the MATLAB-Script in Listing 4.3 by the lines

```
h=diff(tout); tc=0.5*(tout(1:length(tout)-1)+tout(2:length(tout)));
figure, semilogy(tc,h,'ok','MarkerSize',3)
```

will open a new figure and plot the time history of the integration step size h calculated via the MATLAB-Function diff(tout) in a semi-logarithmic graph versus the intermediate value tc of each integration step, Figure 4.8.

The maneuver includes all driving situations, which encompasses standstill, a rolling, a spinning, a braked, and a locked wheel. Despite all that, the time history of integration step size h that is automatically adjusted by the ode23 solver to the

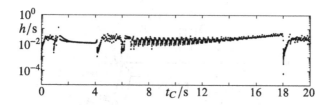

FIGURE 4.8
Integration step size h versus intermediate time t_C.

dynamics of the system is quite smooth. Only at the beginning ($t = 0$) when the solver is in the start-up phase and at $t = 18$ s when the brake is released again and the wheel changes suddenly from locking to rolling, a few smaller step sizes will be needed to maintain the required accuracy.

4.4   Differentials

4.4.1   Classic Design

If a vehicle is driven by a single engine, the drive torque must be transmitted to the wheels by allowing them to rotate at different speeds. This is achieved through differentials. Whereas cars with rear-wheel drive are equipped with one differential only, vehicles with all-wheel drive usually have three differentials. The differentials at the front and the rear axle split the torque equally to the left and to the right. The center differential, however, usually transmits more driving torque to the rear than to the front. This can be achieved by implementing different sized sun gears, Figure 4.9.

Taking no account of the mass and inertia properties of the planetary gears, the differential can be modeled by four rigid bodies that perform rotations only. They are listed in Table 4.1.

TABLE 4.1
Bodies of a differential (inertia of planetary gears neglected)

#	Body	Angular Velocity	Inertia	Applied Torques
1	Output shaft and pinion 1	Ω_1	Θ_1	T_1
2	Output shaft and pinion 2	Ω_2	Θ_2	$T_2 - T_L$
3	Crown wheel and housing	Ω_C	Θ_C	T_L
4	Input shaft and pinion	Ω_0	Θ_0	T_0

The torque T_0 is applied to the input and the torques T_1, T_2 are applied to the output shafts. The torque T_L is generated by friction plates between the output shaft 2 and the housing, and it will model the effects of a locking device. The gear ratio

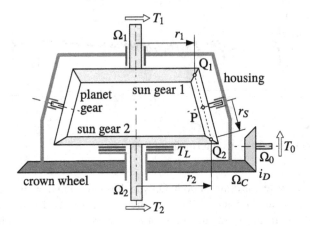

FIGURE 4.9
Differential with different sun gears and locking device.

between the input pinion and the crown wheel, which is fixed to the housing, is given by i_D. Then,

$$\Omega_0 = i_D \, \Omega_C \tag{4.38}$$

represents a first constraint equation. In addition, the rotation of the input shafts are coupled by the planet gears. The velocities in the contact points Q_1 and Q_2 are simply given by

$$v_{Q1} = r_1 \, \Omega_1 \quad \text{and} \quad v_{Q2} = r_2 \, \Omega_2 \tag{4.39}$$

and its mean value

$$v_P = \frac{1}{2} \left(v_{Q1} + v_{Q2} \right) = \frac{1}{2} \left(r_1 \, \Omega_1 + r_2 \, \Omega_2 \right) \tag{4.40}$$

determines the velocity of its center point P. As P is located on the rotation axis of a planet gear that is fixed to the housing,

$$v_P = \frac{r_1 + r_2}{2} \, \Omega_C \tag{4.41}$$

will hold in addition. Combining both relations finally results in a second constraint equation,

$$\Omega_C = \frac{r_1}{r_1 + r_2} \, \Omega_1 + \frac{r_2}{r_1 + r_2} \, \Omega_2 \tag{4.42}$$

which couples the angular velocity of the housing or the crown wheel respectively to the angular velocities of the output shafts. Introducing the internal ratio

$$\varrho = \frac{r_1}{r_1 + r_2} \tag{4.43}$$

the two constraint equations (4.42) and (4.38) will finally result in

$$\Omega_C = \varrho \, \Omega_1 + (1 - \varrho) \, \Omega_2 \tag{4.44}$$

$$\Omega_0 = i_D \left(\varrho \, \Omega_1 + (1 - \varrho) \, \Omega_2 \right) \tag{4.45}$$

The equation of motion will be generated via Jordain's Principle of virtual power. Taking only the angular momentum of the bodies into account, it will read as

$$\delta\Omega_1 \left(\Theta_1 \dot{\Omega}_1 - T_1\right) \quad + \quad \delta\Omega_2 \left(\Theta_2 \dot{\Omega}_2 - (T_2 - T_L)\right)$$
$$+ \quad \delta\Omega_C \left(\Theta_C \dot{\Omega}_C - T_L\right) + \delta\Omega_0 \left(\Theta_0 \dot{\Omega}_0 - T_0\right) = 0 \tag{4.46}$$

Considering the constraint equations yields

$$\delta\Omega_1 \left(\Theta_1 \dot{\Omega}_1 - T_1\right) + \delta\Omega_2 \left(\Theta_2 \dot{\Omega}_2 - (T_2 - T_L)\right)$$
$$+ \left(\varrho\,\delta\Omega_1 + (1-\varrho)\,\delta\Omega_2\right) \left(\Theta_C \left(\varrho\,\dot{\Omega}_1 + (1-\varrho)\dot{\Omega}_2\right) - T_L\right)$$
$$+ i_D \left(\varrho\,\delta\Omega_1 + (1-\varrho)\,\delta\Omega_2\right) \left(\Theta_0\, i_D \left(\varrho\,\dot{\Omega}_1 + (1-\varrho)\,\dot{\Omega}_2\right) - T_0\right) = 0 \tag{4.47}$$

Collecting all terms with $\delta\Omega_1$ and $\delta\Omega_2$ results in two coupled differential equations that in matrix form read as

$$\begin{bmatrix} \Theta_1 + \varrho^2\Theta^* & \varrho\,(1-\varrho)\Theta^* \\ \varrho\,(1-\varrho)\Theta^* & \Theta_2 + (1-\varrho)^2\Theta^* \end{bmatrix} \begin{bmatrix} \dot{\Omega}_1 \\ \dot{\Omega}_2 \end{bmatrix} = \begin{bmatrix} T_1 + \varrho\,T_L + \varrho\,i_D\,T_0 \\ T_2 - \varrho\,T_L + (1-\varrho)i_D\,T_0 \end{bmatrix} \tag{4.48}$$

where the abbreviation $\Theta^* = \Theta_C + i_D^2\Theta_0$ was used to shorten the elements of the mass matrix. A practical layout implies $r_1 > 0$ and $r_2 > 0$. Then, the ratio defined in Equation (4.43) will be limited to the range $0 < \varrho < 1$ which always results in non-vanishing off-diagonal elements in the mass matrix.

The locking torque T_L is generated by friction plates between the sun gear 2 and the crown wheel. Corresponding to the braking torque specified in Equation (4.28), it will be modeled by

$$T_L = T_L^{st} + d_N \left(\Omega_2 - \Omega_C\right) \quad \text{and} \quad |T_L| \leq T_L^{mx} \tag{4.49}$$

where T_L^{mx} denotes the maximum torque that can be transmitted by the friction plates between the sun gear 2 and the crown wheel, and $d_N > 0$ is a fictitious damping parameter that generates an appropriate torque when the angular velocities of the sun gear 2 and the crown wheel differ. Finally, a steady-state locking torque is provided by T_L^{st} even if $\Omega_2 = \Omega_C$ holds. In this particular case, the equations of motion for the differential given in Equation (4.48) deliver two demands on the steady-state locking torque,

$$0 = T_1 + \varrho\,T_L^{st} + \varrho\,i_D\,T_0 \quad \text{and} \quad 0 = T_2 - \varrho\,T_L^{st} + (1-\varrho)i_D\,T_0 \tag{4.50}$$

These two equations $f_1(x) = 0$ and $f_2(x) = 0$ for one unknown $x = T_L^{st}$ can be solved by a least squares approach as best as possible,

$$\tfrac{1}{2}f_1^2(x) + \tfrac{1}{2}f_2^2(x) \longrightarrow \text{Min} \quad \text{or} \quad \frac{d\,f_1}{dx}\,f_1(x) + \frac{d\,f_2}{dx}\,f_2(x) = 0 \tag{4.51}$$

Here, it yields at first

$$\varrho\left(T_1 + \varrho\,T_L^{st} + \varrho\,i_D\,T_0\right) + (-\varrho)\left(T_2 - \varrho\,T_L^{st} + (1-\varrho)i_D\,T_0\right) = 0 \tag{4.52}$$

and will deliver the steady-state locking torque

$$T_L^{st} = \frac{1}{2\varrho}(T_2 - T_1) + \left(\frac{1}{2\varrho} - 1\right)T_0 \tag{4.53}$$

which depends on the internal ratio ϱ of the differential and is adjusted to the torque difference $T_2 - T_1$ at the output shafts associated with the torque T_0 applied to the input shaft.

4.4.2 Active Differentials

Today, advanced drive trains include an electronically active center differential and an active yaw-control rear differential unit. The active center differential (ACD) is an electronically controlled hydraulic multi-plate clutch that distributes torque between the front and rear to improve traction under acceleration out of a corner. It works in conjunction with active yaw control (AYC), which can actively split torque based on input from various sensors in the vehicle measuring longitudinal and lateral acceleration, steering, brakes, and throttle position.

The sport differential supplements the classic differential at the rear axle with gearboxes on the left and right whose drive superposition stages are rotating 10% faster than the half-shafts, $\Omega_{S1} = 1.1\,\Omega_1$ and $\Omega_{S2} = 1.1\,\Omega_2$. The two components can be coupled by a multi-plate clutch running in an oil bath, Figure 4.10.

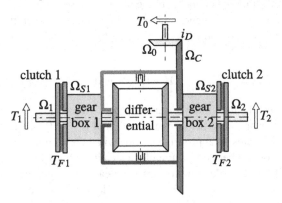

FIGURE 4.10
Enhanced rear axle differential.

When the car is accelerated in a left corner for example, the clutch on the outer side (clutch 2) is engaged, which by friction will generate a torque (T_{F2}) and speed up the corresponding half-shaft ($\Omega_2 \uparrow$) and finally the attached wheel too. As a result, the longitudinal slip at the outer tire is increased, which induces an additional driving force and reduces or completely eliminates the understeer tendency of the car. As the friction torque reacts on the gearbox too, the drive torque transferred to the opposite wheel is decreased accordingly. Thus, this active differential is able to torque the drive torque (torque vectoring) to the left and to the right wheel as needed in specific driving situations. Usually, almost the complete input torque can be diverted to one rear wheel in this way. The torque shift from the inner to the outer wheels also reduces the cornering resistance, in particular in sharp bends [54].

4.5 Generic Drive Train

The subsystem consisting of the drive shafts, the differentials, and the half-shafts interacts on one side with the engine and on the other side with the wheels, Fig. 4.11. Engine, clutch, transmission, wheels, and tires are described separately. Hence, the

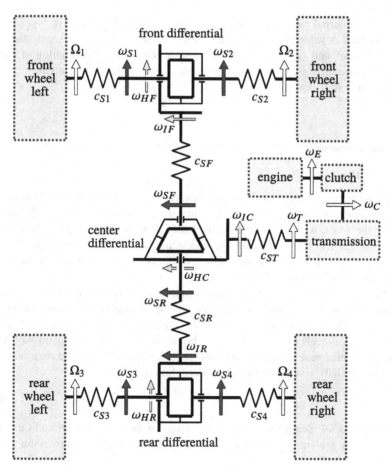

FIGURE 4.11
Drive train model.

angular velocities of the wheels $\Omega_1, \ldots, \Omega_4$, and the engine, or respectively the transmission output angular velocity ω_T, serve as inputs for this subsystem. The angular velocities of the drive shafts ω_{S1}: front left, ω_{S2}: front right, ω_{SF}: front, ω_{SR}: rear, ω_{S3}: rear left, ω_{S4}: rear right specify the generalized coordinates within this generic 4×4 all-wheel drive model. Via the tire forces and torques, the whole drive train is coupled with the steering system and the vehicle framework.

The generic drive train includes three differentials that may include locking or active parts. Their dynamic behavior can be described by adjusting the equations of motion deduced in Section 4.4 appropriately. The internal ratios of the front and rear differential defined in Equation (4.43) amount to $\varrho_F = \varrho_R = \frac{1}{2}$, which means that the driving torque will be distributed equally to the left and right wheels. The effect of active differentials can be modeled by additional torques generated by friction or electric motors.

Via the internal ratio ϱ_C of the center differential, different drive types can be distinguished. A value of $\varrho_C = 1$ means front-wheel drive, $0 < \varrho_C < 1$ stands for all-wheel drive, and $\varrho_C = 0$ will model a rear-wheel drive. The equation of motion for the generic drive train including the modeling of torques and hints for real-time applications can be found in [44].

4.6 Transmission

The transmission or gearbox allows the gear ratio to be adjusted. This is necessary because combustion engines work best if they run at a limited rate of revolutions. By shifting the gears, which can be done manually or automatically, the engine is kept at its most efficient rate while allowing the vehicle to run at a large range of speed. Operating the gear lever of a manual transmission brings a different train of gear wheels into play, Figure 4.12.

The different ratios of teeth on the gear wheels involved produce different speeds. If a gear is selected, the dog teeth lock the required upper gear wheel to the transmission shaft. Then, the transmission goes from the clutch shaft via the counter shaft and the lower gear wheels to the upper gear wheels and finally to the transmission shaft. Selecting reverse gear introduces the idler wheel, which reverses the rotation of the transmission shaft. Usually the gear ratio is defined as

$$r_G = \frac{\omega_T}{\omega_C} \tag{4.54}$$

where ω_T and ω_C denote the angular velocities of the transmission and the clutch shaft. Typical gear ratios are given in Table 4.2. The angular momentum for the transmission shaft results in

$$\Theta_T \dot{\omega}_T = r_G T_C - T_T^{FR} - T_T \tag{4.55}$$

where Θ_T is a generalized inertia that includes all rotating parts of the transmission. That is why it will depend on the gear ratio $\Theta_T = \Theta_T(r_G)$. The friction in the

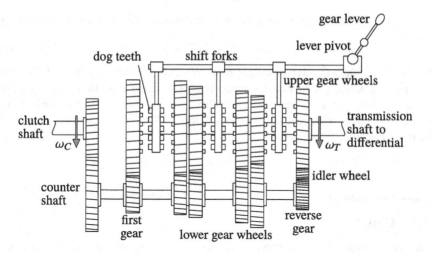

FIGURE 4.12
Manual transmission.

TABLE 4.2
Typical gear ratios for a passenger car

gear	reverse	neutral	first	second	third	fourth	fifth
ratio	−4.181	0	3.818	2.294	1.500	1.133	0.911

transmission is described by T_T^{FR}, and the torque T_T represents the external load that can be modeled by a torsional spring damper model

$$T_T = -c_T \, \Delta\varphi_{T0} - d_T \, (\omega_T - \omega_C) \tag{4.56}$$

where c_T, d_T describe the stiffness and damping properties of the shaft connecting the transmission with the differential, and ω_T, ω_C name the angular velocities of the flexible shaft at the transmission output and the input from the clutch. Finally, the differential equation

$$\frac{d}{dt} \Delta\varphi_{T0} = \omega_T - \omega_C \tag{4.57}$$

defines the torsional angle between the input and the output shaft.

For a gear to be engaged, the different speeds of the rotating parts need to be matched and locked together. The synchromesh uses friction to do this smoothly and quietly. Pushed by the selector fork, the collar slides along the transmission shaft, rotating with it. The collar fits over a cone on the upper gear wheel, making the wheel speed up or slow down until both are moving with the same speed. Then, the dog teeth are engaged, locking the upper gear wheel to the collar and, hence, to the transmission shaft.

The synchromesh mode of action can be approximated by a first-order differential equation

$$H_{syn} \dot{r}_G^D = -r_G^D + r_G \qquad (4.58)$$

where r_G^D denotes the dynamic gear ratio, H_{syn} is the time constant of the synchromesh process, and r_G denotes the static gear ratio. By this differential equation, the jump from one static gear ratio to another will be dynamically smoothed, which comes very close to the real synchromesh process. This dynamic gear ratio will then be used instead of the static one.

4.7 Clutch

The clutch makes use of friction to transmit the rotation of the engine crankshaft to the transmission. When the clutch pedal is released ($p_G = 0$), the clutch spring forces the clutch plate and the flywheel, which is turned by the crankshaft, together, Fig. 4.13. Then, the angular momentum for the clutch plate reads as

$$\Theta_P \dot{\omega}_P = T_C - T_D \qquad (4.59)$$

where Θ_P, ω_P describe the inertia and the angular velocity of the clutch plate. According to the principle "actio" equals "reactio," the torque T_C represents the load

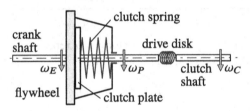

FIGURE 4.13
Clutch model without clutch pedal mechanism.

applied to the engine. The torque in the drive disk can be modeled by a torsional spring damper combination. Assuming linear characteristics, one gets

$$T_D = -c_D \Delta\varphi_{PC} - d_D (\omega_P - \omega_C) \qquad (4.60)$$

where ω_C names the angular velocity of the clutch shaft and c_D, d_D describe the stiffness and damping properties of the drive disk. The differential equation

$$\frac{d}{dt} \Delta\varphi_{PC} = \omega_P - \omega_C \qquad (4.61)$$

defines the torsional angle φ_{PC} in the drive disk. Similar to the brake torque defined in Equation (4.28), the friction-based clutch torque is described by

$$T_C = T_C^{st} + d_N (\omega_E - \omega_P) \quad \text{with} \quad |T_C| \le T_C^{mx} \qquad (4.62)$$

where ω_E denotes the angular velocity of the engine and $d_N > 0$ is a constant. The static part T_C^{st} provides a steady-state locking torque when the angular velocities of the engine ω_E and the clutch plate ω_P are equal. In steady state, and when $\omega_E = \omega_P$ will hold in addition, Equations (4.66) to (4.62) simply yield

$$0 = T_E - T_{FR} - T_C^{st} \quad \text{and} \quad 0 = T_C^{st} - T_D \tag{4.63}$$

These two equations for one unknown are again solved by a least squares approach that here will result in

$$\frac{\partial}{\partial T_C^{st}} \left(\epsilon_1^2 + \epsilon_2^2 \right) = 2 \left(T_E - T_{FR} - T_C^{st} \right) (-1) + 2 \left(T_C^{st} - T_D \right)^2 = 0 \tag{4.64}$$

Thus, the steady-state locking torque

$$T_C^{st} = \frac{1}{2} (T_E - T_{FR} + T_D) \tag{4.65}$$

will be adjusted to the torques $T_E - T_{FR}$ applied at the engine and to the torque in the drive disk T_{DD} as best as possible.

The maximum friction torque T_C^{mx} transmitted by the clutch depends on the position of the clutch pedal p_C, Figure 4.14. Pressing the clutch pedal reduces the normal

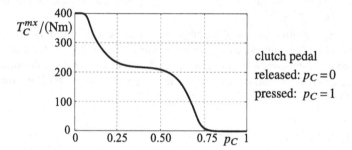

FIGURE 4.14
Example of the maximum friction torque transmitted by a passenger car clutch.

force between the clutch plate and the flywheel and hence reduces the maximum friction torque.

4.8 Power Sources

4.8.1 Combustion Engine

Combustion engines are very common in ground vehicles. In a first approximation, the torque T_E of a combustion engine can be characterized as a function of its angular

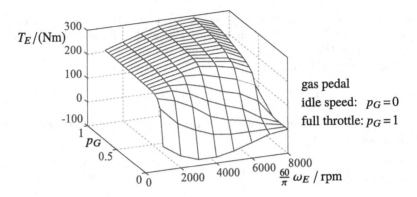

FIGURE 4.15
Example of a combustion engine torque: $T_E = T_E(\omega_E, p_G)$.

velocity ω_E and the gas pedal p_G, Fig. 4.15. Then, the dynamics of the engine can be described by the angular momentum

$$\Theta_E \dot{\omega}_E = T_E - T_{FR} - T_C \qquad (4.66)$$

where Θ_E labels the inertia of the engine, T_{FR} names the friction torque, and T_C is the external load from the clutch.

This simple approach usually is sufficient for vehicle handling and ride analysis. It is even used to design automotive control systems [23]. A sophisticated and real-time capable combustion engine model en-DYNA®, where the air flow, the fuel supply, the torque calculation, and the exhaust system are modeled in detail, is provided by TESIS [11].

4.8.2 Electric Drive

Compared with internal combustion engine cars, electric cars are quieter, have no local CO_2 emissions, and deliver full torque right from the start. A convenient overnight charge of the battery at home or at work is possible if the daily mileage is below 300 km. Otherwise a reliable access to public charging facilities is required.

Purely electric-driven cars require no clutch and no gear shift and are easy to drive. If two or more electric motors are mounted in the car, the drive train becomes partly or completely obsolete in addition.

Electric-driven cars have become more and more popular because the number and variety of electric cars available on the market is rapidly increasing.

4.8.3 Hybrid Drive

The power sources of a hybrid vehicle can be combined in two different ways. In a series hybrid, the combustion engine turns a generator, and the generator can either charge the batteries or power an electric motor that drives the transmission. Thus, the gasoline engine never directly powers the vehicle. A parallel hybrid has a fuel tank that supplies gasoline to the combustion engine and a set of batteries that supplies power to the electric motor. The batteries store energy recovered from braking or generated by the engine. Both the engine and the electric motor can turn the transmission at the same time. Hence, a significant increase in the drive torque is possible, Figure 4.16.

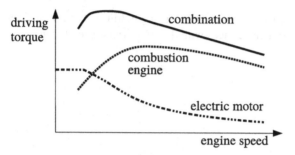

FIGURE 4.16
Hybrid drive torque characteristics.

Usually the combustion engine in a conventional car will be sized for the peak power requirement, whereas the hybrid car may use a much smaller engine, one that is sized closer to the average power requirement than to the peak power. Hence, fuel consumption and pollution may be reduced significantly.

Exercises

4.1 Radius and mass of a passenger car wheel are defined by $r = 0.3$ m and $m = 15$ kg. Estimate the inertia Θ of the wheel about its rotation axis by approximating it as a homogeneous cylinder.

Consider if the real value will be larger or smaller than the estimated one.

4.2 A wheel-tire combination is characterized by the data

$$\Theta = 7.5 \text{ kgm}^2 \qquad r_S \approx r_D = 0.54 \text{ m} \qquad dF_x^0 = 360\,000 \text{ N/m}$$
$$c_x = 680\,000 \text{ N/m} \qquad d_x = 1000 \text{ N/(m/s)} \qquad v_N = 0.02 \text{ m/s}$$

Calculate and plot the eigenvalues, the eigenfrequencies and the damping ratio for different driving velocities $0 \le v \le 25$ m/s by using the MATLAB-Script in Listing 4.1.

Check the influence of the fictitious velocity v_N on the results.

4.3 Use the MATLAB-Script in Listing 4.3 in combination with the functions in Listing 4.2 and Listing 3.3 to perform simulations with the grade angles $\alpha = 0$, $\alpha = 30°$, and $\alpha = 45°$.

Analyze and discuss the simulation results.

Change the lookup tables defined by the vectors `p.t_Drive` and `p.Drive_Torque` as well as `p.t_Brake` and `p.Brake_Torque`, which determine the time histories of the drive and brake torque appropriately so that various driving and braking maneuvers can be simulated.

4.4 Extend the MATLAB-Script in Listing 4.3 in such a way that the vehicle wheel tire model can run on surfaces with different friction coefficients.

5

Suspension System

CONTENTS

5.1 Purpose and Components

The automotive industry uses different kinds of wheel/axle suspension systems. Important criteria are costs, space requirements, kinematic properties, and compliance attributes. The main purposes of a vehicle suspension system are

- carry the car and its weight,
- maintain correct wheel alignment,
- control the vehicle's direction of travel,
- keep the tires in contact with the road, and
- reduce the effect of shock forces.

Vehicle suspension systems consist of the following:

- Guiding elements:
    - Control arms, links
    - Struts
    - Leaf springs

- Force elements:
    - Coil spring, torsion bar, air spring, leaf spring
    - Anti-roll bar, anti-sway bar or stabilizer
    - Damper
    - Bushings, hydro-mounts

- Tires

From a modeling point of view, force elements may be separated into static and dynamic systems. Examples and modeling aspects will be discussed in Chapter 6. Tires are some kind of air springs that support the total weight of the vehicle. The spring action of the tire is very important to the ride quality and safe handling of the vehicle. In addition, the tire must provide the forces and torque that keep the vehicle on track. The tire was already discussed in detail in Chapter 3.

5.2 Some Examples

5.2.1 Multipurpose Systems

The double wishbone suspension, the MacPherson suspension, and the multi-link suspension are multipurpose wheel suspension systems, Figure 5.1. They are used as steered front or nonsteered rear axle suspension systems. These suspension systems are also suitable for driven axles. Usually, the damper is attached to the knuckle,

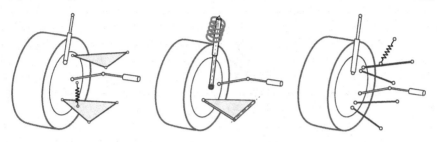

FIGURE 5.1
Double wishbone, MacPherson, and multi-link suspensions.

whereas the coil spring is connected to one control arm of the double wishbone or the multi-link suspension. In a MacPherson suspension however, the spring is attached to the knuckle and will have a well-defined inclination to the strut axis. Thus, bending torques at the strut, which cause high friction forces between the damper piston and the housing, can be reduced.

On pickups, trucks, and buses, solid axles are used often. They are guided either by leaf springs or by rigid links, Figure 5.2. Solid axles tend to tramp on rough

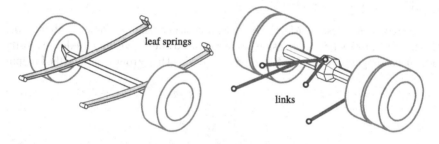

FIGURE 5.2
Solid axles guided by leaf springs and links.

roads. Leaf-spring-guided solid axle suspension systems are very robust. Dry friction between the leaves leads to locking effects in the suspension. Although the leaf springs provide axle guidance on some solid axle suspension systems, additional links in longitudinal and lateral direction are used. Thus, the typical wind-up effect on braking can be avoided. Solid axles suspended by air springs need at least four links for guidance. In addition to good driving comfort, air springs allow level control too.

5.2.2 Specific Systems

The semi-trailing arm, the short-long arm axle (SLA), and the twist beam axle suspension are suitable only for non-steered axles, Figure 5.3. The semi-trailing arm is a simple and cheap design that requires only a small space. It is mostly used for driven rear axles. The short-long arm axle design allows a nearly independent layout

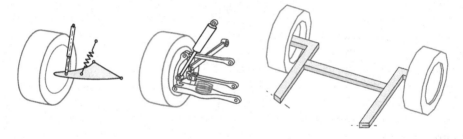

FIGURE 5.3
Semi-trailing arm, short-long arm, and twist beam suspension.

of longitudinal and lateral axle motions. It is similar to the central control arm axle suspension, where the trailing arm is completely rigid and, hence, only two lateral links are needed. The twist beam axle suspension exhibits either a trailing arm or a semi-trailing arm characteristic. It is used for nondriven rear axles only. The twist beam axle provides enough space for spare tire and fuel tank.

5.2.3 Steering Geometry

On steered front axles, the MacPherson-damper strut axis, the double wishbone axis, and the multi-link wheel suspension or the enhanced double wishbone axle are mostly used in passenger cars, Figures 5.4 and 5.5. The wheel body rotates around the kingpin

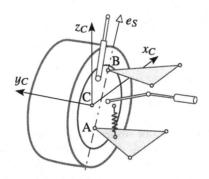

FIGURE 5.4
Double wishbone wheel suspension.

line at steering motions. For the double wishbone axle, the ball joints A and B, which determine the kingpin line, are both fixed to the wheel body. Whereas the ball joint A is still fixed to the wheel body at the standard MacPherson wheel suspension, the top-mount T is now fixed to the vehicle body. On a multi-link axle the kingpin line is no longer defined by real joints. Here, as well as with an enhanced MacPherson wheel

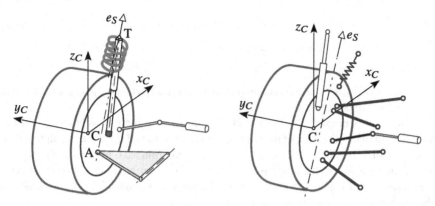

FIGURE 5.5
MacPherson and multi-link wheel suspensions.

suspension, where the A-arm is resolved into two links, the momentary rotation axis serves as the kingpin line. In general, the momentary rotation axis is neither fixed to the wheel body nor to the chassis, and it will change its position during wheel travel and steering motions.

The unit vector e_S describes the direction of the kingpin line. Within the vehicle-fixed reference frame F, it can be fixed by two angles. The caster angle v denotes the angle between the z_F-axis and the projection line of e_S into the x_F-z_F-plane. In a similar way, the projection of e_S into the y_F-z_F-plane results in the kingpin inclination angle σ, Figure 5.6. On many axles the kingpin and caster angle can no

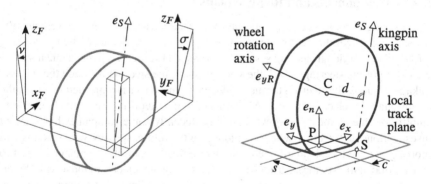

FIGURE 5.6
Kingpin inclination as well as caster and steering offset.

longer be determined directly. Here, the current rotation axis at steering motions, which can be taken from kinematic calculations, will yield a virtual kingpin line. The current values of the caster angle v and the kingpin inclination angle σ can be calculated from the components of the unit vector e_S in the direction of the kingpin axis,

$$\tan v = \frac{-e_{S,F}^{(1)}}{e_{S,F}^{(3)}} \quad \text{and} \quad \tan \sigma = \frac{-e_{S,F}^{(2)}}{e_{S,F}^{(3)}} \tag{5.1}$$

where $e_{S,F}^{(1)}$, $e_{S,F}^{(2)}$, $e_{S,F}^{(3)}$ are the components of the unit vector $e_{S,F}$ expressed in the vehicle-fixed axis system F.

The contact point P, the local track normal e_n, and the unit vectors e_x and e_y, which point in the direction of the longitudinal and the lateral tire force, result from the contact geometry. The axle kinematics define the kingpin axis. In general, the point S where an extension of the kingpin axis meets the road surface does not coincide with the contact point P, Figure 5.6. As both points are located on the local track plane, for the left wheel the vector from S to P can be written as

$$r_{SP} = -c\,e_x + s\,e_y \tag{5.2}$$

where s names the steering offset or scrub radius, and c is the caster offset. Caster and steering offset will be positive if S is located in front of and inwards of P. The kingpin offset d describes the distance between the wheel center C and the king pin axis. It is an important quantity in evaluating the overall steering behavior [29].

5.3 Steering Systems

5.3.1 Components and Requirements

The steering system is a very important interface between driver and vehicle. Via the steering wheel the driver controls the vehicle and gets feedback by the steering torque. The traditional steering system of high-speed vehicles is a mechanical system consisting of the steering wheel, the steering shaft, the steering box, and the steering linkage. Usually the steering torque produced by the driver is amplified by hydraulic or nowadays by electric systems. Modern steering systems use an overriding gear to amplify or change the steering wheel angle. Recently some companies have started investigations on "steer-by-wire" techniques. In the future, steer-by-wire systems will probably be used as standard. Here, an electronically controlled actuator is used to convert the rotation of the steering wheel into steering movements of the wheels. Steer-by-wire systems are based on mechanics, micro-controllers, electric-motors, power electronics, and digital sensors. At present, fail-safe systems with a mechanical backup system are under investigation. Modeling concepts for modern steering systems are discussed in [48].

The steering system must guarantee easy and safe steering of the vehicle. The entirety of the mechanical transmission devices must be able to cope with all loads and stresses occurring in operation. In order to achieve good maneuverability, a maximum steering angle of approx. 40° must be provided at the front wheels of passenger cars.

Depending on the wheel base, buses and trucks need maximum steering angles up to 55° at the front wheels.

5.3.2 Rack-and-Pinion Steering

Rack-and-pinion is the most common steering system of passenger cars, Figure 5.7. The rack may be located either in front of or behind the axle. First, the rotations of the

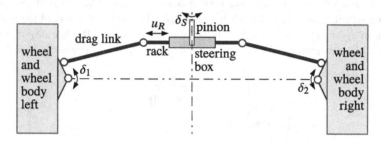

FIGURE 5.7
Rack-and-pinion steering.

steering wheel δ_S are transformed by the steering box to the rack travel $u_R = u_R(\delta_S)$ and then via the drag links transmitted to the wheel rotations $\delta_1 = \delta_1(u_R), \delta_2 = \delta_2(u_R)$. Hence, the overall steering ratio depends on the ratio of the steering box and on the kinematics of the steering linkage.

5.3.3 Lever Arm Steering System

Using a lever arm steering system, Figure 5.8, large steering angles at the wheels are possible. This steering system is used on trucks with large wheel bases and

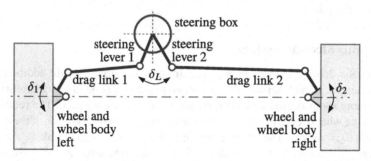

FIGURE 5.8
Lever arm steering system.

independent wheel suspension at the front axle. Here, the steering box can be placed outside the axle center. At first, the rotations of the steering wheel δ_S are transformed by the steering box to the rotation of the steer levers $\delta_L = \delta_L(\delta_S)$. The drag links transmit this rotation to the wheel $\delta_1 = \delta_1(\delta_L)$, $\delta_2 = \delta_2(\delta_L)$. Hence, the overall steering ratio again depends on the ratio of the steering box and on the kinematics of the steering linkage.

5.3.4   Toe Bar Steering System

On solid axles the toe bar steering system is used, Figure 5.9. The rotations of the steering wheel δ_S are transformed by the steering box to the rotation of the steering lever arm $\delta_L = \delta_L(\delta_S)$ and further on to the rotation of the left wheel, $\delta_1 = \delta_1(\delta_L)$. The toe bar transmits the rotation of the left wheel to the right wheel, $\delta_2 = \delta_2(\delta_1)$.

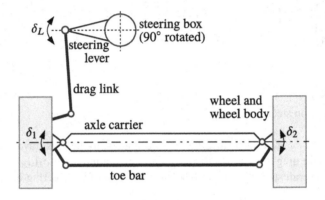

FIGURE 5.9
Toe bar steering system.

The steering ratio is defined by the ratio of the steering box and the kinematics of the steering link. Here, the ratio $\delta_2 = \delta_2(\delta_1)$ given by the kinematics of the toe bar can be changed separately.

5.3.5   Bus Steering System

In buses the driver sits more than $2\,m$ in front of the front axle. In addition, large steering angles at the front wheels are needed to achieve good maneuverability. That is why more sophisticated steering systems are needed, Figure 5.10. The rotations of the steering wheel δ_S are transformed by the steering box to the rotation of the steering lever arm $\delta_L = \delta_L(\delta_S)$. The left lever arm is moved via the steering link $\delta_A = \delta_A(\delta_L)$. This motion is transferred by a coupling link to the right lever arm. Finally, the left and right wheels are rotated via the drag links, $\delta_1 = \delta_1(\delta_A)$ and $\delta_2 = \delta_2(\delta_A)$.

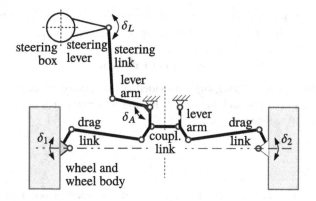

FIGURE 5.10
Typical bus steering system.

5.3.6 Dynamics of a Rack-and-Pinion Steering System

5.3.6.1 Equation of Motion

Modern steering systems include power assistance, and also some overriding gears or even steer-by-wire facilities [48]. The essential parts of a pure mechanical rack-and-pinion steering system are shown in Figure 5.11. The tire forces applied at the contact points P_1, P_2 are denoted by F_{T1}, F_{T2}, and the vectors T_{T1}, T_{T2} summarize

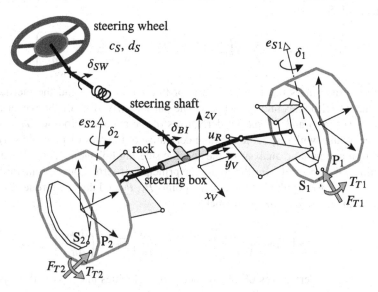

FIGURE 5.11
Double wishbone suspension with a rack-and-pinion steering system.

torques acting in the tire contact patches. The torsional compliance of the steering shaft, modeled by the stiffness c_S and the damping parameter d_S, generates the torque

$$T_S = c_S \left(\delta_{SW} - \delta_{BI} \right) + d_S \left(\dot{\delta}_{SW} - \dot{\delta}_{BI} \right) \tag{5.3}$$

where δ_{SW} denotes the steering angle provided by the rotation of the steering wheel. The maximum steering wheel angle δ_{SW}^{mx} and the maximum rack displacement u_R^{mx} define the ratio of the steering box[1]

$$i_{SB} = \frac{\delta_{SW}^{mx}}{u_R^{mx}} \tag{5.4}$$

Then, the rotation angle of the steering shaft at the steering box input is simply provided by

$$\delta_{BI} = i_{SB} \, u_R \tag{5.5}$$

The steering linkage is considered a pure massless kinematical constraint here. It transmits the rack displacements u_R to the wheels and forces them to rotate with the angles δ_1 and δ_2 around the kingpin axes, marked in Figure 5.11 by the unit vectors e_{S1}, e_{S2}.

Applying Jordain's principle of virtual power, the equation of motion for the rack and the wheels including the wheel bodies will result in

$$m_{RG} \, \ddot{u}_R = F_{TG} + F_{SM} - F_{SF} \tag{5.6}$$

The generalized mass

$$m_{RG} = m_R + \sum_{i=1}^{2} \left\{ \left(\frac{\partial v_{Ci}}{\partial \dot{u}_R} \right)^T m_i \frac{\partial v_{Ci}}{\partial \dot{u}_R} + \left(\frac{\partial \omega_{0i}}{\partial \dot{u}_R} \right)^T \Theta_i \frac{\partial \omega_{0i}}{\partial \dot{u}_R} \right\} \tag{5.7}$$

adds to the rack mass m_R essential parts of the mass m_1, m_2 and the inertias Θ_1, Θ_2 of the wheel bodies and the wheels. The partial derivatives of the wheel center velocities v_{C1}, v_{C2} and the partial derivatives of the angular velocities ω_{01}, ω_{02} of the wheel bodies with respect to the time derivative of the rack displacement $\dot{u}_R$ will result from a kinematic analysis of the specific steering linkage, which will be done in Section 5.4 elaborately. Furthermore, the principle of virtual power transforms the tire forces F_{T1}, F_{T2} applied at the contact points P_1, P_2 and the tire torques T_{T1}, T_{T2} to a generalized force acting at the rack:

$$F_{TG} = \left(\frac{\partial v_{P1}}{\partial \dot{u}_R} \right)^T F_{T1} + \left(\frac{\partial \omega_{01}}{\partial \dot{u}_R} \right)^T T_{T1} + \left(\frac{\partial v_{P2}}{\partial \dot{u}_R} \right)^T F_{T2} + \left(\frac{\partial \omega_{02}}{\partial \dot{u}_R} \right)^T T_{T2} \tag{5.8}$$

Here, the partial derivatives of the contact point velocities v_{P1}, v_{P2} with respect to the rack displacement u_R will also be required.

---

[1]Note: Some steering boxes will generate a nonlinear ratio. Then, the steering box ratio depends on the rack displacement, $i_{SB} = i_{SB}(u_R)$.

The torque T_S acting in the steering shaft provides the steering force

$$F_{SM} = i_{SB} T_{SS} \tag{5.9}$$

Finally, the friction force is approximated similar to the braking torque defined in Equation (4.28) by an enhanced dry friction model. It provides the friction force as

$$F_{SF} = F_{SF}^{st} + d_N \dot{u}_R \quad \text{and} \quad \left| F_{SF} \right| \le F_{SF}^{mx} \tag{5.10}$$

where F_{SF}^{mx} denotes the maximum friction force. In steady state when $\dot{u}_R = 0$ and $\ddot{u}_R = 0$ will hold, the static or locking force

$$F_{SF}^{st} = F_{TG} + F_{SM} \tag{5.11}$$

counteracts all other forces applied to the rack and will prevent movements as long as $\left| F_{SF}^{st} \right| \le F_{SF}^{mx}$ is assured. If the rack is exposed to the steering force F_{SM} only (steering motions in a liftoff situation), the critical damping of free rack oscillations will be given by

$$d_{CR} = 2\sqrt{i_{SB}^2 \, c_{TB} \, m_{RG}} = 2 \, i_{SB} \sqrt{c_{TB} \, m_{RG}} \tag{5.12}$$

By setting

$$d_N = \tfrac{1}{2} d_{CR} = i_{SB} \sqrt{c_{TB} \, m_{RG}} \tag{5.13}$$

the numerical damping in the friction model can be adjusted to the specific properties of the steering system.

5.3.6.2 Steering Forces and Torques

According to Equation (5.6), the force balance

$$0 = F_{TG} + F_{SM} - F_{SF} \tag{5.14}$$

characterizes the steering system at a steady state. Neglecting the friction force $F_{SF} \approx 0$, the steering force F_{SM} that in compliance with Equation (5.9) is generated by the steering torque T_{SS}, must counteract the generalized tire force F_{TG}. Hence, at a steady state, the generalized tire force provides, by

$$T_{SS}^{st} = -\frac{1}{i_{SB}} F_{TG}, \tag{5.15}$$

the largest part of the steering torque.

In general, the partial angular velocities of the wheel bodies, which according to Equation (5.8) are required to calculate the generalized tire force, can be written as

$$\frac{\partial \omega_{0i}}{\partial \dot{u}_R} = e_{Si} \frac{\partial \dot{\delta}_i}{\partial \dot{u}_R} = e_{Si} \, i_{SLi}, \quad i = 1, 2, \tag{5.16}$$

where e_{S1}, e_{S2} denote the unit vectors in the direction of the king pin axes and i_{SL1}, i_{SL2} characterize the ratios of the left and the right steering linkage, which in general are not constant but will depend on the rack displacement. The king pin axis will not change its position and orientation during steer motions on standard double wishbone and MacPherson suspension systems. Then, the points S_1, S_2 where the king pin axis intersects the local road plane stay where they are and the velocity of the contact points P_1 and P_2 induced by steering motions is just given by

$$v_{Pi} = \omega_{0i} \times r_{SiPi} = e_{Si}\, \dot{\delta}_i \times r_{SiPi} = e_{Si} \times r_{SiPi}\, \dot{\delta}_i, \quad i = 1, 2. \quad (5.17)$$

As a consequence, the corresponding partial velocities read as

$$\frac{\partial v_{P1}}{\partial \dot{u}_R} = e_{Si} \times r_{SiPi}\, i_{SLi}, \quad i = 1, 2, \quad (5.18)$$

where according to Equation (5.16) i_{SL1}, i_{SL2} denote the ratios of the steering linkage. Inserting Equations (5.16) and (5.18) into Equation (5.8) results in

$$F_{TG} = i_{SL1}\left[(e_{S1} \times r_{S1P1})^T F_{T1} + e_{S1}^T T_{T1}\right] + i_{SL2}\left[(e_{S2} \times r_{S2P2})^T F_{T2} + e_{S2}^T T_{T2}\right]. \quad (5.19)$$

Taking into account that

$$(a \times b)^T c = (-b \times a)^T c = \left(-\tilde{b}\, a\right)^T c = \left(\tilde{b}^T a\right)^T c = a^T\, \tilde{b}\, c = a^T (b \times c) \quad (5.20)$$

holds for arbitrary vectors a, b, c, Equation (5.19) can be rearranged to

$$F_{TG} = i_{SL1}\, e_{S1}^T\, (r_{S1P1} \times F_{T1} + T_{T1}) + i_{SL2}\, e_{S2}^T\, (r_{S2P2} \times F_{T2} + T_{T2}). \quad (5.21)$$

Referring to Figure 5.6, the orientation of the king pin axis can be described by the kingpin and caster angle. According to Equation (5.1),

$$e_{S1,F} = \frac{1}{\sqrt{\tan^2 \nu + \tan^2 \sigma + 1}} \begin{bmatrix} -\tan \nu \\ -\tan \sigma \\ 1 \end{bmatrix} \quad (5.22)$$

holds for the left wheel and

$$e_{S2,F} = \frac{1}{\sqrt{\tan^2 \nu + \tan^2 \sigma + 1}} \begin{bmatrix} -\tan \nu \\ +\tan \sigma \\ 1 \end{bmatrix} \quad (5.23)$$

for the right wheel, where the left/right symmetry of a standard suspension system was taken into account. In addition, the vectors from S_1 to P_1 and from S_2 to P_2 can be expressed by the caster offset c and steering offset s. According to Equation (5.2),

$$r_{S1P1} = \begin{bmatrix} -c \\ s \\ 0 \end{bmatrix} \quad \text{and} \quad r_{S2P2} = \begin{bmatrix} -c \\ -s \\ 0 \end{bmatrix} \quad (5.24)$$

will hold at the left and the right wheel in design position when symmetry is taken into account again. Combining Equations (5.22), (5.23), (5.24) with Equations (5.21) and (5.15) finally provides the steady-state steering torque in design position as

$$
\begin{aligned}
T_{SS}^{st} = -i_{S1} \begin{bmatrix} -\tan v \\ -\tan \sigma \\ 1 \end{bmatrix}^T \left(\begin{bmatrix} -c \\ +s \\ 0 \end{bmatrix} \times \begin{bmatrix} F_{x1} \\ F_{y1} \\ F_{z1} \end{bmatrix} + \begin{bmatrix} T_{x1} \\ T_{y1} \\ T_{z1} \end{bmatrix} \right) \\
-i_{S2} \begin{bmatrix} -\tan v \\ +\tan \sigma \\ 1 \end{bmatrix}^T \left(\begin{bmatrix} -c \\ -s \\ 0 \end{bmatrix} \times \begin{bmatrix} F_{x2} \\ F_{y2} \\ F_{z2} \end{bmatrix} + \begin{bmatrix} T_{x2} \\ T_{y2} \\ T_{z2} \end{bmatrix} \right),
\end{aligned}
\tag{5.25}
$$

where F_{xi}, F_{yi}, F_{zi}, $i = 1, 2$, are the components of the tire force vectors F_{T1}, F_{T2} expressing the longitudinal, the lateral, and the vertical tire forces or wheel loads, and T_{xi}, T_{yi}, T_{zi}, $i = 1, 2$, name the components of the tire torque vectors T_{T1}, T_{T2} representing the tipping torque, the rolling resistance torque, and the sum of the bore and the self-aligning torque. Furthermore, the abbreviations

$$
i_{Si} = \frac{i_{SLi}}{i_{SB} \sqrt{\tan^2 v + \tan^2 \sigma + 1}}, \quad i = 1, 2
\tag{5.26}
$$

define generalized ratios of the steering system. They will be equal or at least approximately equal to or close to the design position. Taking $i_{S1} = i_{S2} = i_S$ for granted, Equation (5.25) simplifies to

$$
\begin{aligned}
T_{SS}^{st} &= i_S \left[s \left(F_{x1} - F_{x2} \right) + c \left(F_{y1} + F_{y2} \right) + \left(s \tan v + c \tan \sigma \right) \left(F_{z1} - F_{z2} \right) \right. \\
&= \left. \tan v \left(T_{x1} + T_{x2} \right) - \tan \sigma \left(T_{y1} - T_{y2} \right) + T_{z1} + T_{z2} \right].
\end{aligned}
\tag{5.27}
$$

In normal driving situations, $F_{x1} \approx F_{x2}$, $F_{z1} \approx F_{z2}$, $T_{x1} \approx -T_{x2}$, and $T_{y1} \approx T_{y2}$ will hold. Then, the steering torque

$$
T_{SS}^{st} \approx c \left(F_{y1} + F_{y2} \right) + T_{z1} + T_{z2}
\tag{5.28}
$$

is dominated by the lateral forces and the bore and the self-aligning torques. However, during cornering, the term

$$
\left(s \tan v + c \tan \sigma \right) \left(F_{z1} - F_{z2} \right)
\tag{5.29}
$$

will provide an additional steering torque due to the load transfer from the inner to the outer wheel, which will increase according to the lateral acceleration. Modern vehicles are often equipped with an Electronic Stability Program (ESP) that may activate the brake at a single wheel. Then, $F_{x1} \neq F_{x2}$ will be the case and the term

$$
s \left(F_{x1} - F_{x2} \right)
\tag{5.30}
$$

will contribute to the steering torque unless the suspension system exhibits a vanishing or at least a sufficient small steering offset, $s \approx 0$. In parking maneuvers, the steering torque is dominated by the tire bore torques.

5.3.6.3 Parking Effort

One major design task of a steering system is the investigation of the parking effort. During a static parking maneuver, the steering wheel is turned while the vehicle is still at full stop. In this case, the forces and torques applied to the wheels are dominated by the tire bore torque. The motions of the vehicle can be neglected for a basic study. In addition, we assume here that the steering motion of the wheels is a simple rotation around a vertical axis running through the wheel centers C_1, C_2 and the contact points P_1, P_2. Then, the partial velocities needed in Equations (5.7) and (5.8) will simplify to

$$\frac{\partial v_{Ci}}{\partial \dot{u}_R} = \frac{\partial v_{Pi}}{\partial \dot{u}_R} = \begin{bmatrix} 0 \\ 0 \\ 0 \end{bmatrix} \quad \text{and} \quad \frac{\partial \omega_{0i}}{\partial \dot{u}_R} = \begin{bmatrix} 0 \\ 0 \\ 1 \end{bmatrix} i_{SLi}, \quad i = 1, 2, \tag{5.31}$$

where according to Equation (5.16), i_{SL1}, i_{SL2} abbreviate the partial derivatives of the time derivatives of the wheel steering angles δ_1, δ_2 with respect to the time derivative of the rack displacement u_R. Then, the generalized mass and the generalized rack force simplify to

$$m_{RG} = m_R + i_{SL1}^2 \Theta_{z1} + i_{SL2}^2 \Theta_{z2}, \tag{5.32}$$

$$F_{TG} = i_{SL1}(T_{S1} + T_{B1}) + i_{SL1}(T_{S2} + T_{B2}), \tag{5.33}$$

where Θ_{z1}, Θ_{z2} denote the inertias of the left and the right wheel and wheel body around the vertical z-axis. According to Equation (3.1), the tire torques around the z-axis consist of the self-aligning torques T_{S1}, T_{S2} and the bore torques T_{B1}, T_{B2}.

If the steering linkage is designed to satisfy the Ackermann geometry, the steering angles of the left and the right wheel will be provided by

$$\tan \delta_{1,2} = \frac{a \tan(i_{SL} u_R)}{a \mp \frac{1}{2}s \tan(i_{SL} u_R)}, \tag{5.34}$$

where i_{SL} denotes the ratio of the steering linkage at $u_R = 0$, s describes the track width, and a the wheel base of the vehicle. The time derivative of Equation (5.34) results in

$$\frac{1}{\cos^2 \delta_{1,2}} \dot{\delta}_{1,2} = \frac{a^2}{\left(a \mp \frac{1}{2}s \tan(i_{SL} u_R)\right)^2} \frac{1}{\cos^2(i_{SL} u_R)} i_{SL} \dot{u}_R \tag{5.35}$$

and provides the ratios of the steering linkage defined by the partial derivatives

$$i_{SL1,2} = \frac{\partial \dot{\delta}_{1,2}}{\partial \dot{u}_R} = i_{SL} \frac{a^2}{\left(a \mp \frac{1}{2}s \tan(u_R)\right)^2} \frac{\cos^2 \delta_{1,2}}{\cos^2(i_{SL} u_R)} \tag{5.36}$$

as functions of the rack displacement u_R because the corresponding steering angles $\delta_{1,2} = \delta_{1,2}(u_R)$ will be delivered by Equation (5.34).

The performance of the simple steering system model can be studied in a stand-alone simulation, now. The MATLAB-Function provided in Listing 5.1 describes the dynamics of the simple steering system model during a parking maneuver by a set of first-order differential equations.

Listing 5.1

Function fun_05_steering_system: steering system dynamics

```
 1 function [xdot,out] = fun_05_steering_system(t,x,p)
 2 % simple rack and pinion steering system model without steering wheel
 3
 4 ur  = x(1);   urdot = x(2); % rack displacement and velocity
 5 ga1 = x(3);   ga2   = x(4); % left and right tire torsional deflection
 6
 7 % soft periodic steering wheel input over one period
 8 if t <= 1/p.fSW
 9   dsw    =  p.ASW*sin(2*pi*p.fSW*t)^2;
10   dswdot = 2*p.ASW*sin(4*pi*p.fSW*t)*pi*p.fSW;
11   if t > 0.5/p.fSW, dsw=-dsw; dswdot = -dswdot; end
12 else
13   dsw = 0; dswdot = 0;
14 end
15 dm = p.iSL*ur; tand = tan(dm); % mean value of wheel steering angle
16
17 % kinematics of steering linkage (ackermann geometry)
18 d1    = atan2( p.a*tand , p.a-0.5*p.s*tand );
19 dd1   = p.iSL * p.a^2/(p.a-0.5*p.s*tand)^2 * cos(d1)^2/cos(dm)^2;
20 d1dot= dd1*urdot;
21 d2    = atan2( p.a*tand , p.a+0.5*p.s*tand );
22 dd2   = p.iSL * p.a^2/(p.a+0.5*p.s*tand)^2 * cos(d2)^2/cos(dm)^2;
23 d2dot= dd2*urdot;
24
25 % torque in steering column and rack force
26 dbi = p.iSB*ur;
27 tss = p.cS*(dsw-dbi) + p.dS*(dswdot-p.iSB*urdot);
28 fsm = p.iSB*tss;
29
30 % left bore torque (enhanced model @ stand still, vt=vtn)
31 sb = -p.rb*d1dot/p.vtn; sg = abs(sb); % bore slip = generalized slip
32 [~,fos] = tmy_fcombined(sg,p.df0,p.fm,p.sm,p.fs,p.ss);
33 % twist angle dynamics  and bore or turn torque
34 ga1dot=-(p.cbo*ga1*p.vtn+fos*d1dot*p.rb^2)/(p.dbo*p.vtn+fos*p.rb^2);
35 tbo1 = p.cbo*ga1 + p.dbo*ga1dot;
36
37 % right bore torque (enhanced model @ stand still, vt=vtn)
38 sb = -p.rb*d2dot/p.vtn; sg = abs(sb); % bore slip = generalized slip
39 [~,fos] = tmy_fcombined(sg,p.df0,p.fm,p.sm,p.fs,p.ss);
40 % twist angle dynamics and bore or turn torque
41 ga2dot=-(p.cbo*ga2*p.vtn+fos*d2dot*p.rb^2)/(p.dbo*p.vtn+fos*p.rb^2);
42 tbo2 = p.cbo*ga2 + p.dbo*ga2dot ;
43
44 ftg = dd1*tbo1 + dd2*tbo2; % generalized tire torques applied to rack
45
46 % enhanced dry friction model
47 fsfst = ftg + fsm;  % static part
48 mrg = p.mR+(dd1^2+dd2^2)*p.Thz; % general. rack mass
49 dN = p.iSB * sqrt( p.cS * mrg ); % numerical damping adjusted to data
50 fsf = fsfst + dN*urdot;   fsf = sign(fsf)*min(abs(fsf),p.frmx);
51
52 urdotdot = ( ftg+fsm-fsf ) / mrg ; % rack acceleration
53 xdot = [ urdot; urdotdot; ga1dot; ga2dot ]; % state derivatives
54
55 % additional output
56 out.dsw = dsw; out.dbi=dbi; out.d1=d1;    out.d2=d2;
57 out.ftg = ftg; out.fsf=fsf; out.tss=tss;
58
59 end
```

The function assumes a periodic steering input with a soft transition from positive to negative angles achieved by a squared sine function, and it includes a small part of the TMeasy tire model that automatically generates the dynamic bore torques. The MATLAB-Script provided in Listing 5.2 sets the required data, performs a time simulation, and finally plots some results.

Listing 5.2

Script app_05_steering_system.m: simulation of the parking effort

```
 1 par.ASW  = 540/180*pi; % steering wheel amplitude [rad]
 2 par.fSW  = 0.2;    % steering frequency [Hz]
 3 par.a    = 2.7;    % wheel base [m]
 4 par.s    = 1.5;    % track_width [m]
 5 par.mR   = 2.5;    % mass of rack [kg]
 6 par.Thz  = 1.3;    % wheel body & wheel inertia about z-axis [kgm^2]
 7 par.iSB  = 125;    % ratio steering wheel angle / rack displ. [rad/m]
 8 par.iSL  = 9.3;    % ratio wheel steering angle / rack displ. [rad/m]
 9 par.cS   = 573;    % stiffness of steering shaft [Nm/rad]
10 par.dS   = 5;      % damping of steering shaft [Nm/(rad/s)]
11 par.frmx = 120;    % maximum friction force in steering box [N]
12 par.df0  = 90000;  % tire force initial inclination [N/-]
13 par.fm   = 4250;   % maximum tire force [N]
14 par.sm   = 0.18;   % slip where f=f_m [-]
15 par.fs   = 4100;   % sliding force [N]
16 par.ss   = 0.5;    % slip where f=f_s [-]
17 par.vtn  = 0.01;   % fictitious velocity [m/s]
18 par.cbo  = 2400;   % torsional tire stiffness [Nm/rad]
19 par.dbo  = 2;      % torsional tire damping [Nm/(rad/s)]
20 par.rb   = 0.065; % equivalent bore radius [m]
21
22 % perorm simulation over one period of a sinusoidal steer input
23 t0=0; tE=1/par.fSW; x0 = [0; 0; 0; 0 ];
24 [tout,xout]=ode23(@(t,x) fun_05_steering_system(t,x,par),[t0,tE],x0);
25
26 % pre-allocate output vectors to speed up loop and plot some results
27 z=tout; dsw=z; dbi=z; d1=z; d2=z; ftg=z; fsf=z; tss=z;
28 for i=1:length(tout)
29   [~,out]=fun_05_steering_system(tout(i),xout(i,:),par);
30   dsw(i)=out.dsw; dbi(i)=out.dbi; d1(i)=out.d1; d2(i)=out.d2;
31   ftg(i)=out.ftg; fsf(i)=out.fsf; tss(i)=out.tss;
32 end
33 subplot(2,2,1)
34   plot(dsw*180/pi,-tss), grid on
35   title('steering torque [Nm] / steering wheel angle / degree')
36   xy=axis; axis([[-1,1]*par.ASW*180/pi,xy(3),xy(4)])
37 subplot(2,2,2)
38   plot(tout,(dbi-dsw)*180/pi), grid on,
39   title('twist angle of steering shaft / degree')
40   xy=axis; axis([0,tE,xy(3),xy(4)])
41 subplot(2,2,3)
42   plot(tout,[d1,d2,-xout(:,3:4)]*180/pi), grid on
43   legend('\delta_1','\delta_2','tire twist left','tire twist right')
44   xy=axis; axis([0,tE,xy(3),xy(4)])
45 subplot(2,2,4)
46   yyaxis left,  plot(tout,ftg/1000), hold on, ylim([-10,10])
47   yyaxis right, plot(tout,fsf), grid on, ylim([-200,200])
48   legend('rack force / kN','friction force / N')
```

The chosen wheelbase of $a = 2.7$ m and the track width of $s = 1.5$ m match typical mid-sized passenger cars. The rack mass $m_R = 2.5$ kg is roughly estimated. The inertia of wheel and wheel body are simply calculated on the basis of a cylindrical body of mass $m = 50$ kg, radius $r = 0.3$ m, and width $b = 0.2$ m, where a homogeneous mass distribution was assumed in addition. Then, the inertia is given by $\Theta = \frac{1}{12}m\left(b^2 + 3r^2\right) = 1.3$ kg m^2 the ratios of the steering system are sized up by specifying the maximum steering wheel amplitude with $\delta_{SW}^{mx} = 540°$ and estimating the maximum rack displacement with $u_R^{mx} = 0.075$ m. Then, the ratio of the steering wheel angle to rack displacement results in $i_{SB} = 540\,\pi/180/0.075 \approx 125$ rad/m. Assuming an average wheel steering angle of $\delta = 40°$ the ratio of the wheel steering angle to the rack displacement will be obtained as $i_{SL} = 40\,\pi/180/0.075/= 9.3$ rad/m. So, the overall ratio of steering wheel angle to wheel steering angle is given by $i_S = 125/9.3 = 13.44$ rad/rad, which corresponds quite well with mechanical rack-and-pinion steering systems. The torsional stiffness $c_S = 10/(1\,\pi/180) = 573$ Nm/rad would allow the steering shaft to twist up to 1° when the rack is locked and a torque of 10 Nm is applied at the steering wheel. The damping value $d_S = 5$ Nm/(rad/s) is roughly estimated. The tire data represent a typical passenger car tire.

The MATLAB solver ode23, an explicit Runge-Kutta (2,3) pair of Bogacki and Shampine, is slightly more efficient here than the MATLAB standard solver ode45 because the enhanced friction model is based on a simple function with sharp bends. The simulation results are shown in Figure 5.12, where the MATLAB commands yyaxis left and yyaxis right are used to combine different scaled graphs (here the rack force and the friction force) within one plot.

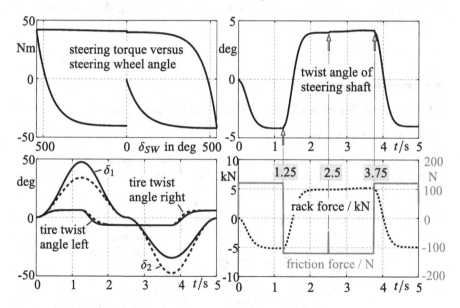

FIGURE 5.12
Stand-alone simulation results for periodic steer input.

The maximum bore torque is obtained by $T_B^{mx} = r_B F^M = 0.065*4250 = 276.25$ Nm at each tire. Then, in a first approximation, the maximum steering torque is given by $T_S^{mx} = 2\,T_B^{mx}/i_S = 2*276.25/13.44 = 41$ Nm, where i_S names the overall ratio of the steering system. This corresponds quite well with the first graph in Figure 5.12, where the steering wheel torque is plotted versus the steering wheel angle δ. The lower left graph shows the time histories of the tire twist angles and the wheel steering angles. As long as the tire patch sticks to the road, the tire twist angles will follow the wheel steering angles. When the tire patches start to slide, the maximum tire twist angles are limited to $\pm 6.4°$, which corresponds to the value $0.065 * 4100/2400 * 180/\pi = 6.4°$, which is calculated with the actual bore radius $r_B = 0.065$ m, the sliding force $F^S = 4100$ N, and the torsional tire stiffness $c_{BO} = 2400$ Nm/rad. The steering motions of the wheels are point-symmetric, ranging from $+47.4°$ to $-34.1°$ or $+34.1°$ to $-47.4°$, respectively. Amplifying the average value by the overall ratio of steering system results in $\delta_M^{mx} = 0.5(\pm 47.4 \pm 34.1)i_S = \pm 40.75 * 13.44 = \pm 547.7°$, which is slightly larger than the amplitude of the periodic steering wheel angle input $\delta_S^{mx} = \pm 540°$.

The friction in the steering box, which is approximated by an enhanced dry friction model where the maximum friction force is limited to $F_F^{mx} = 120$ N, causes the steering shaft to twist, which reduces the effective steering angle at the steering box input. At times 1.25 s and 3.75 *s*, the friction force, which is plotted in the lower right graph of Figure 5.12, jumps from $+120$ N to -120 N and back again. At the time 2.5 s the steering wheel input, provided by a series of two squared sine functions with opposite signs, shows a point of inflection that causes the friction force to drop to zero at this very moment. On closer inspection of the upper right graph in Figure 5.12, discontinuities are noticeable in the time history of the steering shaft twist angle at these particular times.

5.4 Kinematics of a Double Wishbone Suspension

5.4.1 Modeling Aspects

A classical double wishbone suspension consists of two control arms, the wheel carrier, and the wheel, Figure 5.13. An additional subframe, which will often be

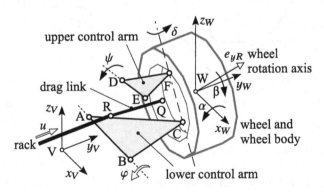

FIGURE 5.13
Double wishbone kinematic model.

present in sophisticated suspension systems is not considered here. Then, the lower and upper control arms will be attached directly to the chassis. In addition, the compliance of the corresponding bushings in A, B and D, E will be neglected. Ball joints in C and F connect the control arms with the wheel body. The rotation with the angle δ around the kingpin axis defined by C and F is controlled by the drag link. Via appropriate joints it is attached at point R to the rack and at point Q to the wheel body. The double wishbone suspension has two degrees of freedom. The hub motion is mainly controlled by the rotation of the lower control arm around the axis A-B and the steer motion is mainly induced by the rack movements. Hence, the angle φ and the displacement u are taken as generalized coordinates.

5.4.2 Position and Orientation

The momentary position and orientation of the wheel body are described with respect to the vehicle-fixed axis system with origin V located in the midst of the axle and where the axes x_V, y_V, and z_V point to the front to the left and upward. The axes x_W, y_W, and z_W are fixed to the wheel body. They will be parallel to the corresponding axes of the vehicle-fixed system in the design position. The vector

$$r_{VW,V} = r_{VA,V} + A_\varphi\, r_{AC,\varphi} + A_{VW}\, r_{CW,W} \tag{5.37}$$

specifies the momentary position of the wheel center W, where the vectors $r_{VA,V}$, $r_{AC,\varphi}$, $r_{CW,W}$ are fixed by the topology of the suspension and A_φ, A_{VW} are matrices describing the rotation of the lower wishbone and the wheel body with respect to the vehicle-fixed axis system. The lower control arm rotates with the angle φ around an axis that is determined by the location of the bushings A and B. Then, the relation

$$A_\varphi = e_{AB,V}\, e_{AB,V}^T + \left(I_{3\times3} - e_{AB,V}\, e_{AB,V}^T \right) \cos\varphi + \tilde{e}_{AB,V} \sin\varphi \tag{5.38}$$

defines the corresponding rotation matrix [30], where $e_{AB,V}$ is a unit vector pointing into the direction A→B, $I_{3\times3}$ is the 3×3 matrix of identity, and $\tilde{e}_{AB,V}$ denotes the skew-symmetric matrix defined by the components of the vector $e_{AB,V}$. The rotation matrix of the wheel body is composed of three rotations,

$$A_{VW} = A_\alpha\, A_\beta\, A_\delta \tag{5.39}$$

where

$$A_\alpha = \begin{bmatrix} 1 & 0 & 0 \\ 0 & \cos\alpha & -\sin\alpha \\ 0 & \sin\alpha & \cos\alpha \end{bmatrix}, \quad A_\beta = \begin{bmatrix} \cos\beta & 0 & \sin\beta \\ 0 & 1 & 0 \\ -\sin\beta & 0 & \cos\beta \end{bmatrix} \tag{5.40}$$

are elementary rotations around the corresponding x- and y-axis and

$$A_\delta = e_{CF,W}\, e_{CF,W}^T + \left(I_{3\times3} - e_{CF,W}\, e_{CF,W}^T \right) \cos\delta + \tilde{e}_{CF,W} \sin\delta \tag{5.41}$$

defines, similar to Equation 5.38, the rotation of the wheel body around the kingpin axis defined by the ball joints C and F.

5.4.3   Constraint Equations

5.4.3.1   Control Arms and Wheel Body

The ball joint F connects the upper control arm to the wheel body. Similar to Equation (5.37), the momentary position of F attached to the wheel body is defined by the vector

$$r_{VF,V} = r_{VA,V} + A_\varphi r_{AC,\varphi} + A_{VW} r_{CF,W} \tag{5.42}$$

where the vector $r_{CF,W}$, measured in the wheel body-fixed axis system, is given by data and describes the position of the upper ball joint F relative to the lower ball joint C. As F is located on the upper control arm too, the vector

$$r_{VF,V} = r_{VD,V} + A_\psi r_{DF,\psi} \tag{5.43}$$

will represent another way to determine the momentary position of the upper ball joint. The vector $r_{VD,V}$ describes the location of bushing D with respect to the vehicle-fixed axis system, and the vector $r_{DF,\psi}$ defines the location of the ball joint F relative to D. Both vectors are defined by the topology of the suspension. Similar to Equation (5.38), the matrix

$$A_\psi = e_{DE,V}\, e_{DE,V}^T + \left(I_{3\times3} - e_{DE,V}\, e_{DE,V}^T\right)\cos\psi + \tilde{e}_{DE,V}\sin\psi \tag{5.44}$$

describes the rotation of the upper control arm with the angle ψ around an axis determined by the location of the bushings D and E. Equation (5.42) with Equation (5.43) results in a first constraint equation

$$r_{VA,V} + A_\varphi r_{AC,\varphi} + A_{VW} r_{CF,W} = r_{VD,V} + A_\psi r_{DF,\psi} \tag{5.45}$$

Using Equation (5.39) and rearranging some terms results in

$$\underbrace{A_\alpha A_\beta A_\delta}_{A_{VW}} r_{CF,W} = \underbrace{r_{VD,V} - \left(r_{VA,V} + A_\varphi r_{AC,\varphi}\right)}_{r_{CD,V}(\varphi)} + A_\psi r_{DF,\psi} \tag{5.46}$$

The rotation matrix A_δ given by Equation (5.41) describes the rotation around the kingpin axis defined by the unit vector $e_{CF,W}$. As the vector $r_{CF,W}$ coincides with the rotation axis $e_{CF,W}$, it will be not affected by this rotation,

$$A_\delta r_{CF,W} = r_{CF,W} \tag{5.47}$$

Then, Equation (5.46) simplifies to

$$A_\alpha A_\beta r_{CF,W} = r_{CD,V}(\varphi) + A_\psi r_{DF,\psi} \tag{5.48}$$

which makes it possible to calculate the angles α, β, and ψ as a function of the rotation angle φ, which was chosen as a generalized coordinate. Although Equation (5.48) results in three nonlinear trigonometric relations, an algebraic solution is possible here.

Squaring[2] both sides of Equation (5.48) results in

$$\left(A_\alpha A_\beta r_{CF,w}\right)^T\left(A_\alpha A_\beta r_{CF,w}\right) = \left(r_{CD,V}+A_\psi r_{DF,\psi}\right)^T\left(r_{CD,V}+A_\psi r_{DF,\psi}\right) \quad (5.49)$$

where $r_{CD,V}(\varphi)$ was shortened by $r_{CD,V}$. Taking the orthogonality, which is defined in Equation (1.8), for the rotation matrices A_α, A_β, and A_ψ into account, one gets

$$\underbrace{r_{CF,w}^T r_{CF,w}}_{CF^2} = \underbrace{r_{CD,V}^T r_{CD,V}}_{CD^2} + 2 r_{CD,V}^T A_\psi r_{DF,\psi} + \underbrace{r_{DF,\psi}^T r_{DF,\psi}}_{DF^2} \quad (5.50)$$

Making use of Equation (5.44) and rearranging some terms yields

$$r_{CD,V}^T \left\{ee^T + \left(I - ee^T\right)\cos\psi + \tilde{e}\sin\psi\right\} r_{DF,\psi} + \tfrac{1}{2}\left(CD^2 + DF^2 - CF^2\right) = 0 \quad (5.51)$$

where the subscripts of $e_{DE,V}$ and $I_{3\times3}$ were just omitted. By using the abbreviations

$$a = r_{CD,V}^T\left(I - ee^T\right)r_{DF,\psi}, \quad b = r_{CD,V}^T\,\tilde{e}\,r_{DF,\psi} = r_{CD,V}^T\,e\times r_{DF,\psi} \quad (5.52)$$

and

$$c = -\left[r_{CD,V}^T\,ee^T\,r_{DF,\psi} + \tfrac{1}{2}\left(CD^2 + DF^2 - CF^2\right)\right] \quad (5.53)$$

Equation (5.51) can be written as

$$a\cos\psi + b\sin\psi = c \quad (5.54)$$

Interpreting the terms $a/\sqrt{a^2 + b^2}$ and $b/\sqrt{a^2 + b^2}$ as the sine and cosine of a fictitious angle ϑ, Equation (5.54) will read as

$$\sin\vartheta\,\cos\psi + \cos\vartheta\,\sin\psi = \frac{c}{\sqrt{a^2 + b^2}} \quad (5.55)$$

Using the addition theorem $\sin\vartheta\,\cos\psi + \cos\vartheta\,\sin\psi = \sin(\vartheta+\psi)$ and taking into account that the fictitious angle is defined by the relationship $\tan\vartheta = a/b$, one finally gets

$$\sin(\vartheta+\psi) = \frac{c}{\sqrt{a^2 + b^2}} \quad \text{or} \quad \psi = \arcsin\frac{c}{\sqrt{a^2 + b^2}} - \arctan\frac{a}{b} \quad (5.56)$$

Note that practical results will be achieved only if the parameter b is not negative. In this case, multiplication of Equation (5.54) with -1 will change the sign appropriately. Equation (5.56) provides the rotation angle of the upper control arm as a function of the lower one, $\psi = \psi(\varphi)$. Then, the corresponding rotation matrix is determined via Equation (5.44) and the right-hand side of Equation (5.48) can be summarized in the vector

$$r_{CF,V}(\varphi) = r_{CD,V}(\varphi) + A_\psi\left(\psi(\varphi)\right)r_{DF,\psi} \quad (5.57)$$

---

[2]Note: Squaring a vector is equivalent to its scalar product, $r^2 = r^T r$.

Multiplying the constraint equation (5.48) by A_α^T and taking Equations (5.57) and (5.40) into consideration results in

$$\begin{bmatrix} \cos\beta & 0 & \sin\beta \\ 0 & 1 & 0 \\ -\sin\beta & 0 & \cos\beta \end{bmatrix} \begin{bmatrix} r_{CF,W}^{(1)} \\ r_{CF,W}^{(2)} \\ r_{CF,W}^{(3)} \end{bmatrix} = \begin{bmatrix} 1 & 0 & 0 \\ 0 & \cos\alpha & \sin\alpha \\ 0 & -\sin\alpha & \cos\alpha \end{bmatrix} \begin{bmatrix} r_{CF,V}^{(1)} \\ r_{CF,V}^{(2)} \\ r_{CF,V}^{(3)} \end{bmatrix} \tag{5.58}$$

which immediately will deliver two[3] equations,

$$r_{CF,W}^{(1)} \cos\beta + r_{CF,W}^{(3)} \sin\beta = r_{CF,V}^{(1)}, \tag{5.59}$$

$$r_{CF,V}^{(2)} \cos\alpha + r_{CF,V}^{(3)} \sin\alpha = r_{CF,W}^{(2)}. \tag{5.60}$$

These equations of the type of Equation (5.54) are solved similar to Equation (5.56) and will provide the angles $\alpha = \alpha(\varphi)$, $\beta = \beta(\varphi)$ needed to compose the elementary rotation matrices A_α and A_β.

5.4.3.2   Steering Motion

The drag link is attached in R to the rack and in Q to the wheel body. The momentary positions of these points are determined by

$$r_{VR,V} = r_{VR,V}^D + \begin{bmatrix} 0 \\ u \\ 0 \end{bmatrix} \tag{5.61}$$

and

$$r_{VQ,V} = r_{VA,V} + A_\varphi r_{AC,\varphi} + A_\alpha A_\beta A_\delta r_{CQ,W}, \tag{5.62}$$

where the vectors $r_{VR,V}^D$, $r_{VA,V}$, $r_{AC,\varphi}$, $r_{CQ,W}$ are given by the topology of the suspension and a pure lateral movement of the rack was assumed. The drag link is supposed to be rigid, which results in a second constraint equation

$$\left(r_{VQ,V} - r_{VR,V} \right)^T \left(r_{VQ,V} - r_{VR,V} \right) = \ell_{RQ}^2 \tag{5.63}$$

where ℓ_{RQ} denotes the length of the drag link. Introducing the abbreviation

$$r_{RC,V} = r_{VA,V} + A_\varphi r_{AC,\varphi} - r_{VR,V} \tag{5.64}$$

the constraint equation will read as

$$\left(r_{RC,V} + A_\alpha A_\beta A_\delta r_{CQ,W} \right)^T \left(r_{RC,V} + A_\alpha A_\beta A_\delta r_{CQ,W} \right) = \ell_{RQ}^2 \tag{5.65}$$

which results in

$$\underbrace{r_{RC,V}^T r_{RC,V}}_{RC^2} + 2 \underbrace{r_{RC,V}^T A_\alpha A_\beta A_\delta r_{CQ,W}}_{r_{RC,H}^T} + \underbrace{r_{CQ,W}^T r_{CQ,W}}_{CQ^2} = \ell_{RQ}^2 \tag{5.66}$$

---

[3]Note: The third equation corresponds with Equation (5.49) and thus can provide no additional information.

Inserting the rotation matrix A_δ defined in Equation (5.41) finally yields

$$r_{RC,H}^T \left(I - e e^T\right) r_{CQ,W} \, \cos \delta + r_{RC,H}^T \tilde{e} \, r_{CQ,W} \, \sin \delta =$$
$$-\left[r_{RC,H}^T e e^T r_{CQ,W} + \tfrac{1}{2}\left(RC^2 + CQ^2 - \ell_{RQ}^2\right)\right] \tag{5.67}$$

where the subscripts of the identity matrix $I_{3\times3}$ and the unit vector $e_{CF,W}$ were omitted. Equation (5.67) is again of the type of Equation (5.54) and can be solved accordingly. As the movements of points R and Q depend on the rack displacement and on the rotation of the lower control arm, the steering angle δ will be a function of u and φ. The corresponding rotation matrix A_δ is defined by Equation (5.41).

Now, the momentary position of the wheel center W relative to the vehicle-fixed axis system can be computed. According to Equation (5.37), one gets

$$r_{VW,V}(\varphi, u) = r_{VA,V} + A_\varphi(\varphi) r_{AC,\varphi} + A_{VW}(\varphi, u) r_{CW,W} \tag{5.68}$$

where the rotation matrix

$$A_{VW}(\varphi, u) = A_\alpha(\varphi) A_\beta(\varphi) A_\delta(\varphi, u) \tag{5.69}$$

describes the orientation of the wheel body-fixed axis system relative to the vehicle-fixed axis system. Within the wheel body-fixed axis system, the unit vector in the direction of the wheel rotation axis may be determined by the toe angle δ_0 and the camber angle $\hat{\gamma}_0$. According to Equation (1.2), one gets

$$e_{yR,V} = \frac{1}{\sqrt{\tan^2 \delta_0 + 1 + \tan^2 \hat{\gamma}_0}} \begin{bmatrix} \tan \delta_0 \\ 1 \\ -\tan \hat{\gamma}_0 \end{bmatrix} \tag{5.70}$$

Then, the orientation of the wheel rotation axis with respect to the vehicle-fixed axis system is simply given by

$$e_{yR,V} = A_{VW} \, e_{yR,V} \tag{5.71}$$

A standard MacPherson suspension is modeled in [59]. Besides an analytic solution, the multibody approach with commercial software packages is presented too.

5.4.4 Velocities

The time derivative of Equation (5.68) provides the velocity of the wheel center with respect to the vehicle-fixed axis system V,

$$\dot{r}_{VW,V} = \omega_{\varphi,V} \times \underbrace{A_\varphi(\varphi) r_{AC,\varphi}}_{r_{AC,V}} + \omega_{VW,V} \times \underbrace{A_{VW}(\varphi, u) r_{CW,W}}_{r_{CW,V}} \tag{5.72}$$

where $r_{VA,V} = $ const. was already taken into account. The rotation matrix A_φ defined in Equation (5.38) describes the orientation of the lower control arm around an axis that is defined by the unit vector $e_{AB,V}$. Then, the angular velocity of the lower control arm is simply given by

$$\omega_{\varphi,V} = e_{AB,V}\,\dot{\varphi} \tag{5.73}$$

According to Equation (5.39), the rotation matrix A_{VW} describing the orientation of the wheel body W relative to the vehicle-fixed axis system V is composed of a series of rotations which are defined in Equations (5.40) and (5.41). Then, the corresponding angular velocity is given by

$$\omega_{VW,V} = e_{xV,V}\,\dot{\alpha} + \underbrace{A_\alpha\,e_{y_\alpha,\alpha}}_{e_{y_\alpha,V}}\,\dot{\beta} + \underbrace{A_\alpha\,A_\beta\,e_{CF,W}}_{e_{CF,V}}\,\dot{\delta} \tag{5.74}$$

where

$$e_{xV,V} = \begin{bmatrix} 1 \\ 0 \\ 0 \end{bmatrix} \quad \text{and} \quad e_{y_\alpha,\alpha} = \begin{bmatrix} 0 \\ 1 \\ 0 \end{bmatrix} \tag{5.75}$$

denote unit vectors in the direction of the corresponding rotation axis. The time derivative of the first constraint equation (5.45) yields

$$\omega_{\varphi,V} \times r_{AC,V} + \omega_{VW,V} \times r_{CF,V} = \omega_{\psi,V} \times \underbrace{A_\psi\,r_{DF,\psi}}_{r_{DF,V}} \tag{5.76}$$

where the position vectors $r_{AC,V}$, $r_{CW,V}$ are defined in Equation (5.72). Inserting the angular velocities provided by Equations (5.73) and (5.74) results in

$$e_{AB,V} \times r_{AC,V}\dot{\varphi} + \left(e_{xV,V}\dot{\alpha} + e_{y_\alpha,V}\dot{\beta} + e_{CF,V}\dot{\delta}\right) \times r_{CF,V} = e_{DE,V} \times r_{DF,V}\dot{\psi} \tag{5.77}$$

where the angular velocity of the upper control arm is given by

$$\omega_{\psi,V} = e_{DE,V}\,\dot{\psi} \tag{5.78}$$

and $e_{DE,V}$ denotes the unit vector in the direction of the rotation axis. The unit vector $e_{CF,V}$ points in the direction of the vector $r_{CF,V}$. Then, the corresponding cross-product vanishes, $e_{CF,V} \times r_{CF,V} = 0$, and Equation (5.77) simplifies to

$$e_{AB,V} \times r_{AC,V}\dot{\varphi} + e_{xV,V} \times r_{CF,V}\dot{\alpha} + e_{y_\alpha,V} \times r_{CF,V}\dot{\beta} = e_{DE,V} \times r_{DF,V}\dot{\psi} \tag{5.79}$$

The kinematics of the double wishbone suspension is fully determined by the generalized coordinates φ and u, which represent the rotation of the lower control arm and the rack displacement. The constraint equation (5.48) provides the angles α and β as well as the rotation angle of the upper control arm ψ as functions of φ,

$$\psi = \psi(\varphi), \quad \alpha = \alpha(\varphi), \quad \beta = \beta(\varphi) \tag{5.80}$$

Whereas according to the constraint equation (5.67), the angle δ depends on φ and u,

$$\delta = \delta(\varphi,\,u) \tag{5.81}$$

then the time derivatives of the corresponding angles are obtained as

$$\dot{\alpha} = \frac{\partial \alpha}{\partial \varphi} \dot{\varphi}, \quad \dot{\beta} = \frac{\partial \beta}{\partial \varphi} \dot{\varphi}, \quad \dot{\psi} = \frac{\partial \psi}{\partial \varphi} \dot{\varphi}, \quad \dot{\delta} = \frac{\partial \delta}{\partial \varphi} \dot{\varphi} + \frac{\partial \delta}{\partial u} \dot{u} \tag{5.82}$$

Although just the angle δ depends on both generalized coordinates, the partial derivatives were used here for all angles in order to keep the calculations more general.

Inserting Equation (5.82) into Equation (5.79), rearranging some terms, and canceling the angular velocity of the lower control arm $\dot{\varphi}$ finally results in

$$e_{DE,V} \times r_{DF,V} \frac{\partial \psi}{\partial \varphi} - e_{xv,V} \times r_{CF,V} \frac{\partial \alpha}{\partial \varphi} - e_{y\alpha,V} \times r_{CF,V} \frac{\partial \beta}{\partial \varphi} = e_{AB,V} \times r_{AC,V} \tag{5.83}$$

which finally represents, for the three unknown partial derivatives $\partial \psi / \partial \varphi$, $\partial \alpha / \partial \varphi$, and $\partial \beta / \partial \varphi$, the corresponding scalar equations.

The time derivative of the second constraint equation (5.63) results in

$$2 \underbrace{\left(r_{VQ,V} - r_{VR,V} \right)^T}_{r_{RQ,V}} \underbrace{\left(\dot{r}_{VQ,V} - \dot{r}_{VR,V} \right)}_{\dot{r}_{RQ,V}} = 0 \tag{5.84}$$

where the position vectors $r_{VR,V}$, $r_{VQ,V}$ are defined in Equations (5.61), (5.62), and their time derivatives are given by

$$\dot{r}_{VR,V} = \begin{bmatrix} 0 \\ \dot{u} \\ 0 \end{bmatrix} = \begin{bmatrix} 0 \\ 1 \\ 0 \end{bmatrix} \dot{u} = e_{yv,V} \dot{u} \tag{5.85}$$

$$\dot{r}_{VQ,V} = \omega_{\varphi,V} \times r_{AC,V} + \omega_{VW,V} \times r_{CQ,V} \tag{5.86}$$

The angular velocities $\omega_{\varphi,V}$ and $\omega_{VW,V}$ are defined in Equations (5.73) and (5.74). Taking Equation (5.82) into account provides the time derivative of the vector $r_{VQ,V}$ as a function of the angular velocity of the lower control arm $\dot{\varphi}$ and the rack velocity $\dot{u}$

$$\dot{r}_{VQ,V} = \frac{\partial \dot{r}_{VQ,V}}{\partial \dot{\varphi}} \dot{\varphi} + \frac{\partial \dot{r}_{VQ,V}}{\partial \dot{u}} \dot{u} \tag{5.87}$$

where the corresponding partial derivatives are given by

$$\frac{\partial \dot{r}_{VQ,V}}{\partial \dot{\varphi}} = e_{AB,V} \times r_{AC,V} + \left(e_{xv,V} \frac{\partial \alpha}{\partial \varphi} + e_{y\alpha,V} \frac{\partial \beta}{\partial \varphi} + e_{CF,V} \frac{\partial \delta}{\partial \varphi} \right) \times r_{CQ,V} \tag{5.88}$$

$$\frac{\partial \dot{r}_{VQ,V}}{\partial \dot{u}} = e_{CF,V} \frac{\partial \delta}{\partial u} \times r_{CQ,V} \tag{5.89}$$

Now, Equation (5.84) reads as

$$r_{RQ,V}^T \left(\frac{\partial \dot{r}_{VQ,V}}{\partial \dot{\varphi}} \dot{\varphi} + \frac{\partial \dot{r}_{VQ,V}}{\partial \dot{u}} \dot{u} - e_{yv,V} \dot{u} \right) = 0 \tag{5.90}$$

where the factor 2 was simply canceled. The time derivatives of the generalized coordinates $\dot{\varphi}$ and $\dot{u}$ are completely independent. Then, Equation (5.90) requires that

$$r_{RQ,V}^T \frac{\partial \dot{r}_{VQ,V}}{\partial \dot{\varphi}} = 0 \quad \text{and} \quad r_{RQ,V}^T \left(\frac{\partial \dot{r}_{VQ,V}}{\partial \dot{u}} - e_{yV,V} \right) = 0 \tag{5.91}$$

holds separately. By using Equations (5.88) and (5.89), the demands result in

$$r_{RQ,V}^T \left(e_{AB,V} \times r_{AC,V} + \left[e_{xV,V} \frac{\partial \alpha}{\partial \varphi} + e_{ya,V} \frac{\partial \beta}{\partial \varphi} \right] \times r_{CQ,V} + e_{CF,V} \times r_{CQ,V} \frac{\partial \delta}{\partial \varphi} \right) = 0 \tag{5.92}$$

$$r_{RQ,V}^T \left(e_{CF,V} \times r_{CQ,V} \frac{\partial \delta}{\partial u} - e_{yV,V} \right) = 0 \tag{5.93}$$

which will deliver the partial derivatives $\partial \delta / \partial \varphi$ and $\partial \delta / \partial u$ in a few steps.

Now, the angular velocity of the wheel body-fixed axis system W with respect to the vehicle-fixed axis system V defined by Equation (5.74) can be written as

$$\omega_{VW,V} = \frac{\partial \omega_{VW,V}}{\partial \dot{\varphi}} \dot{\varphi} + \frac{\partial \omega_{VW,V}}{\partial \dot{u}} \dot{u} \tag{5.94}$$

where the corresponding partial derivatives are given by

$$\frac{\partial \omega_{VW,V}}{\partial \dot{\varphi}} = e_{xV,V} \frac{\partial \alpha}{\partial \varphi} + e_{ya,V} \frac{\partial \beta}{\partial \varphi} + e_{CF,V} \frac{\partial \delta}{\partial \varphi} \quad \text{and} \quad \frac{\partial \omega_{VW,V}}{\partial \dot{u}} = e_{CF,V} \frac{\partial \delta}{\partial u} \tag{5.95}$$

Finally, the velocity of the wheel center with respect to the vehicle-fixed axis system V defined by Equation (5.72) is just obtained as

$$\dot{r}_{VW,V} = \frac{\partial \dot{r}_{VW,V}}{\partial \dot{\varphi}} \dot{\varphi} + \frac{\partial \dot{r}_{VW,V}}{\partial \dot{u}} \dot{u} \tag{5.96}$$

where the corresponding partial derivatives are provided by

$$\frac{\partial \dot{r}_{VW,V}}{\partial \dot{\varphi}} = e_{AB,V} \times r_{AC,V} + \frac{\partial \omega_{VW,V}}{\partial \dot{\varphi}} \times r_{CW,V} \quad \text{and} \quad \frac{\partial \dot{r}_{VW,V}}{\partial \dot{u}} = \frac{\partial \omega_{VW,V}}{\partial \dot{u}} \times r_{CW,V} \tag{5.97}$$

The partial derivatives of the angular velocity and the velocity with respect to the generalized velocities, which here equals the time derivatives of the generalized coordinates, form the basis of Jourdain's principle. They are simply called partial velocities in Section 1.5.

5.4.5 Acceleration

The time derivatives of Equations (5.94) and (5.96) provide the angular acceleration of the wheel body-fixed axis system W and the acceleration of the wheel center with respect to the vehicle-fixed axis system V. Although the kinematics of nearly all suspension systems is nonlinear, the partial derivatives of the angular velocity and the velocity of the wheel center defined by Equations (5.95) and (5.97) do not change very much in the normal range of motion. Then, the accelerations can be approximated by

$$\dot{\omega}_{VW,V} \approx \frac{\partial \omega_{VW,V}}{\partial \dot{\varphi}} \ddot{\varphi} + \frac{\partial \omega_{VW,V}}{\partial \dot{u}} \ddot{u} \tag{5.98}$$

$$\ddot{r}_{VW,V} \approx \frac{\partial \dot{r}_{VW,V}}{\partial \dot{\varphi}} \ddot{\varphi} + \frac{\partial \dot{r}_{VW,V}}{\partial \dot{u}} \ddot{u} \qquad (5.99)$$

Neglecting the time derivatives of the partial derivatives saves a lot of computing time and still generates very accurate results [45].

5.4.6 Kinematic Analysis

The function fun_05_dblwb_kin provided by Listing 5.3 calculates the kinematics of a double wishbone suspension system by computing the corresponding equations derived in the previous subsections. It uses the functions uty_skewsym and uty_trigon, which are given in Listings 5.4 and 5.5, to generate a skew-symmetric matrix according to Equation (1.13) and to solve trigonometric equations of the type of Equation (5.54).

Listing 5.3

Function fun_05_dblwb_kin: double wishbone axle kinematics

```
1  function [ avw ...      % rotation matrix wheel rim / ref-sys
2           , rvwv ...     % actual position of wheel center
3           , del ...      % rotation angle arround king pin
4           , pv ...       % partial velocities
5           ] = ...
6  fun_05_dblwb_kin ...    % kinematics of a double wishbone suspension
7           ( phi ...      % rotation angle of lower control arm
8           , u ...        % rack displacement
9           , p ...        % structure of model parameter
10          )
11
12 % lower wishbone
13   rab = p.rvbk-p.rvak; eab = rab/norm(rab); eabs = uty_skewsym(eab);
14   eabeab = eab*eab.'; eyeab = eye(3,3)-eabeab;
15   aphi = eabeab + eyeab*cos(phi) + eabs*sin(phi);
16
17 % upper wishbone
18   rcfk = p.rvfk-p.rvck; rdfk = p.rvfk-p.rvdk;
19   racv=aphi*(p.rvck-p.rvak); rcdv=p.rvdk-(p.rvak+racv);
20   rde=p.rvek-p.rvdk;
21   ede=rde/norm(rde); edeede=ede*ede.'; edes=uty_skewsym(ede);
22   a = rcdv.'*(eye(3,3)-edeede)*rdfk;  b = rcdv.'*cross(ede,rdfk);
23   c = -rcdv.'*edeede*rdfk-0.5*(rdfk.'*rdfk+rcdv.'*rcdv-rcfk.'*rcfk);
24   psi = uty_trigon(a,b,c);
25   apsi = edeede + (eye(3,3)-edeede)*cos(psi) + edes*sin(psi);
26
27 % orientation of wheel body due to control arm motion
28   rvcv = p.rvak + racv;
29   rdfv = apsi*(p.rvfk-p.rvdk); rvfv = p.rvdk + rdfv; rcfv = rvfv-rvcv;
30   be = uty_trigon(rcfk(1),rcfk(3),rcfv(1));
31   abe = [ cos(be)  0  sin(be) ; ...
32              0     1    0     ; ...
33          -sin(be)  0  cos(be) ];
34   al = uty_trigon(rcfv(2),rcfv(3),rcfk(2));
35   aal = [ 1     0        0     ; ...
36           0  cos(al) -sin(al) ; ...
37           0  sin(al)  cos(al) ] ;
38
39 % rotation arround king pin
40   rvrv = p.rvrk + [ 0; u; 0 ]; rrcv = rvcv-rvrv; rrcht=rrcv.'*aal*abe;
41   rrqk = p.rvqk-p.rvrk; rcqk = p.rvqk-p.rvck; rcfk = p.rvfk-p.rvck;
```

```
42   ecf = rcfk/norm(rcfk); ecfecf = ecf*ecf.'; ecfs = uty_skewsym(ecf);
43   a = rrcht*(eye(3,3)-ecfecf)*rcqk;  b = rrcht*cross(ecf,rcqk);
44   c = -(rrcht*ecfecf*rcqk+0.5*(rrcv.'*rrcv+rcqk.'*rcqk-rrqk.'*rrqk));
45   del=uty_trigon(a,b,c);
46   adel=ecfecf+(eye(3,3)-ecfecf)*cos(del)+ecfs*sin(del);
47
48 % wheel body orientation and position
49   avw = aal*abe*adel;  rcwv = avw*(p.rvwk-p.rvck); rvwv = rvcv + rcwv;
50
51 % partial derivatives: dpsi/dphi, dal/dphi, dbe/dphi
52   exvv=[1;0;0]; eyalv=aal*[0;1;0];
53   a = [ cross(ede,rdfv)  -cross(exvv,rcfv)  -cross(eyalv,rcfv) ];
54   c = a \ cross(eab,racv); dpsidphi=c(1); daldphi=c(2); dbedphi=c(3);
55
56 % partial derivatives: ddel/dphi, ddel/du
57   eyvv=[0;1;0]; ecfv=rcfv/norm(rcfv);
58   rcqv=avw*rcqk; rvqv=rvcv+rcqv; rrqv=rvqv-rvrv;
59   a = exvv*daldphi + eyalv*dbedphi; b = cross(eab,racv)+cross(a,rcqv);
60   c = rrqv.'*cross(ecfv,rcqv);
61   ddeldphi = -rrqv.'*b/c; ddeldu = rrqv.'*eyvv/c;
62
63 % partial angular and partial velocities of wheel carrier
64   dodphi=exvv*daldphi+eyalv*dbedphi+ecfv*ddeldphi;  dodu=ecfv*ddeldu;
65   dvdphi=cross(eab,racv)+cross(dodphi,rcwv); dvdu = cross(dodu,rcwv);
66   pv = [ dodphi dodu dvdphi dvdu ];
67
68 end
```

Listing 5.4
Function `uty_skewsym`: generate skew-symmetric matrix

```
1 function matrix = uty_skewsym( vector )
2   matrix = [    0      -vector(3)   vector(2) ; ...
3             vector(3)       0      -vector(1) ; ...
4            -vector(2)   vector(1)       0     ];
5 end
```

Listing 5.5
Function `uty_trigon`: solve trigonometric equation

```
1 function psi = uty_trigon(a,b,c)  % solve   a*cos(psi) + b*sin(psi) = c
2   if b<0
3     psi = asin(-c/sqrt(a^2+b^2)) - atan2(-a,-b);
4   else
5     psi = asin( c/sqrt(a^2+b^2)) - atan2( a, b);
6   end
7 end
```

The MATLAB-Script in Listing 5.6 provides the data for a typical passenger car front suspension and performs a kinematic analysis of the double wishbone suspension. At first, some relevant properties are computed in the design position and displayed via the MATLAB-Function disp. The calculation of kingpin and caster angle is straightforward because its orientation is determined by the ball joints C and F which are defined by the substructures dt.rvck and dt.rvfk in the design position. The caster offset c and the scrub radius s are visualized in Figure 5.6 and defined by Equation 5.2. Here, the vector from the point S where an extension of the kingpin axis intersects the road surface to the contact point P is given by

$$\underbrace{-c\,e_x + s\,e_y}_{r_{SP}} = \underbrace{\lambda\,e_{CF}}_{r_{SC}} + \underbrace{r_{CW} + r_{WP}}_{r_{CP}}, \tag{5.100}$$

where the caster offset c, the scrub radius s, and the parameter λ are yet unknown. The unit vector e_{CF} in the direction of the kingpin axis is defined by the ball joints C and F, and the vectors r_{CW}, r_{WP} are given by data in the design position. The unit vectors e_x, e_y in the longitudinal and the lateral direction are perpendicular to each other perpendicular to the road normal, which is defined by the unit vector e_n. Then, scalar multiplication of Equation 5.100 results in

$$0 = \lambda e_n^T e_{CF} + e_n^T r_{CP}, \qquad (5.101)$$

which immediately delivers the parameter λ or the vector

$$r_{SC} = \lambda e_{CF} = -\frac{e_n^T r_{CP}}{e_n^T e_{CF}} e_{CF} \qquad (5.102)$$

Finally, the caster offset and the scrub radius,

$$c = -e_x^T (r_{SC} + r_{CP}) \quad \text{and} \quad s = e_y^T (r_{SC} + r_{CP}) \qquad (5.103)$$

are obtained by scalar multiplication of Equation 5.100 with e_x and e_y, respectively. Running the MATLAB-Script will provide the following results: kingpin angle = 10.5182, caster angle = 7.0561, caster offset = 0.030326, and scrub radius = 0.0010327, which are typical values for modern passenger car front suspension systems. In addition, the script generates some plots, which are shown in Figure 5.14 and Figure 5.15.

Listing 5.6
MATLAB-Script app_05_dblwb_kin.m: double wishbone kinematic analysis

```
1  par.umx  = 0.0745;  % max. rack displacement
2  par.phmx = 10/180*pi;  % max rotation of lower arm
3  par.rvwk=[ 0.000;  0.768;  0.000]; % W wheel rim center
4  par.rvak=[-0.251;  0.320; -0.080]; % A lower arm @ chassis rear
5  par.rvbk=[ 0.148;  0.320; -0.094]; % B lower arm @ chassis front
6  par.rvck=[ 0.013;  0.737; -0.145]; % C lower arm @ wheel body
7  par.rvdk=[-0.105;  0.435;  0.196]; % D upper arm @ body rear
8  par.rvek=[ 0.122;  0.435;  0.230]; % E upper arm @ body front
9  par.rvfk=[-0.025;  0.680;  0.162]; % F upper arm @ wheel body
10 par.rvrk=[-0.150;  0.380; -0.038]; % R drag link @ rack
11 par.rvqk=[-0.137;  0.690; -0.088]; % Q drag link @ wheel body
12 % additional data
13 toe0  = 0.0000/180*pi; % initial toe angle (ISO def)
14 camb0= 0.8000/180*pi; % initial camber angle (ISO def)
15 rs    = 0.2850; % steady state tire radius
16 en0   = [ 0; 0; 1 ]; % road normal (flat horizontal road)
17
18 % wheel/tire orientation in design position
19 eyrk = [toe0; 1; -camb0]; eyrk = eyrk/norm(eyrk); % wheel rot. axis
20 exk = cross(eyrk,en0); exk=exk/norm(exk);  % longitudinal direction
21 eyk = cross(en0,exk); % lateral direction
22 ezk = cross(exk,eyrk); % radial direction
23 rwpk= -rs*ezk; % wheel center W --> P (contact in design pos.)
24
25 % kingpin and caster angle in design position
26 rcfk = par.rvfk-par.rvck; ecfk=rcfk/norm(rcfk); % kingpin orientation
27 si=atan2(-ecfk(2),ecfk(3)); disp(['sigma = ',num2str(si*180/pi)])
28 nu=atan2(-ecfk(1),ecfk(3)); disp(['nue = ',num2str(nu*180/pi)])
```

```
29
30 % caster offset and scrub radius in design position
31 rcpk = par.rvwk+rwpk-par.rvck; rsck = -(en0.'*rcpk)/(en0.'*ecfk)*ecfk;
32 co = -exk'*(rsck+rcpk); disp(['caster offset= ',num2str(co)])
33 sr =  eyk'*(rsck+rcpk); disp(['scrub radius= ',num2str(sr)])
34
35 % range of motion (rotation of lower arm and rack displacement)
36 n=11; phi=linspace(-1,1,n)*par.phmx; m=15; u=linspace(-1,1,m)*par.umx;
37
38 % pre-allocate vars to speed up loop and compute suspension kinematics
39 xw=zeros(n,m);yw=xw;zw=xw; xp=xw;yp=xw;zp=xw;
40 del=xw; toe=xw; camb=xw; ddel=xw;
41 for i=1:n
42   for j=1:m
43     [avw,rvwv,del(i,j),pd] = fun_05_dblwb_kin(phi(i),u(j),par);
44     eyrv = avw*eyrk; % actual orientation of wheel rotation axis
45     rvpv = rvwv + avw*rwpk; % actual position of contact point P
46     xw(i,j)=rvwv(1);yw(i,j)=rvwv(2);zw(i,j)=rvwv(3); % wheel center
47     xp(i,j)=rvpv(1);yp(i,j)=rvpv(2);zp(i,j)=rvpv(3); % ref. point
48     toe(i,j) = atan2(-eyrv(1), eyrv(2)); % toe angle (+ rot. z-axis)
49     camb(i,j)= atan2( eyrv(3), eyrv(2)); % camber angle (+rot. x-axis)
50     ddel(i,j)=norm(pd(:,2)); % partial derivative ddel/du
51   end
52 end
53
54 n0=round(n/2); m0=round(m/2);  rvpk = par.rvwk + rwpk ;
55 axes('position',[0.05,0.05,0.20,0.90])
56 hold on, axis equal,grid on,title('xz')
57 plot(xw(:,m0),zw(:,m0)),plot(xp(:,m0),zp(:,m0),'--')
58 plot(par.rvwk(1),par.rvwk(3),'ok'), plot(rvpk(1),rvpk(3),'ok')
59 axes('position',[0.30,0.05,0.20,0.90])
60 hold on, axis equal, grid on, title('yz')
61 plot(yw(:,m0),zw(:,m0)), plot(yp(:,m0),zp(:,m0),'--')
62 plot(par.rvwk(2),par.rvwk(3),'ok'),plot(rvpk(2),rvpk(3),'ok')
63 axes('position',[0.60,0.05,0.35,0.40]), colormap('white')
64 surf(u,phi*180/pi,toe*180/pi), grid on, view(-40,10), title('toe')
65 axes('position',[0.60,0.55,0.35,0.40]), colormap('white')
66 surf(u,phi*180/pi,camb*180/pi), grid on, view(-40,10), title('camber')
67
68 figure
69 d1 = del(n0,:); d2 = -d1(m:-1:1); % wheel steering angles in des. pos.
70 a=2.7; s=2*par.rvwk(2); d2a=atan2(a*tan(d1),a+s*tan(d1)); % ackermann
71 axes('position',[0.05,0.55,0.40,0.30]), title('d1(u), d2(u)')
72 plot(u,[d1;d2]*180/pi), grid on, legend('left','right')
73 axes('position',[0.05,0.15,0.40,0.30]), title('d\delta/du')
74 plot(u,ddel(n0,:)), grid on
75 axes('position',[0.55,0.30,0.40,0.40]), axis equal, title('d2(d1)')
76 plot(d1*180/pi,[d2;d2a]*180/pi),grid on, legend('kin','ackermann')
```

At first, the motion of the wheel center W and the wheel body-fixed reference point
P, which in design position coincides with the contact point, are plotted in the x-z-
and the y-z-plane. By fixing the second index in the corresponding matrices xw, yw,
zw, and xp, yp, zp to m0, the motions induced by varying the rotation angle φ are
considered only. As typical for most passenger car front axle suspension systems, the
reference point moves slightly forward when the wheel goes upward. At the same
time, the wheel center travels backward here. This is achieved by a different inclined
rotation axis of the lower and the upper control arms. The MATLAB-Function surf is
used to plot the toe and the camber angle, which define the momentary orientation of
the wheel rotation axis as two-dimensional functions of the rack displacement u and

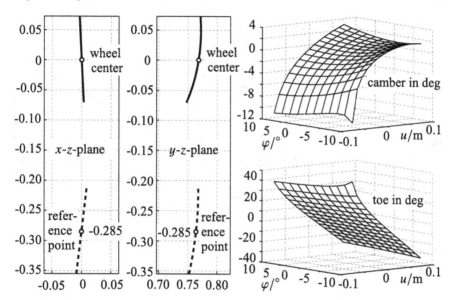

FIGURE 5.14
Double wishbone kinematics.

the rotation angle φ of the lower control arm. Positive signs in the toe and camber angle indicate here positive rotations of the left wheel body about the z- and x-axis, which does not match the definitions in Sections 1.3.3 and 1.3.4 but corresponds with the multibody system approach. Due to different-sized control arms, the wheel body rotates around the x-axis when the wheel is moved up and down. As the kingpin is fixed to the wheel body, it will change its orientation at the same time. Thus the camber angle depends on the generalized coordinates u and φ. The toe angle, however, is mainly influenced by the rack displacement u. A potential dependency on the hub motion (rotation angle φ), which becomes visible here on extreme hub and steer motions, will influence the driving behavior of the vehicle and should therefore be designed properly.

The function fun_05_dblwb_kin provides the rotation angle δ around the kingpin too. The MATLAB command d1 = del(n0,:) extracts the angle of the left wheel in design position as a function of the rack displacement u only. Taking a left/right symmetry for granted, the MATLAB command d2 = -d1(m:-1:1) generates the corresponding angle at the right wheel. Both angles are plotted versus the rack displacement u in the upper left graph of Figure 5.15. The MATLAB-Script given in Listing 5.2 is based on a simple steering system model. A constant ratio of the steering linkage of $i_{SL} = 9.3$ was assumed. As can be seen in the lower left graph of Figure 5.15, the ratio is quite nonlinear here. In a wide range it is close to the mean value of 8.8 rad/m. If the rack displacement approaches its left maximum at $u = -u_{max} = -0.0745$ m, the ratio goes up rather rapidly to 18.1 rad/m, which indicates that the steering linkage comes close to a kinematical singularity. Assuming

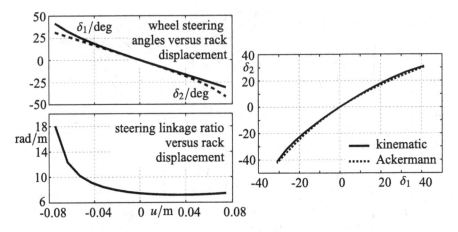

FIGURE 5.15
Double wishbone steering kinematics.

a wheel base of $a = 2.7$ m and assigning the track width via s = 2*par.rvwk(2) to the y-coordinate of the wheel center, the suspension kinematics can be compared with the Ackermann steering geometry. Similar to Equation (5.34), the relation

$$\tan \delta_2^A = \frac{a \tan \delta_1}{a + s \tan \delta_1} \tag{5.104}$$

provides the Ackermann wheel steering angle at the second wheel δ_2^A as a function of the steering angle at the first wheel δ_1. As can be seen from the graph on the right of Figure 5.15 this double wishbone suspension system is designed to match the Ackermann steering geometry quite closely.

Note that the kinematics of a double wishbone suspension combined with a rack-and-pinion steering system are sensitive to the geometric data and may come to singular positions where the kinematics is not defined anymore. That is why the maximum rotation angle of the lower control arm and the maximum rack displacement are specified here via the parameter par.umx and par.phmx. Increasing these values or changing the geometric data too much or improperly may cause MATLAB to generate warnings or error messages.

5.5 Design Kinematics

5.5.1 General Approach

The kinematics of a wheel/axle suspension can be described by two generalized coordinates. In a more general approach the position of the wheel center W and the orientation of the knuckle with respect to the vehicle fixed reference frame V are then defined by

$$
r_{VW,V} = r_{VW,V}^D + \begin{bmatrix} x(u,h) \\ y(u,h) \\ z(u,h) \end{bmatrix} \quad \text{and} \quad A_{VW} = A_{\gamma(u,h)} \, A_{\alpha(u,h)} \, A_{\beta(u,h)} \quad (5.105)
$$

where $r_{VW,V}^D$ defines the design position of the wheel center. The displacements x, y, z as well as the angles α, β, γ, which describe elementary rotations about the x-, y-, and z-axis, depend on the generalized coordinates u and h. In the case of the double wishbone suspension, u represents the rack displacement and h is substituted by the rotation angle φ of the lower control arm. The rotation sequence γ, α, β corresponds with the steer, the camber, and the pitch motion of the knuckle.

The design kinematics approach, proposed in [57], makes it possible to describe any kind of suspension kinematics very accurately and efficiently. In this approach the displacements x, y, z and the elementary rotations α, β, γ are each approximated by a 2-dimensional analytic function in the form of

$$
\begin{aligned}
f(u,y) = {} & a_u \, u + b_u \, u^2 + c_u \, u^3 + a_h \, h + b_h \, h^2 + c_h \, h^3 \\
& + a_{uh} \, u \, h + b_{uh} \, u \, h^2 + c_{uh} \, u^2 \, h + d_{uh} \, u^2 \, h^2
\end{aligned} \quad (5.106)
$$

where $f(u = 0, h = 0) = 0$ was taken for granted. The 10 coefficients of the design kinematics approach are determined by the 2 initial inclinations $df/du|_0$, $df/dh|_0$, the 4 center points f_{P0}, f_{N0}, f_{0P}, f_{0N}, and the 4 corner points f_{PP}, f_{PN}, f_{NP}, f_{NN}, Figure 5.16.

The design kinematics approach defined on four patches of size $\Delta u \times \Delta h$ provides a smooth and accurate inter- and extrapolation. It is superior to commonly used lookup tables because it is comparatively fast and provides not only the function values $f = f(u,h)$, but also continuous derivatives df/du and df/dh which are required for the equations of motion. The Listing 5.7 provides the MATLAB®-Function that performs the corresponding computations.

Listing 5.7

Function `uty_axk_design`: design kinematics approach

```
1 function [f,dfdu,dfdh] = uty_axk_design(u,h,p)
2 % 2-dimensional function f = f(u,h) defined on 4 patches
3
4 % normalized coordinates
5   un = u/p.du;   dxndx = 1/p.du;   hn = h/p.dh;   dyndy = 1/p.dh;
6
7 % skeleton based on independent approximation of 2 cubic functions
8 % defined by dfdu0, and fP0, fN0 as well as dfdh0, and f0P, f0N
9
10 % fuh0 = f(u,0) and dfdu
11   a1n0 = p.dfdu0*p.du;
12   a2n0 =      0.5*( p.fP0  + p.fN0  ) ;
13   a3n0 = ( 0.5*( p.fP0  - p.fN0  ) - a1n0 ) ;
14   fuh0 = ( a1n0 + ( a2n0 + (a3n0*un) ) * un ) * un ;
15   dfdu = ( a1n0 + ( 2*a2n0 + 3*(a3n0*un) ) * un ) * dxndx ;
16
17 % fu0h = f(0,h) and dfdh
```

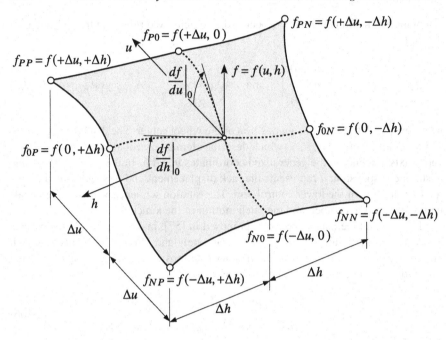

FIGURE 5.16
Definition of the design kinematics approximation.

```
18    a1n0  = p.dfdh0*p.dh;
19    a2n0  =   0.5*( p.f0P + p.f0N ) ;
20    a3n0  = ( 0.5*( p.f0P - p.f0N )  - a1n0 ) ;
21    fu0h  = ( a1n0 + ( a2n0 + (a3n0*hn) ) * hn ) * hn ;
22    dfdh  = ( a1n0 + ( 2*a2n0 + 3*(a3n0*hn) ) * hn ) * dyndy ;
23
24 % compute deviations at corner-points fPP, fPN, fNN, FNP
25    del_fPP = p.fPP - ( p.f0P + p.fP0 );
26    del_fPN = p.fPN - ( p.f0N + p.fP0 );
27    del_fNN = p.fNN - ( p.f0N + p.fN0 );
28    del_fNP = p.fNP - ( p.f0P + p.fN0 );
29
30 % correct remaining deviations by a bi-quadratic function
31    a1 = del_fPP - del_fPN - del_fNP + del_fNN;
32    a2 = del_fPP + del_fPN - del_fNP - del_fNN;
33    a3 = del_fPP - del_fPN + del_fNP - del_fNN;
34    a4 = del_fPP + del_fPN + del_fNP + del_fNN;
35
36 % complete function value and derivatives
37    f    = fuh0 + fu0h + ( (a1+a2*hn) + (a3+a4*hn)*un ) * (un*hn) * 0.25;
38    dfdu = dfdu + ( (a1+a2*hn) + 2*un*(a3+a4*hn) )*hn*dxndx*0.25;
39    dfdh = dfdh + ( (a1+a3*un) + 2*hn*(a2+a4*un) )*un*dyndy*0.25;
40
41 end
```

The design kinematics approach can be used to substitute complex kinematic computations, to gain the kinematic properties of suspension systems by fitting measurements obtained from Kinematic and Compliance (KnC) test rigs, and to approximate statically over-determined suspension systems [57, 60]. The latter feature of the design kinematics will be demonstrated in the following on the example of a twist beam axle suspension.

5.5.2 Example Twist Beam Axle Suspension

In a lumped mass model approach the twist beam is equally separated into a left and a right half, Figure 5.17. Then the left and the right trailing arms together with

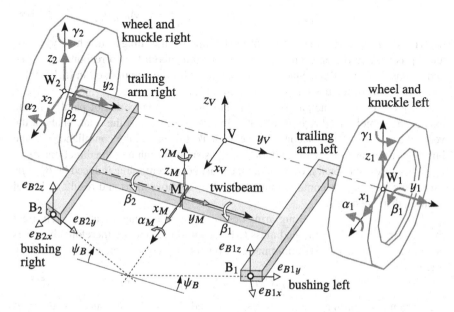

FIGURE 5.17
Lumped mass model of a twist beam suspension supported by elastic bushings.

the corresponding half of the twist beam can be considered as two rigid bodies that are attached to the knuckles at one end and supported in the bushings B_1 and B_2 at the other end. The torsional compliance of the twist beam makes it possible that the left and right trailing arms can perform independent rotations β_1 and β_2 about the twist beam center line fixed by the displacements x_M, y_M, z_M of the point M and the rotation angles α_M and γ_M of the twist beam. Hence, the lumped mass model has $f = 7$ degrees of freedom if the bushings in B_1 and B_2 are modeled by three-dimensional force elements. The simple and straightforward approach, where the bushings are modeled as ball joints, results in an over-determined system, because the two ball joints generate a redundancy in the direction of the line B_1 and B_2. However, the fully elastically suspended lumped mass twist beam model with $f = 7$ degrees of freedom can be reduced to an equivalent kinematic model with $f = 2$ degrees of freedom by a quasi-static analysis. The results of these analyses may then be processed by lookup tables or much better by the design kinematic approach.

The momentary position of point M on the twist beam center line with respect to the vehicle fixed reference frame V is defined by the vector

$$r_{VM,V} = r_{VM,V}^D + \begin{bmatrix} x_M \\ y_M \\ z_M \end{bmatrix} \tag{5.107}$$

where $r_{VM,V}^D$ denotes the design position. The rotation matrices

$$A_{V1} = A_{\gamma_M} A_{\alpha_M} A_{\beta_1} \quad \text{and} \quad A_{V2} = A_{\gamma_M} A_{\alpha_M} A_{\beta_2} \tag{5.108}$$

describe the momentary orientation of the left and right trailing arm, where γ_M, α_M, and β_1 as well as β_2 represent elementary rotations about the corresponding z-, x-, and y-axis. This rotation sequence corresponds with the one defined in (5.105) that forms the basis of the design kinematics approach.

In practice, the principal axis of the bushing compliance is not parallel to the axis of the vehicle fixed reference frame. In this example, just the rotation about the vertical axis is taken into account by the angle ψ. In general, the compliance into the bushing directions e_{B1x}, e_{B2x} and e_{B1z}, e_{B22} is smaller than into the directions e_{B1x}, e_{B2x}, because a torsional motion about the twist beam center line induced by $\beta_1 \neq \beta_2$ causes the bushing attachment points B_1 and B_2 to move apart and results in lateral bushing displacements.

The rotations of the trailing arms described by the angles β_1 and β_2 represent the main motions of the lumped mass twist beam model and are therefore chosen as generalized coordinates. By minimizing the potential energy generated by the bushing forces

$$E_{pot} = u_{B1}^T C_B u_{B1} + u_{B2}^T C_B u_{B2} \rightarrow \text{Min.,} \tag{5.109}$$

the remaining coordinates x_M, y_M, z_M, α_M, and γ_M can be computed as functions of β_1 and β_2 in a quasi-static approach. The bushing displacements into the directions of its principal stiffness axis are provided by

$$u_{Bi,V} = \begin{bmatrix} e_{Bix,V} & e_{Biy,V} & e_{Bix,V} \end{bmatrix} \left(r_{VBi,V} - r_{VBi,V}^D \right), \quad i = 1, 2 \tag{5.110}$$

where $r_{VB1,V}^D$ and $r_{VB2,V}^D$ determine the design position of the left ($i = 1$) and right ($i = 2$) bushing. The momentary position of the bushing attachments are given by

$$\begin{aligned} r_{VB1,V} &= r_{VM,V} + A_{V1} \left(r_{VB1,V}^D - r_{VM,V}^D \right) \\ r_{VB2,V} &= r_{VM,V} + A_{V2} \left(r_{VB2,V}^D - r_{VM,V}^D \right) \end{aligned} \tag{5.111}$$

where (5.107) defines the vector $r_{VM,V}$ and (5.108) provides the rotation matrices A_{V1}, A_{V2}. The same stiffness properties for the left and right bushing were assumed in (5.109). Then, the stiffness matrix is simply defined by

$$C_B = \text{diag} \begin{bmatrix} c_r & c_a & c_r \end{bmatrix} \tag{5.112}$$

where according to the typical design of common twist beam suspension systems hinge joint like bushings with the radial stiffness c_r and the axial stiffness c_a are modeled hereby.

The MATLAB-Script in Listing 5.8 defines the layout of a typical twist beam suspension system. The vectors be1 and be2 sample the motion range of the trailing arms by $n = m = 11$ elements, where dbe1=dbe2=20/180*pi=0.3491 specifies the limits. Then, the quasi-static coordinates x1q, y1q, z1q, al1q, ga1q, and x2q, y2q, z2q, al2q, ga2q that define position and orientation of the knuckle-fixed reference frames are computed as two-dimensional functions of the trailing arm rotations be1 and be2 by minimizing the potential energy stored in the bushing forces.

Listing 5.8
Script app_05_twistbeam.m: twist beam kinematics

```
1 % wheel center
2 p.rVW1D = [ 0.000;  0.750; 0.000 ]; % wheel center left
3 p.rVW2D = [ 0.000; -0.750; 0.000 ]; % wheel center right
4
5 % hardpoints twist beam suspension
6 p.rVMD  = [ 0.150;  0.000; 0.000 ]; % point on twistbeam center line
7 p.rVB1D = [ 0.400;  0.500; 0.000 ]; % bushing left
8 p.rVB2D = [ 0.400; -0.500; 0.000 ]; % bushing right
9
10 % bushing stiffness and principal axis
11 p.cr = 9e5;  p.ca = 3e5;  % radial and axial stiffness in N/m
12 psi = 20/180*pi; % angular position of axial direction: degree --> rad
13 p.eB1x=[ cos(psi);  sin(psi); 0 ]; p.eB2x=[ cos(psi); -sin(psi); 0 ];
14 p.eB1y=[ -sin(psi);  cos(psi); 0 ]; p.eB2y=[ sin(psi);  cos(psi); 0 ];
15 p.eB1z=[    0;         0;      1 ]; p.eB2z=[    0;         0;      1 ];
16
17 n=11; dbe1=20/180*pi; be1=linspace(-1,1,n)*dbe1; % range of beta 1
18 m=11; dbe2=20/180*pi; be2=linspace(-1,1,m)*dbe2; % range of beta 2
19
20 opts = optimset('TolFun',1e-9,'TolX',1e-9); % options for fminsearch
21 x0 = [0; 0; 0; 0; 0 ]; % simple initial conditions and pre-allocations
22 x1q=zeros(n,m); y1q=x1q; z1q=x1q; al1q=x1q; ga1q=x1q;
23 x2q=zeros(n,m); y2q=x2q; z2q=x2q; al2q=x2q; ga2q=x2q;
24 rB1x=x1q; rB1y=x1q; rB1z=x1q; rB2x=x2q; rB2y=x2q; rB2z=x2q;
25 for i=1:n
26   for j=1:m
27     x=fminsearch(@(x)fun_05_twistbeam(x,be1(i),be2(j),p),x0,opts);
28     % get rotation matrices and positions of B1, B2, and W1, W2
29     [~,o] = fun_05_twistbeam(x,be1(i),be2(j),p);
30     rB1x(i,j)=o.rVB1V(1); rB1y(i,j)=o.rVB1V(2); rB1z(i,j)=o.rVB1V(3);
31     rB2x(i,j)=o.rVB2V(1); rB2y(i,j)=o.rVB2V(2); rB2z(i,j)=o.rVB2V(3);
32     % roll-, yaw-angle, x-y-z-displacements of left, right knuckle
33     al1q(i,j)= x(4); ga1q(i,j)=x(5);  al2q(i,j)=x(4); ga2q(i,j)=x(5);
34     x1q(i,j)=o.rVW1V(1); y1q(i,j)=o.rVW1V(2); z1q(i,j)=o.rVW1V(3);
35     x2q(i,j)=o.rVW2V(1); y2q(i,j)=o.rVW2V(2); z2q(i,j)=o.rVW2V(3);
36   end
37 end
```

To achieve an appropriate accuracy, the default options of the MATLAB-Function fminsearch were changed accordingly. The potential energy epot of the bushing forces is computed in the MATLAB-Function provided in Listing 5.9.

Listing 5.9
Function fun_05_twistbeam: potential energy stored in twist beam bushings

```
1 function [epot,out] = fun_05_twistbeam(x,be1,be2,p)
2
```

```
 3 % get displacements as well as roll and yaw angle of twist beam
 4 xM = x(1); yM = x(2); zM = x(3); alM = x(4); gaM = x(5);
 5
 6 % position and orientation of twist beam and trailing arms
 7 Aga = [ cos(gaM) -sin(gaM)  0;  sin(gaM)  cos(gaM)  0;  0 0 1 ];
 8 Aal = [ 1 0 0;  0  cos(alM) -sin(alM);  0  sin(alM)  cos(alM) ];
 9 rVMV = p.rVMD + [ xM; yM; zM ];     A = Aga*Aal;
10 out.AV1 = A*[ cos(be1)  0  sin(be1); 0 1 0; -sin(be1)  0  cos(be1) ];
11 out.AV2 = A*[ cos(be2)  0  sin(be2); 0 1 0; -sin(be2)  0  cos(be2) ];
12
13 % momentary position of bushing attachments B1 and B2
14 out.rVB1V = rVMV + out.AV1*(p.rVB1D-p.rVMD);
15 out.rVB2V = rVMV + out.AV2*(p.rVB2D-p.rVMD);
16
17 % bushing displacements in principal axis system
18 uB1 = [ p.eB1x  p.eB1y  p.eB1z ].'*( out.rVB1V - p.rVB1D );
19 uB2 = [ p.eB2x  p.eB2y  p.eB2z ].'*( out.rVB2V - p.rVB2D );
20
21 % potential energy of bushing forces
22 cB = diag([ p.cr  p.ca   p.cr ]);  % stiffness matrix
23 epot = 0.5*( uB1.'*cB*uB1 + uB2.'*cB*uB2 );
24
25 if nargout > 1  % relative displacement of knuckle centers
26    out.rVW1V = rVMV + out.AV1*(p.rVW1D-p.rVMD) - p.rVW1D;
27    out.rVW2V = rVMV + out.AV2*(p.rVW2D-p.rVMD) - p.rVW2D;
28 end
29
30 end
```

The additional output structure out collects the rotation matrices of the trailing arms, the momentary position of the bushing attachment points, and the displacements of the wheel centers relative to the design positions. The latter are only computed if the additional output is actually requested.

Two three-dimensional plots generated by the MATLAB-Commands

```
subplot(1,2,1); plot3(rB2x,rB2y,rB2z,'ok','LineWidth',1), grid on
title('Bushing attachment right'), axis equal, view(105,30)
subplot(1,2,2); plot3(rB1x,rB1y,rB1z,'ok','LineWidth',1), grid on
title('Bushing attachment left'),  axis equal, view(105,30)
```

illustrate the momentary positions of the bushing attachment points B_1 and B_2 while the trailing arms are moved up and down, Figure 5.18.

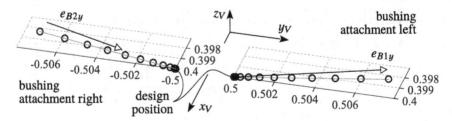

FIGURE 5.18
Twist beam bushing attachment point movements.

They start at the design positions p.rVB1D = [0.400; 0.500; 0.000] and p.rVB2D = [0.400; -0.500; 0.000] and move outwards along straight lines that deviate slightly from the axial directions e_{B1y} and e_{B2y} due to the radial compliance of the hinge joints in B_1 and B_2.

In general, the quasi-static solution is suitable just for offline computations, because it is cumbersome and time consuming. However, the point-by-point results obtained by executing the commands in Listing 5.8 can be replaced by the design kinematics approach. According to Figure 5.16, each of the two-dimensional functions x1q(be1,be2) ... ga1q(be1,be2) and x2q(be1,be2) ... ga2q(be1,be2) can be approximated by the design kinematic approach $f = f(u, h)$.

At first the patch sizes used in Listing 5.8 to define the motion range of the trailing arm rotation angles are assigned by the MATLAB-Commands

```
dk1(5).du=0; dk1(5).dh=0;  dk2(5).du=0; dk2(5).dh=0;
for i=1:5
  dk1(i).du = dbe1;  dk1(i).dh = dbe2;
  dk2(i).du = dbe1;  dk2(i).dh = dbe2;
end
```

to the corresponding values of structure fields dk1(i) and dk2(i), $i = 1(1)5$ that will collect the characteristic parameter required for the design kinematics approaches.

The calculation of the corner and center points is quite simple. The MATLAB-commands

```
dk1(5).fPP=0; dk2(5).fPP=0; u = +dbe1; h = +dbe2;
x = fminsearch(@(x) fun_05_twistbeam(x,u,h,p),x0,opts);
[~,o] = fun_05_twistbeam(x,u,h,p);
f1 = [ o.rVW1V(1) o.rVW1V(2) o.rVW1V(3)   x(4)   x(5) ];
f2 = [ o.rVW2V(1) o.rVW2V(2) o.rVW2V(3)   x(4)   x(5) ];
for i=1:5; dk1(i).fPP=f1(i);  dk2(i).fPP=f2(i); end
```

for example, compute one of the corner points (fPP at u = +dbe1 and h = +dbe2) and assign the results to corresponding values of the design kinematic structure fields. The placeholders f1 and f2 collect the quasi-static coordinates of the left (x_1, y_1, z_1, α_1, γ_1) and right wheel (x_2, y_2, z_2, α_2, γ_2) hereby.

Then, the initial inclinations into the direction of the variables $u = \beta_1$ and $h = \beta_2$ are computed via central differential quotients for the left and right wheel. The MATLAB-Commands

```
eps=1.e-3; dk1(5).dfdu0=0; dk2(5).dfdu0=0;
u = [ -1 +1 ]*eps*dbe1; h = [ 0 0 ]; f1=zeros(2,5); f2=f1;
for k=1:2
  x = fminsearch(@(x) fun_05_twistbeam(x,u(k),h(k),p),x0,opts);
  [~,o] = fun_05_twistbeam(x,u(k),h(k),p);
  f1(k,:) = [ o.rVW1V(1) o.rVW1V(2) o.rVW1V(3)   x(4)   x(5) ];
  f2(k,:) = [ o.rVW2V(1) o.rVW2V(2) o.rVW2V(3)   x(4)   x(5) ];
end
for i=1:5
  dk1(i).dfdu0 = ( f1(2,i) - f1(1,i) ) / ( 2*eps*dbe1 );
  dk2(i).dfdu0 = ( f2(2,i) - f2(1,i) ) / ( 2*eps*dbe1 );
end
```

for example, will deliver all the initial inclinations into the direction of the variable $u = \beta_1$ and assign the results to the corresponding values of the design kinematics structure fields.

The complete design kinematics structure fields dk1(i) and dk2(i), $i = 1(1)5$ will represent the "fingerprint" of the twist beam suspension defined in Listing 5.8. The values of the structure fields related to the axle kinematics are listed in Table 5.1 and Table 5.2. The MATLAB-Function provided in Listing 5.7 takes the design

TABLE 5.1
Twist beam design kinematics parameter of left wheel body ($\Delta u = \Delta h = 0.3491$)

$u = \beta_1 / h = \beta_2$	x_1	y_1	z_1	α_1	γ_1
df/du	-0.00000	-0.00000	+0.46250	+0.25000	+0.00000
df/dh	-0.00000	+0.00000	-0.06250	-0.25000	-0.00000
f_{PP}	+0.02412	-0.00000	+0.13681	-0.00000	+0.00000
f_{P0}	+0.02729	-0.00506	+0.15761	+0.08530	-0.01502
f_{PN}	+0.02244	-0.01938	+0.17699	+0.16937	+0.00000
f_{0P}	-0.00413	-0.00506	-0.02130	-0.08530	+0.01502
f_{0N}	-0.00413	-0.00506	+0.02130	+0.08530	+0.01502
f_{NP}	+0.02244	-0.01938	-0.17699	-0.16937	-0.00000
f_{N0}	+0.02729	-0.00506	-0.15761	-0.08530	-0.01502
f_{NN}	+0.02412	+0.00000	-0.13681	+0.00000	-0.00000

TABLE 5.2
Twist beam design kinematics parameter of right wheel body ($\Delta u = \Delta h = 0.3491$)

$u = \beta_1 / h = \beta_2$	x_2	y_2	z_2	α_2	γ_2
df/du	+0.00000	-0.00000	-0.06250	+0.25000	+0.00000
df/dh	-0.00000	+0.00000	+0.46250	-0.25000	-0.00000
f_{PP}	+0.02412	-0.00000	+0.13681	-0.00000	+0.00000
f_{P0}	-0.00413	+0.00506	-0.02130	+0.08530	-0.01502
f_{PN}	+0.02244	+0.01938	-0.17699	+0.16937	+0.00000
f_{0P}	+0.02729	+0.00506	+0.15761	-0.08530	+0.01502
f_{0N}	+0.02729	+0.00506	-0.15761	+0.08530	+0.01502
f_{NP}	+0.02244	+0.01938	+0.17699	-0.16937	-0.00000
f_{N0}	-0.00413	+0.00506	+0.02130	-0.08530	-0.01502
f_{NN}	+0.02412	+0.00000	-0.13681	+0.00000	-0.00000

kinematics structures as input and delivers the function value and optionally the derivatives at arbitrary values of the trailing arm rotation angles. The MATLAB commands for instance

```
x2D = zeros(n,m);
for i=1:n, for j=1:m
   x2D(i,j) = uty_axk_design(be1(i),be2(j),dk2(1));
end, end
```

will compute the two-dimensional array x1D that collects the values of the longitudinal displacements of the right wheel center approximated by the design kinematics in the trailing arm motion ranges defined by the vectors be1 and be2. The MATLAB commands

```
subplot(3,2,1), grid on, hold on, view(-45,30)
surf(be1,be2,x2D.','FaceColor',[0.9,0.9,0.9]),
xlabel('be_1'), ylabel('be_2'), zlabel('x_2')
for i=1:n, for j=1:m
   plot3(be1(i),be2(j),x2q(i,j),'ok','Linewidth',1,'Markersize',2)
end, end
```

will generate a surface plot[4] colored in light gray and add the results of the point-by-point quasi-static computation as single points.

---

[4]Note: within the MATLAB-Command surf(x,y,Z) the input x corresponds to the columns of Z and the input y to the rows. That is why, the transposed matrix provided by x2D.' is used as input Z in this case.

The complete results[5] of the point-by-point quasi-static approaches obtained from Listing 5.8 and the design kinematics approximations are plotted in Figures 5.19 and 5.20. The gray surfaces which represent the design kinematics approaches

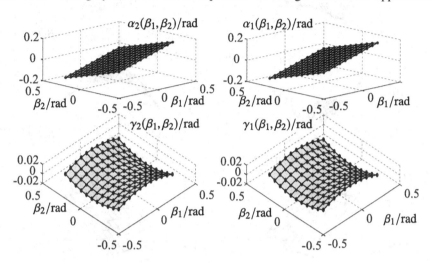

FIGURE 5.19

Twist beam kinematics: rotations of right and left knuckles or wheel bodies.

coincide perfectly with the single dots which illustrate the results of the quasi-static computation.

The roll and yaw motions of the right and left knuckles are identical, which is confirmed by the values in the columns α_1, γ_1 and α_2, γ_2 of Table 5.1 and Table 5.2. The roll motion $\alpha_1 = \alpha_2$ provides a camber compensation which is a characteristic feature of the twist beam suspension system. It may be reduced to a simple function that just depends on $\beta_1 - \beta_2$ and delivers a zero value for $\beta_1 = \beta_2$. The amount of the camber compensation depends on the location of the twist beam center line. The maximum camber compensation is achieved if point M is located in between the wheel centers W_1 and W_2. Then, the twist beam suspension corresponds with a rigid axle guided by two trailing arms. Such a layout represents the standard axle suspension system of a heavy-duty semi-trailer. If point M is located in between the bushings B_1 and B_2, the twist beam suspension degenerates to a simple trailing arm suspension where no camber compensation is present at all.

The yaw-motion $\gamma_1 = \gamma_2$ which corresponds with the steer motion of the axle or the knuckles, respectively, is rather complex. At the center points, where $\beta_1 = \pm\Delta\beta_1$ and $\beta_2 = 0$ or $\beta_1 = 0$ and $\beta_2 = \pm\Delta\beta_2$ will hold, it amounts to $\gamma_{1/2}^{max} = \pm0.01502 = \pm0.86°$. This steering motion, although comparatively small, is the main disadvantage of a twist beam suspension system, because it generates unwanted and nearly unpredictable

---

[5]The MATLAB-Script app_05_twistbeam.m, which is available on the book's home page, includes all necessary commands.

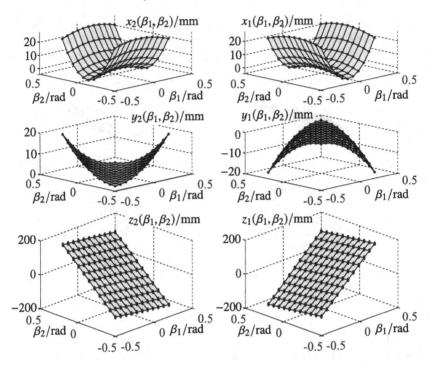

FIGURE 5.20
Twist beam kinematics: displacements of right and left wheel centers.

steering effects when a vehicle equipped with a rear axle twist beam suspension is driving straight ahead on rough roads.

The longitudinal x_2, x_1 and the vertical z_2, z_1 displacements of the right and left knuckle are symmetric. The x- and z-displacements on the right depend mainly on β_2 and the corresponding displacements on the left depend mainly on β_1.

The lateral displacements y_2 and y_1 act in opposite directions and vanish completely if $\beta_1 = \beta_2$ holds. The lateral displacements at the wheel centers W_2 and W_1 correspond with the movements of the bushings B_2 and B_1 which are illustrated in Figure 5.18.

5.6   Race Car Suspension System

5.6.1   General Layout

Race car suspension systems differ from the standard layout. The kinematics is entirely tuned to optimize the tire/road interaction and the suspension has to get along with huge down forces generated by specific wings mounted at the chassis. Usually, push or pull rods are used to operate standard spring/damper elements, the anti-roll bar, as well as an additional (third or center) spring. Figure 5.21 shows the typical layout of a race car suspension, where rocker elements are marked by light-grey triangles.

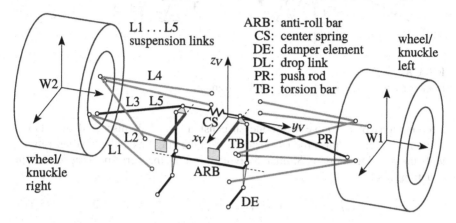

FIGURE 5.21
Race car suspension with push rod mechanism.

In this particular example, the links 1 and 2 as well as the links 3 and 4 represent the lower and upper A-arms of a double wishbone suspension system. Link 5 serves as drag link which is fixed at one end to the rack or just the chassis if not steered. The push rod PR is attached at one end at the knuckle and at the other end at the main rocker. In this particular case, the main rockers on the left and right side are mounted at torsion bars TB representing the standard suspension spring. The rotations of the main rockers are transferred to drop links DL and produce displacements of the attachment points of the center spring CS. The drop links cause additional rockers to rotate and operate the anti-roll bar ARB and the damper elements (DE) in addition.

Some axle layouts substitute the push rod, which exhibits a buckling risk, by a pull rod. Others replace the damper by a spring damper combination, which reduces the torsion bars to simple swivel joints. In standard axle layouts, the spring is usually attached to one link or control arm, which requires a rather heavy component instead of a lightweight lean link. In addition, race car suspension systems install all force elements inside the chassis and move heavier parts like the dampers to the lowest possible position. This improves the aerodynamic properties of a race car suspension system and lowers the height of the center of gravity in addition. This rather sophisticated type of a suspension system is also employed on high-performance cars [58].

5.6.2 Kinematics

Figure 5.22 shows the push rod rocker combination with the center spring in the y_V-z_V-plane of the vehicle fixed reference frame. The kinematics is similar to the

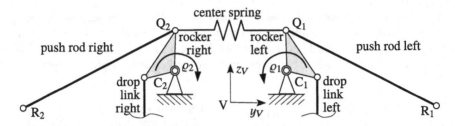

FIGURE 5.22
Rocker rotation generated by push rod movements.

steer motions of a double wishbone suspension, which is discussed in detail in Section 5.4.3.2. The push rod corresponds with the drag link and the rocker rotation with the steering motion. Similar to Equation 5.63 the corresponding constraint equations read as

$$\left(r_{VQi,V} - r_{VRi,V}\right)^T \left(r_{VQi,V} - r_{VRi,V}\right) = \ell_{PR}^2, \quad i = 1, 2 \tag{5.113}$$

where ℓ_{PR} denotes the length of each push rod. The momentary positions of the push rod attachment points at the knuckles R_1, R_2 are defined by

$$r_{VR_1,V} = r_{VR_1,V}(z_1, u_1) \quad \text{and} \quad r_{VR_2,V} = r_{VR_2,V}(z_2, u_2) \tag{5.114}$$

where z_1, z_2 and u_1, u_2 denote the hub motions and the steer inputs at the left and right wheel. The positions of the push rod attachments at the rockers Q_1, Q_2 are defined by

$$r_{VQ_1,V} = r_{VC_1,V} + A_{\varrho_1} r_{C_1Q_1,D} \quad \text{and} \quad r_{VQ_2,V} = r_{VC_2,V} + A_{\varrho_2} r_{C_2Q_2,D} \tag{5.115}$$

where the vectors $r_{VC_1,V}$, $r_{VC_2,V}$ and $r_{C_1Q_1,D}$, $r_{C_2Q_2,D}$ are given by data and the matrices $A_1(\varrho_1)$, $A_2(\varrho_2)$ describe the rocker rotations. In general, the rocker rotation axes will not just point into the direction of the vehicle fixed x_V-axis. Then, similar to Equation (5.38), the rotation matrices are defined by

$$A_{\varrho i} = e_{Ri,V} e_{Ri,V}^T + \left(I_{3\times 3} - e_{Ri,V} e_{Ri,V}^T\right) \cos \varrho_i + \tilde{e}_{Ri,V} \sin \varrho_i, \quad i = 1, 2 \tag{5.116}$$

where the unit vectors $e_{R1,V}$ and $e_{R2,V}$ determine the rocker rotation axes. As described in Section 5.4.3.2 the constraint equations can be transferred to trigonometric equations and solved analytically for the rocker rotation angles. According to Equation (5.114) the rotation angles will depend in general on the hub and steer motions of the wheels

$$\varrho_1 = \varrho_1(z_1, u_1) \quad \text{and} \quad \varrho_2 = \varrho_2(z_2, u_2) \tag{5.117}$$

A dependency on the steer motions u_1 and u_2 is surely not intended, because then steer inputs have to act against the resistance of suspension forces. The vectors $r_{VR_1,V}$ and $r_{VR_2,V}$ and hence, the rocker rotation angles ϱ_1 and ϱ_2 will not or just slightly depend on the steer motions, if the push rods are attached at the knuckles or at least close to the kingpin axes.

The dependency of ϱ_1 and ϱ_2 on u_1 and u_2 vanishes also, if the movements of the attachment points R_1, R_2 due to the steer motions u_1 and u_2 are performed perpendicular to the momentary orientations $r_{Q_iR_i,V} = r_{VQ_i,V} - r_{VR_i,V}$ of the push rods, $i = 1, 2$.

The main rocker rotations are then transmitted via the drop links to the lower rockers, the anti-roll bar and the damper or spring/damper elements, respectively. The drop link / lower rocker kinematics corresponds to the push rod / main rocker arrangement and can be solved by a similar approach.

Due to the symmetric push rod / rocker layout, the center spring employed to compensate the down forces is practically not activated at vehicle roll motions, where $z_2 = -z_1$ will hold. Hence, the center spring does not affect the overall roll stiffness of the axle and as a consequence the steering tendency of the vehicle, which as discussed in Chapter 9, depends on the ratio of the overall roll stiffness of the front to the rear axle.

Exercises

5.1 Use part of the MATLAB-Script in Listing 5.6 to study the influence of the initial toe and camber angle set by the variables toe0 and camb0 on the caster offset and scrub radius.

5.2 A passenger car is equipped with double wishbone suspension systems at both axles. The specific data for the front and rear axle are given by

```
% suspension kinematics (front left)
dt.umx  = 0.0745; % max. rack displacement
dt.phimx= 10/180*pi; % max rotation of lower control arm
dt.rvwk= [ 0.0000;  0.7680;  0.0000]; % W: wheel rim center
dt.rvak= [-0.2510;  0.3200; -0.0800]; % A: lower control arm @ chassis rear
dt.rvbk= [ 0.1480;  0.3200; -0.0940]; % B: lower control arm @ chassis front
dt.rvck= [ 0.0130;  0.7370; -0.1450]; % C: lower control arm @ wheel body
dt.rvdk= [-0.1050;  0.4350;  0.1960]; % D: upper control arm @ body rear
dt.rvek= [ 0.1220;  0.4350;  0.2300]; % E: upper control arm @ body front
dt.rvfk= [-0.0250;  0.6800;  0.1620]; % F: upper control arm @ wheel body
dt.rvrk= [-0.1500;  0.3800; -0.0381]; % R: drag link @ rack
dt.rvqk= [-0.1370;  0.6900; -0.0879]; % Q: drag link @ wheel body


% suspension kinematics (rear left)
dt.umx  = 0.0; % max. rack displacement
dt.phimx= 12/180*pi; % max rotation of lower control arm
dt.rvwk= [ 0.0000;  0.7675;  0.0000]; % W: wheel rim center
dt.rvak= [-0.2900;  0.3200; -0.0890]; % A: lower control arm @ chassis rear
dt.rvbk= [ 0.0500;  0.3200; -0.0725]; % B: lower control arm @ chassis front
dt.rvck= [-0.0260;  0.7370; -0.1435]; % C: lower control arm @ wheel body
dt.rvdk= [-0.2220;  0.4530;  0.2245]; % D: upper control arm @ body rear
dt.rvek= [ 0.0370;  0.4530;  0.2020]; % E: upper control arm @ body front
dt.rvfk= [ 0.0200;  0.6800;  0.1630]; % F: upper control arm @ wheel body
dt.rvrk= [-0.1540;  0.3160;  0.0120]; % R: drag link @ rack
```

```
dt.rvqk= [-0.1550;  0.6900; -0.0470]; % Q: drag link @ wheel body
```

The data are based on local vehicle-fixed coordinate systems that are located in the corresponding axle center. Both axles are characterized by the additional data

```
toe0 = 0.00/180*pi; % initial toe angle (sign accord. rot. around z-axis)
camb0=-0.50/180*pi; % initial camber angle (sign accord. rot. around x-axis)
rs   = 0.3450;      % steady-state tire radius
en0  = [ 0; 0; 1 ]; % road normal (flat horizontal road)
```

Modify the MATLAB-Script given in Listing 5.6 appropriately so that the kinematics of both axles can be analyzed. Note that the rear axle is not steered, which is simply realized by setting the maximum rack displacement to zero. Compare the results of the movements of the wheel center W and the reference point *P* in particular. Adjust the inner loop and the plotting commands appropriately.

5.3 The figure, where the underlying grid is based in all directions on a distance of 0.05 m, shows the layout of a simple double wishbone suspension for the left rear wheel of a passenger car.

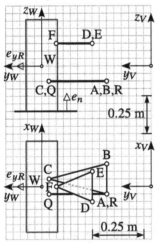

The absence of a steering system is taken into account by a vanishing maximum rack displacement, dt.umx=0. In addition, the wheel is mounted on the wheel body without an initial toe and camber angle, toe0=0 and camb0=0. Furthermore, a flat horizontal road can be assumed, en0 = [0; 0; 1].

Complete the data set needed to analyze the double wishbone suspension by inspecting the sketch. Estimate the maximum possible rotation angle of the lower control arm.

Modify the MATLAB-Script given in Listing 5.6 appropriately to restrict the kinematic analysis to the hub motion of the wheel induced by the rotation of the lower control arm.

5.4 The MATLAB-Script given in Listing 5.8 is available in a full version on the book's home page.

Modify this script such that the point M, which defines the location of the twist beam center line, is located in between

     a)   the wheels: p.rVMD = [0.000; 0.000; 0.000];
     b)   the bushings: p.rVMD = [0.400; 0.000; 0.000];

Analyze and discuss the suspension kinematics for these two modifications and compare the results with the original layout by inspecting the three-dimensional plots as well as the design kinematics parameter.

Note: the MATLAB-Functions fun_05_twistbeam and uty_axk_design required hereby are also provided on the book's home page.

6

Force Elements

CONTENTS

6.1   Standard Force Elements

6.1.1   Springs in General

Springs support the weight of the vehicle. In vehicle suspensions, coil springs, air springs, torsion bars, and leaf springs are used, Figure 6.1.

Coil springs, torsion bars, and leaf springs absorb additional load by compressing. Thus, the ride height depends on the loading condition. A linear coil spring may be characterized by its free length L_F and the spring stiffness c, Figure 6.2. The force acting on the spring is then given by

$$F_S = c \left(L_F - L \right) = c\,\Delta L \tag{6.1}$$

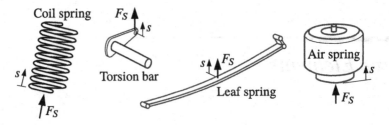

FIGURE 6.1
Vehicle suspension springs.

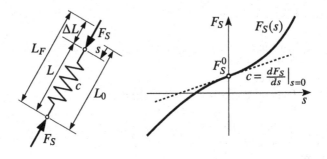

FIGURE 6.2
Linear coil spring and general spring characteristics.

where L denotes the actual length and ΔL the overall deflection of the spring. Mounted in a vehicle suspension, the spring has to support the corresponding chassis weight. Hence, the spring will be compressed to the configuration length $L_0 < L_F$. Now, Equation (6.1) can be written as

$$F_S = c\left(L_F - (L_0 - s)\right) = c\left(L_F - L_0\right) + cs = F_S^0 + cs \qquad (6.2)$$

where F_S^0 denotes the spring preload and $s = L_0 - L$ describes the spring displacement measured from the spring's configuration length. Note, $s > 0$ indicates compression. In general, the spring force F_S can be defined by a nonlinear function of the spring displacement s,

$$F_S = F_S(s) = F_S(L_0 - L) \qquad (6.3)$$

Now, arbitrary spring characteristics can be linearized or approximated by elementary functions, like polynomials, or by tables which are then inter- and extrapolated by linear functions or cubic splines.

The complex behavior of leaf springs can only be approximated by simple nonlinear spring characteristics, $F_S = F_S(s)$. For detailed investigations, sophisticated models must be used [51].

6.1.2 Air Springs

Air springs are rubber cylinders filled with compressed air. They are becoming more popular on passenger cars, light trucks, and heavy trucks because here the correct vehicle ride height can be maintained regardless of the loading condition by adjusting the air pressure.

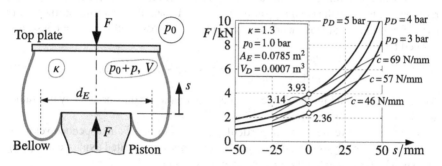

FIGURE 6.3
Layout of a standard air spring and typical force characteristics.

A standard air spring consists of the piston, the bellow, and the top plate, Figure 6.3. The force provided by the air spring is given by

$$F = p\,A_E \tag{6.4}$$

where p denotes the excessive pressure and $A_E = \frac{\pi}{4}d_E^2$ represents an effective cross section. The effective cross section A_E of an air spring or the effective diameter d_E, respectively, may depend on the spring travel s in some layouts. The polytropic state change

$$p_0 + p = \left(p_0 + p_D\right)\left(\frac{V_D}{V_D - A_E s}\right)^{\kappa} \tag{6.5}$$

delivers the excessive pressure $p = p(s)$ in a quasi-static approach, where a potential stretching of the bellow is neglected. The spring travel s measures the displacement of the piston relative to the top plate. The design position, where the state of the air spring is characterized by the excessive pressure p_D as well as the volume V_D, is defined by $s = 0$. Under normal operating conditions the atmospheric pressure is given by $p_0 = 1$ bar and the value $\kappa = 1.3$ may be used for the polytropic exponent.

In practice, a bump stop included in the air spring keeps the actual air spring volume $V = V_D - A_E s$ above a minimum value by limiting the spring travel to $s < V_D/A_E$.

In the design position the air spring generates the force $F_D = p_D A_E$ whilst it is exposed to the static load F_0 which depends on the corresponding weight of vehicle and load. Hence, an adjusted or controlled design pressure of $p_D = F_0/A_E$ will maintain the ride height or the design position respectively regardless of the loading conditions.

As illustrated in the right plot of Figure 6.3 the polytropic state change in (6.5) formally generates the excessive pressure p and via (6.4) the force too as progressive functions of the spring travel s. That is why, some air springs are designed such that the effective area A_E depends in a degressive manner on the spring travel s, thus generating less progressive spring characteristics.

A linear approximation to the air spring characteristics in the form $F \approx F_D + c\,s$ requires the force in the design position $F_D = p_D A_E$ and the air spring stiffness c. In case of constant or at least nearly constant effective cross sections $A_E = $ const. it is defined by

$$c = \left.\frac{dF}{ds}\right|_{s=0} = \left.\frac{dp}{ds}\right|_{s=0} A_E = \kappa\left(p_0 + p_D\right)\frac{A_E}{V_D}A_E \qquad (6.6)$$

As can be seen from Figure 6.3, the linear approximation is only practically in a rather small range of the spring travel.

For detailed investigations, in particular on the comfort of air suspended vehicles, the quasi-static analysis should be extended to more sophisticated dynamic air spring models which include the thermal balance in addition [10].

6.1.3   Anti-Roll Bar

The anti-roll or anti-sway bar or stabilizer is used to reduce the roll angle during cornering and to provide additional stability. Usually, it is simply a U-shaped metal rod connected to both of the lower control arms, Figure 6.4. Thus, the two wheels of an axle are interconnected by a torsion bar spring. This affects each one-sided bouncing. The axle with the stronger stabilizer is rather inclined to breaking out, in order to reduce the roll angle.

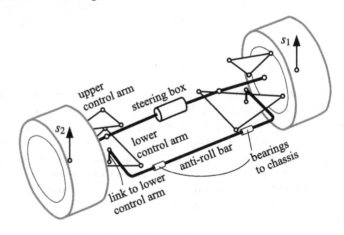

FIGURE 6.4
Axle with anti-roll bar attached to lower control arms.

When the suspension at one wheel moves up and on the other down, the anti-roll bar generates a force acting in the opposite direction at each wheel. In a good approximation, this force is given by

$$F_{arb} = \pm c_{arb} \, (s_1 - s_2) \tag{6.7}$$

where s_1, s_2 denote the vertical suspension motions of the left and the right wheel center, and c_{arb}^{W} in N/m names the stiffness of the anti-roll bar with respect to the vertical suspension motions of the wheel centers.

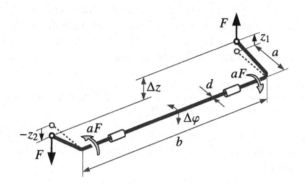

FIGURE 6.5
Anti-roll bar loaded by vertical forces.

Assuming a simple U-shaped anti-roll bar, its stiffness is defined by the geometry and material properties. Vertical forces with the magnitude F applied in the opposite direction at both ends of the anti-roll bar, result in the vertical displacement Δz measured between both ends of the anti-roll bar, Figure 6.5. The stiffness of the anti-roll bar itself is then defined by

$$c = \frac{F}{\Delta z} \tag{6.8}$$

Neglecting all bending effects and taking small deflections for granted, one gets

$$\Delta z = a \, \Delta \varphi = a \, \frac{aF \, b}{G \, \frac{\pi}{32} \, d^4} \tag{6.9}$$

where G denotes the modulus of shear and the distances a, b are defined in Figure 6.5. Hence, the stiffness of the anti-roll bar is given by

$$c = \frac{\pi}{32} \frac{G \, d^4}{a^2 \, b} \tag{6.10}$$

Depending on the axle design, the ends of the anti-roll bar are attached via links to the knuckle or, as shown in Figure 6.4, to the lower control arm. In both cases, the displacement of the anti-roll bar end is given as a function of the vertical suspension motion of the wheel center. For small displacements, one gets

$$z_1 = i_{arb} \, s_1 \quad \text{and} \quad z_2 = i_{arb} \, s_2 \tag{6.11}$$

where i_{arb} denotes the ratio of the vertical motions of the wheel centers s_1, s_2 and the anti-roll bar ends z_1, z_2. Now, the stiffness of the anti-roll bar with respect to the vertical suspension motions of the wheel centers is given by

$$c_{arb} = i_{arb}^2 \frac{\pi}{32} \frac{G \, d^4}{a^2 \, b} \tag{6.12}$$

The stiffness strongly depends (fourth power) on the diameter d of the anti-roll bar.

For a typical passenger car, the layout of an anti-roll bar is defined by $a = 230$ mm, $b = 725$ mm, $d = 20$ mm and $i_{arb} = 2/3$. The shear modulus of steel is given by $G = 85\,000$ N/mm^2. Equation (6.12) delivers then

$$c_{arb} = \left(\frac{2}{3}\right)^2 \frac{\pi}{32} \frac{85\,000 \text{ N/mm}^2 \, (20 \text{ mm})^4}{(230 \text{ mm})^2 \, 725 \text{ mm}} = 15.5 \text{ N/mm} = 15\,500 \text{ N/m} \tag{6.13}$$

This simple calculation will not produce the real stiffness exactly, because bending effects and compliances in the anti-roll bar bearings will reduce the stiffness of the anti-roll bar in practice.

6.1.4   Damper

Dampers are basically oil pumps. As the suspension travels up and down, the hydraulic fluid is forced by a piston through tiny holes, called orifices. This slows down the suspension motion. Today, twin-tube and mono-tube dampers are used in vehicle suspension systems, Figure 6.6. Whereas the twin-tube layout stores the oil that is displaced by the piston rod when entering the cylinder in a remote oil chamber, the mono-tube type uses a floating piston that operates against a gas chamber and makes the capacity of the cylinder adaptable.

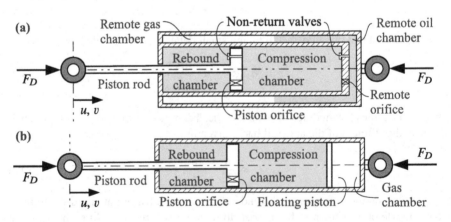

FIGURE 6.6
Types of suspension dampers: (a) twin-tube and (b) mono-tube.

In standard vehicle dynamics applications, simple characteristics

$$F_D = F_D(v) \tag{6.14}$$

are used to describe the damper force F_D as a function of the damper velocity $v = \dot{u}$. To obtain these characteristics, the damper is excited with a sinusoidal displacement signal $u = u_0 \sin 2\pi f t$. By varying the frequency in several steps from $f = f_0$ to $f = f_E$, different force displacement curves $F_D = F_D(u)$ are obtained, Figure 6.7.

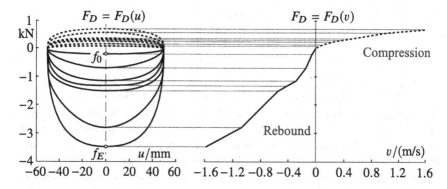

FIGURE 6.7
Damper characteristics generated from measurements [21].

By taking the peak values of the damper force at the displacement $u = 0$, which corresponds with the velocity $v(u = 0) = \pm 2\pi f u_0$, the characteristics $F_D = F_D(v)$ is generated now. Here, the compression cycle is associated with positive and the rebound cycle with negative damper velocities. Typical passenger car or truck dampers will have more resistance during the rebound cycle than the compression cycle.

Usually, nonlinear damper characteristics are simply provided by lookup tables. A simple linear or a smoother spline inter- or extrapolation will provide the damping forces. The MATLAB-Script provided in Listing 6.1 provides the lookup table for the damper characteristic $F_D = F_D(v)$, computes the damping force generated by a sine excitation with different frequencies, and plots the damping characteristic as well as the force displacement curves.

Listing 6.1
MATLAB-Script app_06_damper_char.m: nonlinear damper characteristic

```
1 % nonlin damper char F=F(v) provided by lookup table [ v/(m/s)  F/N ]
2 damper_table = ...
3 [ -1.5080   -3500; ...
4   -1.1310   -2800; ...
5   -0.5655   -1500; ...
6   -0.4524   -1250; ...
7   -0.3016   -1000; ...
8   -0.1508    -650; ...
9   -0.0377    -200; ...
10   0.0000       0; ...
11   0.0377     100; ...
12   0.1508     150; ...
```

```
13      0.3016      200; ...
14      0.4524      250; ...
15      0.5655      300; ...
16      1.1310      500; ...
17      1.5080      600 ];
18
19 subplot(1,2,1), hold on % generate F=F(u) via sine excitation
20 u0=0.06; freq=[0.1 0.4 0.8 1.2 1.5 3.0 4.0]; % amps in m & freqs in Hz
21 for i=1:length(freq)
22    t=linspace(0,1/freq(i),101); % time intervals
23    u=zeros(size(t)); F=u; % pre-allocate vars to speed up inner loop
24    for j=1:length(t)
25       u(j)=u0*sin(2*pi*freq(i)*t(j));
26       vij=2*pi*freq(i)*u0*cos(2*pi*freq(i)*t(j));
27       F(j)=interp1q(damper_table(:,1),damper_table(:,2),vij);
28    end
29    plot(u,F), grid on % force displacement curve at freq(i)
30 end
31 subplot(1,2,2) % plot characteristics F=F(v) and mark given points
32 plot(damper_table(:,1),damper_table(:,2),'-ok','MarkerSize',5),grid on
```

The lookup table was generated by simply inspecting the damper characteristics $F_D = F_D(v)$ provided in Figure 6.7. According to [21], the damper was excited with a sinusoidal displacement signal at different frequencies and an amplitude of approximately half the maximum stroke of the damper. Estimating the amplitude with $u = 0.06$ m and detecting the velocity values in the damper characteristics with $v \approx [0.04\, 0.15\, 0.30\, 0.45\, 0.57\, 1.13\, 1.51]$ m/s results via the relationship $v = 2\pi f u$ in the excitation frequencies defined in Listing 6.1. The MATLAB-Function interp1q provides the actual force for each velocity in between the range defined by the lookup table by a simple linear interpolation. Note that this function is fast, but does not extrapolate.

The results generated by the script just copy the force velocity characteristic and generate force displacement characteristics that are quite similar to the ones in Figure 6.7. However, the specific shape of the curves cannot be reproduced exactly by this simple model approach.

Dynamic damper models, like the one presented in [1], calculate the damper force via the fluid pressure applied to each side of the piston. The dynamic changes of the fluid and gas pressures in the compression and rebound chambers as well as in the gas chamber are calculated by applying physical principles.

6.1.5   General Point-to-Point Force Element

6.1.5.1   Generalized Forces

Usually, the mounts that connect springs, dampers, and actuators to different bodies can be regarded as ball joints, Figure 6.8. Then, the action line of the force generated by the element that is mounted between the points P and Q is defined by the unit vector

$$e_{PQ,0} = \frac{r_{PQ,0}}{|r_{PQ,0}|} = \frac{r_{0Q,0} - r_{0P,0}}{|r_{0Q,0} - r_{0P,0}|} \tag{6.15}$$

where

$$L = |r_{PQ,0}| = \sqrt{r_{PQ,0}^T r_{PQ,0}} \tag{6.16}$$

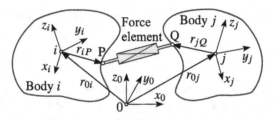

FIGURE 6.8
A point-to-point force element attached between two bodies.

defines the actual length of the force element in addition. The point P is attached to body i and point Q to body j. The vectors $r_{0i,0}$, $r_{0j,0}$ and the rotation matrices A_{0i}, A_{0j} describe the momentary position and orientation of axis systems fixed in the center of each body with respect to the earth-fixed reference frame 0. Then, the position of the attachment points is defined by

$$r_{0P,0} = \underbrace{r_{0i,0} + A_{0i}\, r_{iP,i}}_{r_{iP,0}} \quad \text{and} \quad r_{0Q,0} = \underbrace{r_{0j,0} + A_{0j}\, r_{jQ,j}}_{r_{jQ,0}} \tag{6.17}$$

where the vectors $r_{iP,i}$, $r_{jQ,j}$ characterize the position of P and Q with respect to the corresponding body-fixed axis systems. The deflection of the force element is just given by

$$s = L_0 - L \tag{6.18}$$

where L_0 denotes the length of the force element in the design position and the actual length L is defined by Equation (6.16).

According to Equation (6.3), spring forces can be described by appropriate force displacement characteristics, $F_S = F_S(s)$. In the approach of Equation (6.14), the force provided by a hydraulic damper can be described as a function of the damper velocity $F_D = F_D(v)$. Corresponding to the nonlinear characteristic plotted in Figure 6.7, $v > 0$ will indicate compression, which is in conformity with the sign definition of the spring deflection s too. The time derivative of the force element deflection s given in Equation (6.18) delivers the damper velocity at first as

$$v = \dot{s} = \frac{d}{dt}(L_0 - L) = -\dot{L} \tag{6.19}$$

By using Equation (6.16) it results in

$$v = -\frac{2\, r_{PQ,0}^T \dot{r}_{PQ,0}}{2\sqrt{r_{PQ,0}^T r_{PQ,0}}} = -\frac{r_{PQ,0}^T}{|r_{PQ,0}|}\, \dot{r}_{PQ,0} = -e_{PQ,0}^T\left(\dot{r}_{0Q,0} - \dot{r}_{0P,0}\right) \tag{6.20}$$

where Equation (6.15) was used to reinsert the unit vector e_{PQ} and to put the time derivative of the vector r_{PQ} down to the time derivatives of the position vectors r_{0Q}

and r_{0P}. According to Equation (6.17), the time derivatives of the position vectors are given by

$$\dot{r}_{0P,0} = v_{0i,0} + \omega_{0i,0} \times r_{iP,0} \quad \text{and} \quad \dot{r}_{0Q,0} = v_{0j,0} + \omega_{0j,0} \times r_{jQ,0} \tag{6.21}$$

where $v_{0i,0} = \dot{r}_{0i,0}$, $v_{0j,0} = \dot{r}_{0j,0}$ name the absolute velocities of the body centers and $\omega_{0i,0}$, $\omega_{0j,0}$ denote the absolute angular velocities of the bodies. Finally, the force acting in an arbitrary point-to-point force element can be described by

$$F = F(s, v, u, x), \tag{6.22}$$

where the dependency on a control signal u will include actuators and the vector x collects the internal states that are needed to model dynamic force elements too.

The equations of motion for different vehicle models are generated by applying Jourdain's principle of virtual power throughout this textbook. The forces and torques applied to the bodies are converted via partial velocities to generalized forces hereby. Similar to the appropriate terms in Equation (1.28), the force F acting in a point-to-point element is transformed by

$$q_F = \left(\frac{\partial v}{\partial z}\right)^T F(s, v, u, x) \tag{6.23}$$

to the corresponding parts of the generalized force vector q. According to Equation (6.20), the partial derivatives of the force element deflection velocity v with respect to the vector of the generalized velocities z are given by

$$\frac{\partial v}{\partial z} = -e_{PQ,0}^T \left(\frac{\partial \dot{r}_{0Q,0}}{\partial z} - \frac{\partial \dot{r}_{0P,0}}{\partial z}\right) = e_{PQ,0}^T \frac{\partial \dot{r}_{0P,0}}{\partial z} - e_{PQ,0}^T \frac{\partial \dot{r}_{0Q,0}}{\partial z} \tag{6.24}$$

which, by using Equation (6.21), results in

$$\frac{\partial v}{\partial z} = e_{PQ}^T \left(\frac{\partial v_{0i}}{\partial z} + \frac{\partial \omega_{0i}}{\partial z} \times r_{iP}\right) - e_{PQ}^T \left(\frac{\partial v_{0j}}{\partial z} + \frac{\partial \omega_{0j}}{\partial z} \times r_{jQ}\right) \tag{6.25}$$

where the comma-separated subscript 0, which indicates that all vectors are expressed in the earth-fixed axis system, was omitted. Then, the part q_F of the generalized force vector q related to the force F that acts in a point-to-point force element is given by

$$q_F = \left(\frac{\partial v_{0i}}{\partial z} + \tilde{r}_{iP}^T \frac{\partial \omega_{0i}}{\partial z}\right)^T e_{PQ} F - \left(\frac{\partial v_{0j}}{\partial z} + \tilde{r}_{jQ}^T \frac{\partial \omega_{0j}}{\partial z}\right)^T e_{PQ} F \tag{6.26}$$

where the cross-products in the partial velocities were replaced by the corresponding skew-symmetric matrix vector multiplications via the relationship $\omega \times r = -r \times \omega = -\tilde{r}\, \omega = \tilde{r}^T \omega$. Rearranging some terms and reintroducing the cross-product notation finally results in

$$q_F = \frac{\partial v_{0i}^T}{\partial z} F_{PQ} + \frac{\partial \omega_{0i}^T}{\partial z} r_{iP} \times F_{PQ} - \frac{\partial v_{0j}^T}{\partial z} F_{PQ} - \frac{\partial \omega_{0j}^T}{\partial z} r_{jQ} \times F_{PQ} \tag{6.27}$$

where $F_{PQ} = e_{PQ} F$ simply defines the vector that vectorially describes the magnitude and the orientation of the force acting in the point-to-point element. If bodies i and j are moving without kinematical constraints, the components of the vectors v_{0i}, v_{0j}, ω_{0i}, ω_{0j} can be used as generalized velocities. Then, the partial derivatives are trivial and Equation (6.27) simply yields

$$
q_F = \begin{bmatrix} F_{PQ} \\ r_{iP} \times F_{PQ} \\ -F_{PQ} \\ -r_{jQ} \times F_{PQ} \end{bmatrix}
\begin{array}{l} \leftarrow \text{ force applied to body } i \\ \leftarrow \text{ torque applied to body } i \\ \leftarrow \text{ force applied to body } j \\ \leftarrow \text{ torque applied to body } j \end{array}
\tag{6.28}
$$

which satisfies the "actio = reactio" principle and produces, via appropriate cross-products, the torques applied to bodies i and j automatically.

6.1.5.2 Example

For the sake of simplicity, the quarter car model in Section 1.6 was equipped just with a torsional spring damper combination in a first model approach. Now, the torsional damper will be replaced by a point-to-point damper element that is attached to the knuckle in point D and to the chassis in point E, Figure 6.9. Similar to Equation (6.15),

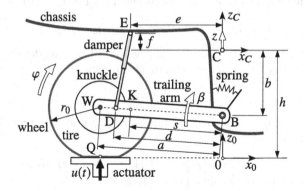

FIGURE 6.9
Quarter car model with a point-to-point damper element.

the action line of the damper force is defined in the chassis-fixed axis system by the unit vector

$$
e_{DE,C} = \frac{r_{DE,C}}{|r_{DE,C}|} = \frac{r_{CE,C} - r_{CD,C}}{|r_{CE,C} - r_{CD,C}|}
\tag{6.29}
$$

where C denotes the center of the chassis mass and

$$
L_{DE} = |r_{DE,C}| = \sqrt{r_{DE,C}^T r_{DE,C}}
\tag{6.30}
$$

defines the actual length of the damper element according to Equation (6.16). Here, the attachment point E is fixed to the chassis, which is common for most suspension systems. Its position relative to the chassis is simply defined by data and is given here by

$$r_{CE,C} = \begin{bmatrix} -e & 0 & f \end{bmatrix}^T \tag{6.31}$$

The attachment point D is fixed to the knuckle. In general, its momentary position may be described via the knuckle center, which here results in

$$r_{CD,C} = r_{CK,C} + A_{CK}\, r_{KD,K} \tag{6.32}$$

where $r_{CK,C}$ describes the momentary position of the knuckle center K and A_{CK} the rotation matrix of the knuckle that is defined by Equation (1.33). The vector $r_{KD,K}$ defines the position of D relative to K. It is expressed in the knuckle-fixed axis system and simply given by data. In this particular case, where it was assumed that the attachment point D is located on a line with the knuckle center K and the revolute joint in B, one gets

$$r_{CD,C} = \begin{bmatrix} -s\cos\beta \\ 0 \\ -b+s\sin\beta \end{bmatrix} + \begin{bmatrix} -(d-s)\cos\beta \\ 0 \\ (d-s)\sin\beta \end{bmatrix} = \begin{bmatrix} -d\cos\beta \\ 0 \\ -b+d\sin\beta \end{bmatrix} \tag{6.33}$$

The damper force F_D depends on the damper velocity v, which is defined as the time derivative of the element deflection s. According to Equation (6.20), one gets

$$v = \dot{s} = -\dot{L}_{DE} = -e^T_{DE,C}\left(\dot{r}_{CE,C} - \dot{r}_{CD,C}\right) \tag{6.34}$$

Here, the time derivative of the vector $r_{CE,C}$ will vanish because the attachment point E is fixed to the chassis. Then, Equation (6.34) simplifies to

$$v = e^T_{DE,C}\, \dot{r}_{CD,C} \tag{6.35}$$

The unit vector $e_{DE,C}$ is defined by Equation (6.29) and the time derivative of Equation (6.33) results in

$$\dot{r}_{CD,C} = \underbrace{\begin{bmatrix} s\sin\beta \\ 0 \\ s\cos\beta \end{bmatrix}\dot{\beta}}_{\dot{r}_{CK,C}} + \underbrace{\begin{bmatrix} 0 \\ 1 \\ 0 \end{bmatrix}\dot{\beta}}_{\omega_{CK,C}} \times \underbrace{\begin{bmatrix} -(d-s)\cos\beta \\ 0 \\ (d-s)\sin\beta \end{bmatrix}}_{r_{KD,C}} \tag{6.36}$$

In this particular case, Equation (6.36) simplifies to

$$\dot{r}_{CD,C} = \frac{\partial r_{CD,C}}{\partial \beta}\dot{\beta} = \begin{bmatrix} d\sin\beta \\ 0 \\ d\cos\beta \end{bmatrix}\dot{\beta} \tag{6.37}$$

The use of the partial derivative should indicate that in general the momentary position of the point D, where the damper is attached to the knuckle, may depend on more than one generalized coordinate. Hence, the damper velocity is given by

$$v = e^T_{DE,C}\frac{\partial r_{CD,C}}{\partial \beta}\dot{\beta} = \frac{\partial v}{\partial \dot{\beta}}\dot{\beta} = \dot{s} = \frac{\partial s}{\partial \beta}\dot{\beta} \tag{6.38}$$

where $\dot{\beta}$, the time derivative of the generalized coordinate β, serves as the trivial generalized velocity. Then, the relationship

$$\frac{\partial v}{\partial \dot{\beta}} = \frac{\partial s}{\partial \beta} \tag{6.39}$$

which can be deduced from Equation (6.38) by simply inspecting the corresponding terms, will hold in general. As mentioned in Section 6.1.4 the damper force F_D is provided as a function of the damper velocity v in standard applications. According to Equation (6.23), the contribution to the generalized force related to the knuckle motion β is obtained finally as

$$q_{F_D}^{\beta} = \left(\frac{\partial v}{\partial \dot{\beta}}\right)^T F_D(v) \tag{6.40}$$

where the partial derivative of the damper velocity v with respect to the angular velocity of the knuckle $\dot{\beta}$ is provided in Equation (6.38).

A simple quarter car model with a trailing arm suspension and a linear torsional spring damper combination acting in the revolute joint was presented in Section 1.6. Replacing the torsional damper by a nonlinear point-to-point damper model requires just a few changes in the corresponding MATLAB programs. At first, the MATLAB-Script given in Listing 1.3 is replaced by the one provided in Listing 6.2.

Listing 6.2
Script app_06_qcm_p2p.m: step input to quarter car model with nonlinear damper

```
1 p = uty_par_qcmta();  % quarter car model (basic parameter)
2
3 % add nonlinear point-to-point damper element
4 p.d = 0.425;   % distance joint B damper attachment at knuckle in m
5 p.e = 0.420;   % damper attach. at chassis (horiz. pos. C-E) in m
6 p.f = 0.020;   % damper attach. at chassis (vert. pos. C-E) in m
7 p.damper_table = ...  % F=F(v) as lookup table [ v/(m/s)   F/N ]
8 [ -1.5080   -3500; ...
9   -1.1310   -2800; ...
10  -0.5655   -1500; ...
11  -0.4524   -1250; ...
12  -0.3016   -1000; ...
13  -0.1508    -650; ...
14  -0.0377    -200; ...
15   0.0000       0; ...
16   0.0377     100; ...
17   0.1508     150; ...
18   0.3016     200; ...
19   0.4524     250; ...
20   0.5655     300; ...
21   1.1310     500; ...
22   1.5080     600 ];
23
24 % define step input
25 tstep=0.75; % [s]   step at @ t=tstep
26 ustep=0.05; % [m]   actuator step value
27
28 % initial states
29 x0 = [ 0; 0; 0; 0; 0; 0];
30
```

```
31 % time simulation
32 t0=0; tE=1.5;
33 [to,xo] = ode45(@(t,x) fun_06_qcm_p2p(t,x,p,tstep,ustep),[t0,tE],x0);
34
35 % get chassis acceleration and plot results
36 zdd=zeros(size(to));
37 for i=1:length(to)
38   xdot=fun_06_qcm_p2p(to(i),xo(i,:).',p,tstep,ustep); zdd(i)=xdot(4);
39 end
40
41 subplot(3,1,1), plot(to,xo(:,1)), grid on
42                 xlabel('t [s]'), ylabel('z in m')
43 subplot(3,1,2), plot(to,xo(:,2)*180/pi), grid on
44                 xlabel('t [s]'), ylabel('\beta in deg')
45 subplot(3,1,3), plot(to,zdd/9.81), grid on
46                 xlabel('t [s]'), ylabel('chassis acceleration in g')
```

The additional parameter structure elements p.d, p.e, p.f provide the damper attachment points according to Figure 6.9 and the nonlinear damper characteristic, defined by a lookup table, is stored in the structure element p.damper_table. Then, the function given in Listing 1.4 is extended and slightly modified to the function provided by Listing 6.3.

Listing 6.3
Function fun_06_qcm_p2p: quarter car model with nonlinear damper element

```
1 function xdot = fun_06_qcm_p2p(t,x,p,tstep,ustep)
2
3 % state variables and shortcuts
4 z  = x(1); be  = x(2); phi  = x(3);   sbe=sin(be); cbe=cos(be);
5 zd = x(4); bed = x(5); phid = x(6);
6
7 % step input to actuator @ t = tstep
8 if t < tstep, u = 0; else, u = ustep; end
9
10 % torque in revolute joint (spring only)
11 Ts = - ( p.Ts0 + p.cs*be );
12
13 % point to point damper element with non linear characteristic
14 rCDC=[-p.d*cbe;0;-p.b+p.d*sbe]; rCEC=[-p.e;0;p.f];
15 rDEC = rCEC-rCDC; LDE=norm(rDEC); eDEC = rDEC/LDE; % action line
16 dvdbed=eDEC.'*[p.d*sbe;0;p.d*cbe]; v = dvdbed*bed; % velocity
17 FD=-uty_interp11(p.damper_table,v); qFDbe=dvdbed*FD; % general. force
18
19 % tire deflection (static tire radius)
20 rS = p.h + z - p.b + p.a*sbe - u ;
21
22 % longitudinal tire force (adhesion assumed)
23 Fx = - p.cx *( p.a*(1-cbe) - rS*phi  ) ...
24      - p.dx *( p.a*sbe*bed - rS*phid ) ;
25
26 % vertical tire force (contact assumed)
27 Fz =  p.cz *( p.r0 - rS );
28
29 % mass matrix
30 Massma=[ p.mC+p.mK+p.mW              (p.s*p.mK+p.a*p.mW)*cbe    0 ; ...
31   (p.s*p.mK+p.a*p.mW)*cbe p.ThetaK+p.s^2*p.mK+p.a^2*p.mW   0 ; ...
32               0                              0               p.ThetaW ];
33
34 % vector of generalized forces and torques
35 q=[Fz-(p.mC+p.mK+p.mW)*p.g+(p.s*p.mK+p.a*p.mW)*sbe*bed^2; ...
36   Ts-(p.s*p.mK+p.a*p.mW)*cbe*p.g+p.a*(Fx*sbe+Fz*cbe)+qFDbe; ...
```

```
37    -rS*Fx ];
38
39 % state derivatives
40 xdot = [ zd;   bed;   phid;   Massma\q ];
41
42 end
```

Here, the torque Ts in the revolute joint, calculated in line 11, is reduced to a preloaded linear torsional spring. The calculations for the point-to-point damper element are done in lines 14 to 17 where the function uty_interp11, provided in Listing 6.4, calls the MATLAB-Function interp1q for a fast linear interpolation and performs a linear extrapolation of the lookup table for values of xi that exceed the range of the x-values defined in the first column of the table. In addition, short cuts for the trigonometric functions $\sin(\beta)$ and $\cos(\beta)$ are defined and the generalized damper force assigned to the variable qFDbe is appropriately added in line 36 to the generalized force vector.

Listing 6.4

Function uty_interp11: evaluation of a lookup table with extrapolation

```
1 function fi = uty_interp11( table, xi )
2 % linear inter- and extrapolation in lookup table:  table = [x, f]
3
4 [n,~] = size(table); % number of table pairs
5
6 if xi >= table(1,1) && xi <= table(n,1) % intermediate points
7    fi = interp1q(table(:,1),table(:,2),xi);
8    return
9 end
10 if xi < table(1,1) % linear extrapolation to the left
11    dfdx = ( table(2,2) - table(1,2) ) / ( table(2,1) - table(1,1) );
12    fi = table(1,2) + dfdx * ( xi - table(1,1) );
13    return
14 end
15 if xi > table(n,1) % linear extrapolation to the right
16    dfdx = ( table(n,2) - table(n-1,2) ) / ( table(n,1) - table(n-1,1) );
17    fi = table(n,2) + dfdx * ( xi - table(n,1) );
18    return
19 end
20
21 end
```

The simulation results of a step input are shown in Figure 6.10. As done in Section 1.6.5, the quarter car model is exposed at $t = 0.75 \, s$ to a step input. The simulation results of the model extended to a point-to-point damper element with a nonlinear characteristic are compared to those generated with the model equipped with a simple linear torsional damper. The time histories of the chassis displacements z and the knuckle rotation β, as well as the vertical chassis acceleration $\ddot{z}$, are quite different although the overall damping behavior is rather the same. Due to the fact that the nonlinear damper has less resistance during compression than rebound, the knuckle rotation induced by the step input is faster and much larger in the nonlinear case. As a consequence, the impact to the chassis mass is reduced, which results in a significantly less maximum acceleration (2.76 g instead of 3.59 g). Thus, the ride comfort can be improved using appropriate nonlinear damper characteristics.

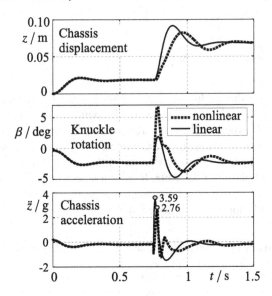

FIGURE 6.10
Simulation results with linear and nonlinear damper.

6.1.6   Rubber Elements

Force elements made of natural rubber or urethane compounds are used in many locations on the vehicle suspension system, Figure 6.11. Those elements require no

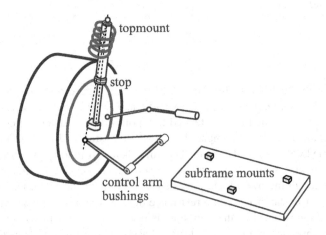

FIGURE 6.11
Rubber elements in vehicle suspension.

lubrication, isolate minor vibration, reduce transmitted road shock, operate noise-free, offer high load-carrying capabilities, and are very durable.

During suspension travel, the control arm bushings provide a pivot point for the control arm. They also maintain the exact wheel alignment by fixing the lateral and vertical location of the control arm pivot points. During suspension travel, the rubber portion of the bushing must twist to allow control arm motion. Thus, an additional resistance to suspension motion is generated.

Bump and rebound stops limit the suspension travel. The compliance of the topmount avoids the transfer of large shock forces to the chassis. The subframe mounts isolate the suspension system from the chassis and allow elasto-kinematic steering effects of the whole axle.

It turns out that those elastic elements can hardly be described by simple spring and damper characteristics, $F_S = F_S(u)$ and $F_D = F_D(v)$, because their stiffness and damping properties change with the frequency of the motion. Here, more sophisticated dynamic models are needed.

6.2 Dynamic Force Elements

6.2.1 Testing and Evaluating Procedures

6.2.1.1 Simple Approach

The effect of dynamic force elements is usually evaluated in the frequency domain. For this, on test rigs or in a simulation, the force element is excited by sine waves

$$x_e(t) = A \sin(2\pi f t) = A \sin\left(\frac{2\pi}{T} t\right) \tag{6.41}$$

with different frequencies $f_0 \leq f \leq f_E$ and amplitudes $A_{min} \leq A \leq A_{max}$, where $T = 1/f$ denotes the time period of the excitation. Starting at $t = 0$, the force element will be in steady state after several periods $t \geq nT$ where the number of cycles $n = 2, 3, \ldots$ must be chosen appropriately. The system response is periodic $F(t + T) = F(T)$ due to nonlinearities; it will be not harmonic, however. That is why the measured or calculated force signal $F = F(t)$ will be approximated by harmonic functions as much as possible. The first harmonic approximation is defined by

$$\underbrace{F(t)}_{\substack{\text{measured/}\\\text{calculated}}} \approx \underbrace{F_0 + \alpha \sin(\tfrac{2\pi}{T} t) + \beta \cos(\tfrac{2\pi}{T} t)}_{\text{first harmonic approximation}} \tag{6.42}$$

A least squares approach over one period from nT to $(n+1)T$

$$\frac{1}{2} \int_{nT}^{(n+1)T} \left\{ F_0 + \alpha \sin\left(\tfrac{2\pi}{T} t\right) + \beta \cos\left(\tfrac{2\pi}{T} t\right) - F(t) \right\}^2 dt \longrightarrow \text{Minimum} \tag{6.43}$$

grants an approximation that fits the original at best. The differentiation of Equation (6.43) with respect to the coefficients F_0, α, and β provides the necessary conditions for a minimum. It results in a system of linear equations

$$\int \left(F_0 + \alpha \sin\left(\tfrac{2\pi}{T}t\right) + \beta \cos\left(\tfrac{2\pi}{T}t\right) - F(t) \right) dt = 0$$

$$\int \left(F_0 + \alpha \sin\left(\tfrac{2\pi}{T}t\right) + \beta \cos\left(\tfrac{2\pi}{T}t\right) - F(t) \right) \sin\left(\tfrac{2\pi}{T}t\right) dt = 0 \qquad (6.44)$$

$$\int \left(F_0 + \alpha \sin\left(\tfrac{2\pi}{T}t\right) + \beta \cos\left(\tfrac{2\pi}{T}t\right) - F(t) \right) \cos\left(\tfrac{2\pi}{T}t\right) dt = 0$$

where the integral limits from nT to $(n+1)T$ were just omitted. As the integration is performed here exactly over one period from nT to $(n+1)T$, the integrals in Equation (6.44) will simplify to

$$\int dt = T, \quad \int \sin dt = \int \cos dt = \int \sin\cos dt = 0, \quad \int \sin^2 dt = \int \cos^2 dt = \tfrac{1}{2}T \qquad (6.45)$$

which immediately deliver the coefficients as

$$F_0 = \tfrac{1}{T} \int F\, dt, \quad \alpha = \tfrac{2}{T} \int F \sin dt, \quad \beta = \tfrac{2}{T} \int F \cos dt \qquad (6.46)$$

However, these are exactly the first coefficients of a Fourier approximation. The first-order harmonic approximation in Equation (6.42) can now be written as

$$F(t) = F_0 + \hat{F} \sin\left(\tfrac{2\pi}{T}t + \Psi\right) = F_0 + \hat{F} \sin(2\pi f t + \Psi) \qquad (6.47)$$

where the amplitude and the phase angle of the sine function are defined by

$$\hat{F} = \sqrt{\alpha^2 + \beta^2} \quad \text{and} \quad \tan\Psi = \frac{\beta}{\alpha} \qquad (6.48)$$

Figure 6.12 shows a simple nondynamic force element consisting of a spring with

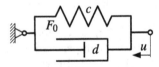

FIGURE 6.12
Linear preloaded spring and linear damper in parallel.

the stiffness c that is preloaded by the force F_0 and a damper with the constant d in parallel. It responds to the periodic excitation $u = A \sin(2\pi f t) = A \sin\left(\tfrac{2\pi}{T}t\right)$ with the force

$$F(t) = F_0 + cu + d\dot{u} = F_0 + \underbrace{cA}_{\alpha} \sin(2\pi f t) + \underbrace{d\,2\pi f\, A}_{\beta} \cos(2\pi f t) \qquad (6.49)$$

which represents exactly the first harmonic signal defined in Equation (6.42). Using Equation (6.48), it can be written in the form of Equation (6.47) where the amplitude and the phase angle are simply given by

$$\hat{F} = A\sqrt{c^2 + (2\pi f d)^2} \quad \text{and} \quad \tan\Psi = \frac{d\,2\pi f\, A}{c\, A} = 2\pi f\, \frac{d}{c} \qquad (6.50)$$

Hence, the response of a pure linear spring element ($c \neq 0$ and $d = 0$) is characterized by a force signal with the amplitude $\hat{F} = A c$ and a vanishing phase angle $\Psi = 0$, or respectively, $\tan \Psi = 0$. Whereas, a pure linear damper element ($c = 0$ and $d \neq 0$) will respond with a force signal of the amplitude $\hat{F} = 2\pi f dA$ and a maximum phase angle of $\Psi = 90°$, which corresponds to $\tan \Psi \rightarrow \infty$. As a consequence, the stiffness and damping properties of general force elements (nonlinear and/or dynamic) will be characterized in the sense of a first harmonic approximation by the dynamic stiffness

$$c_{dyn} = \frac{\hat{F}}{A} \tag{6.51}$$

and the phase angle Ψ, which is also called the dissipation angle.

6.2.1.2 Sweep Sine Excitation

In practice, the frequency response of a system is not determined punctually, but continuously. For this, the system is excited by a sweep sine. In analogy to the simple sine function

$$x_e(t) = A \sin(2\pi f t) \tag{6.52}$$

where the period $T = 1/f$ appears as prefactor at differentiation

$$\dot{x}_e(t) = A 2\pi f \cos(2\pi f t) = \frac{2\pi}{T} A \cos(2\pi f t). \tag{6.53}$$

A generalized sine function can be constructed now. Starting with

$$x_e(t) = A \sin(2\pi h(t)) \tag{6.54}$$

the time derivative results in

$$\dot{x}_e(t) = A 2\pi \dot{h}(t) \cos(2\pi h(t)) \tag{6.55}$$

In the following we demand that the function $h(t)$ generates periods fading linearly in time

$$\dot{h}(t) = \frac{1}{T(t)} = \frac{1}{p - qt} \tag{6.56}$$

where $p > 0$ and $q > 0$ are constants yet to determine. Equation (6.56) yields

$$h(t) = -\frac{1}{q} \ln(p - qt) + C \tag{6.57}$$

The initial condition $h(t = 0) = 0$ fixes the integration constant to

$$C = \frac{1}{q} \ln p \tag{6.58}$$

With Equations (6.58) and (6.57), Equation (6.54) results in a sine-like function

$$x_e(t) = A \sin\left(\frac{2\pi}{q} \ln \frac{p}{p - qt}\right) \tag{6.59}$$

which is characterized by linear fading periods. The important zero values for determining the period duration lie at

$$\frac{1}{q} \ln \frac{p}{p - q\, t_n} = 0, 1, 2, \cdots \quad \text{or} \quad \frac{p}{p - q\, t_n} = e^{nq}, \text{ with } n = 0, 1, 2, \cdots \quad (6.60)$$

and

$$t_n = \frac{p}{q}\left(1 - e^{-nq}\right), \; n = 0, 1, 2, \cdots \quad (6.61)$$

The time difference between two zero values yields the period

$$\begin{aligned} T_n &= t_{n+1} - t_n = \frac{p}{q}\left(1 - e^{-(n+1)q} - 1 + e^{-nq}\right) \\ &= \frac{p}{q}\left(-e^{-nq}\, e^{-q} + e^{-nq}\right) = \frac{p}{q}\, e^{-nq}\left(1 - e^{-q}\right) \end{aligned}, \; n = 0, 1, 2, \cdots \quad (6.62)$$

For the first ($n = 0$) and last ($n = N$) period, one finds

$$\begin{aligned} T_0 &= \frac{1}{f_0} = \frac{p}{q}\left(1 - e^{-q}\right), \\ T_N &= \frac{1}{f_e} = \frac{p}{q}\left(1 - e^{-q}\right) e^{-Nq} = T_0\, e^{-Nq} = \frac{1}{f_0}\, e^{-Nq} \end{aligned} \quad (6.63)$$

With the frequency range to investigate, given by the initial f_0 and the final frequency f_E, the parameter q and the ratio q/p can be calculated from Equation (6.63)

$$q = \frac{1}{N} \ln \frac{f_E}{f_0} \quad \text{and} \quad \frac{q}{p} = f_0\left\{1 - \left[\frac{f_0}{f_E}\right]^{\frac{1}{N}}\right\} \quad (6.64)$$

with $N + 1$ fixing the number of cycles. The passing of the whole frequency range will then take the time

$$t_{N+1} = \frac{p}{q}\left(1 - e^{-(N+1)q}\right) \quad (6.65)$$

Even if the period of the cycles is changed rather rapidly, the sweep sine generated by Equation (6.59) is close to pure sine functions calculated with the periods of the corresponding cycles, Figure 6.13. In this simple example the sine-like function

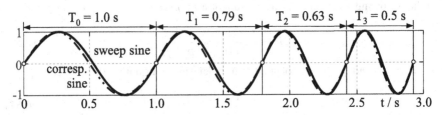

FIGURE 6.13
Sweep sine example.

sweeps from $f_0 = 1$ Hz to $f_e = 2$ Hz in $N + 1 = 4$ cycles. The parameter $q = 0.2310$ and the ratio $q/p = 0.2063$ are determined by Equation (6.64) and the duration $t_{N+1} = 2.9237$ s is provided by Equation (6.65). This method is very efficient in practical applications because it allows sweeping through large frequency ranges in a rather short time.

6.2.2 Spring Damper in Series

6.2.2.1 Modeling Aspects

Usually, suspension dampers are attached at the chassis via rubber topmounts. This combination represents a dynamic force element where a spring is arranged in series to a damper, Figure 6.14. The coordinate s describes the displacements of the force

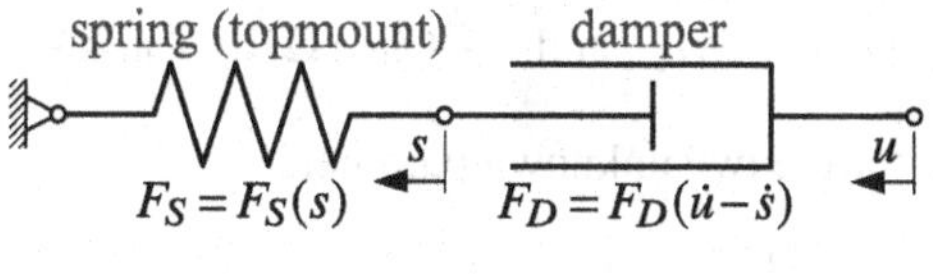

FIGURE 6.14
Spring and damper in series.

element and u denotes the overall deflection of the force element. Then, the forces acting in the spring and the damper are modeled as

$$F_S = F_S(s) \quad \text{and} \quad F_D = F_D(v) = F_D(\dot{u} - \dot{s}) \qquad (6.66)$$

where $v = \dot{u} - \dot{s}$ denotes the time derivative of the damper displacement. For this massless element, the force balance

$$F_D(\dot{u} - \dot{s}) = F_S(s) \qquad (6.67)$$

must hold. By introducing F_D^{-1} as the inverse damper characteristic, Equation (6.67) converts into a nonlinear first-order differential equation for the force displacement s,

$$\dot{s} = -F_D^{-1}(F_S(s)) + \dot{u} \qquad (6.68)$$

which is driven by the time derivative $\dot{u}$ of the overall element deflection and characterizes the dynamic behavior of this force element.

6.2.2.2 Linear Characteristics

If the spring and the damper are modeled by linear characteristics, the forces in the spring and the damper are simply given by $F_S = cs$ and $F_D = d(\dot{u} - \dot{s})$ where c names the spring stiffness and d denotes the damping constant. Now, the force balance $F_S = F_D$ delivers a linear first-order differential equation for the spring displacement s,

$$d(\dot{u} - \dot{s}) = cs \quad \text{or} \quad \dot{s} = -\frac{c}{d}s + \dot{u} \quad \text{or} \quad \frac{d}{c}\dot{s} = -s + \frac{d}{c}\dot{u} \qquad (6.69)$$

where the ratio between the damping coefficient d and the spring stiffness c acts as time constant $T = d/c$ hereby. In this simple case, the steady-state response to the purely harmonic excitation $u(t) = u_0 \sin \Omega t$ and $\dot{u} = u_0 \Omega \cos \Omega t$ respectively can be calculated quite easily. The steady-state response will be of the same type as the excitation. Inserting

$$s_\infty(t) = u_0 (a \sin \Omega t + b \cos \Omega t) \qquad (6.70)$$

into Equation (6.69) results in

$$\underbrace{\frac{d}{c}\, u_0\, (a\Omega \cos \Omega t - b\Omega \sin \Omega t)}_{\dot{s}_\infty} = \underbrace{-u_0\, (a \sin \Omega t + b \cos \Omega t)}_{s_\infty} + \underbrace{\frac{d}{c}\, u_0 \Omega \cos \Omega t}_{\dot{u}} \quad (6.71)$$

Collecting all sine and cosine terms, we obtain two equations,

$$-\frac{d}{c}\, u_0\, b\Omega = -u_0\, a \quad \text{and} \quad \frac{d}{c}\, u_0\, a\Omega = -u_0\, b + \frac{d}{c}\, u_0 \Omega \quad (6.72)$$

which can be solved for the two unknown parameters

$$a = \frac{\Omega^2}{\Omega^2 + (c/d)^2} \quad \text{and} \quad b = \frac{c}{d}\, \frac{\Omega}{\Omega^2 + (c/d)^2} \quad (6.73)$$

Then, the steady-state force response reads as

$$F_S = c\, s_\infty = c\, u_0\, \frac{\Omega}{\Omega^2 + (c/d)^2} \left(\Omega \sin \Omega t + \frac{c}{d} \cos \Omega t \right) \quad (6.74)$$

which can be transformed to $F_S = \hat{F}_S \sin (\Omega t + \Psi)$ where the force magnitude $\hat{F}_S$ and the dissipation angle Ψ are given by

$$\hat{F}_S = \frac{c\, u_0\, \Omega}{\Omega^2 + (c/d)^2} \sqrt{\Omega^2 + (c/d)^2} = \frac{c\, u_0\, \Omega}{\sqrt{\Omega^2 + (c/d)^2}} \quad \text{and} \quad \Psi = \arctan \frac{c/d}{\Omega} \quad (6.75)$$

The dynamic stiffness $c_{dyn} = \hat{F}_S / u_0$ and the dissipation angle Ψ are plotted in Figure 6.15 for different damping values versus the excitation frequency $f = \Omega/(2\pi)$. With increasing frequency, the spring damper combination changes from a pure

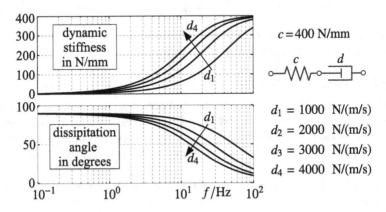

FIGURE 6.15
Frequency response of a linear spring damper combination.

damper performance ($c_{dyn} \to 0$ and $\Psi \approx 90°$) to a pure spring behavior ($c_{dyn} \approx c$ and $\Psi \to 0$). The value of the damping constant d controls the frequency range, where the element provides stiffness and/or damping.

6.2.2.3 Nonlinear Damper Topmount Combination

In Listing 6.1, a nonlinear damper characteristic $F_D = F_D(v)$ is described by a lookup table, which can easily be converted to the inverse characteristic. For the sake of simplicity, the compliance of the topmount will be described by a linear spring here. The function fun_06_damper_topmount in Listing 6.5 provides the sweep sine excitation and computes the dynamics of the nonlinear damper topmount combination.

Listing 6.5

Function fun_06_damper_topmount: nonlinear damper topmount dynamics

```
1 function sdot = fun_06_damper_topmount(t,s,p)
2
3 % sweep excitation ud = d/dt(u)
4    ud = 2*pi*p.amp/(p.p-p.q*t)*cos(2*pi/p.q*log(p.p/(p.p-p.q*t)));
5 % spring force (linear)
6    fs = p.c*s;
7 % force element dynamics
8    sdot = -uty_interp11(p.damper_inv,fs)  +  ud ;
9
10 end
```

The corresponding MATLAB-Script app_06_damper_topmount provided in Listing 6.6 assigns the relevant system parameter to the structure p, runs the simulation in the frequency range from $f_0 = 0.1\,Hz$ to $f_E = 25\,Hz$, performs the least squares approximation in each cycle by solving the appropriate overdetermined system of linear equations, computes the dynamic stiffness as well as the dissipation angle, and finally plots the results.

Listing 6.6

Script app_damper_topmount.m: nonlinear damper topmount frequency response

```
1 % damper char F_D=F_D(v) as lookup table [ v/(m/s)  F_D/N ]
2 ffr = 100;  % extend damper lookup table with dry friction force
3 damper_table = ...
4 [ -1.5080    -3500-ffr; ...
5   -1.1310    -2800-ffr; ...
6   -0.5655    -1500-ffr; ...
7   -0.4524    -1250-ffr; ...
8   -0.3016    -1000-ffr; ...
9   -0.1508     -650-ffr; ...
10  -0.0377     -200-ffr; ...
11   0.0000        0-ffr; ...
12   0.0000        0+ffr; ...
13   0.0377      100+ffr; ...
14   0.1508      150+ffr; ...
15   0.3016      200+ffr; ...
16   0.4524      250+ffr; ...
17   0.5655      300+ffr; ...
18   1.1310      500+ffr; ...
19   1.5080      600+ffr ];
20
21 % inverse damper characteristic as lookup table
22 p.damper_inv = [damper_table(:,2),damper_table(:,1)];
23
24 % topmount stiffness in N/m
25 p.c = 400000;
26
27 % define sweep sine excitation
28 f0  =  0.1;   % lowest frequency in Hz
29 fe  =  25;    % highest frequency in Hz
```

```
30 n    =    50;    % number of frequency interval
31 p.amp=  0.01;   % amplitude in m
32 p.q  = 1/n*log(fe/f0);   p.p  = p.q / ( f0*( 1 - (f0/fe)^(1/n) ) ) ;
33
34 % pre-allocate vars to speed up loop
35 freq=zeros(n+1,1); f0=freq; fs=freq; fc=freq;
36
37 t0=0; s0=0; x0=s0;% initial conditions for first cycle
38 for ifreq = 1:n+1
39
40 % Integrate from t0 to tE results in freq=1/(tE-t0)
41    tE = p.p/p.q*(1-exp(-ifreq*p.q));
42    freq(ifreq) = 1/(tE-t0); disp(['f=',num2str(freq(ifreq))])
43    [t,sout] = ode23s(@(t,s) fun_06_damper_topmount(t,s,p),[t0,tE],x0);
44    t0=tE; x0=sout(length(t),:).'; % initial conditions for next cycle
45
46 % sine and cosine parts of sweep excitation
47    arg = 2*pi/p.q*log(p.p./(p.p-p.q*t)); se=sin(arg); ce=cos(arg);
48
49 % curve fit to force (1. harmonic): f = f0 + fs*sin(arg) + fc*cos(arg)
50    f = p.c*sout(:,1);    A = [ ones(size(arg))   se   ce ];   coeff = A\f;
51    f0(ifreq) = coeff(1);  % offset
52    fs(ifreq) = coeff(2);  % sine part
53    fc(ifreq) = coeff(3);  % cosine part
54
55 end
56
57 % calculate dynamic stiffness and phase angle and plot results
58 cdyn = sqrt(fs.^2+fc.^2)/p.amp;  psi = atan2(fc,fs)*180/pi;
59 subplot(3,1,1), plot(freq,cdyn,'k','Linewidth',1), grid on
60 subplot(3,1,2), plot(freq, psi,'k','Linewidth',1), grid on
61 subplot(3,1,3), plot(freq,  f0,'k','Linewidth',1), grid on
```

Note that the nonlinear damper characteristic is extended to a dry friction component. This will cause the damper force F_D to jump from $-F_{Fr}$ to $+F_{Fr}$ at vanishing damper velocities $v = 0$, which results in a discontinuous damper characteristic. However, the dynamics of the damper topmount combination is computed via the inverse damper characteristic, which is still nonlinear and may have sharp bends but nevertheless is a continuous function, Figure 6.16. Because the stiffness of the

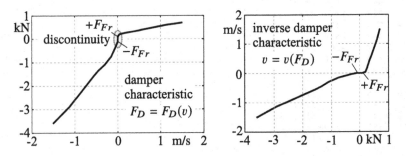

FIGURE 6.16
Nonlinear damper characteristic with discontinuous dry friction and its inverse.

topmount is considerably large in addition the low-order and implicit solver ode23s is used here for the time integration.

The frequency response of the damper topmount combination is shown in Figure 6.17. Besides the results for an excitation amplitude of $A = 10$ mm that will be generated by the MATLAB-Script in Listing 6.6, the results for the excitation amplitude of $A = 2$ mm and $A = 20$ mm are plotted too.

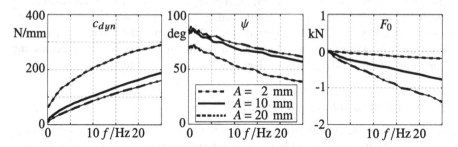

FIGURE 6.17
Frequency response of a damper topmount combination.

The overall behavior is similar to the linear spring damper element in series shown in Figure 6.15. Again, the damping properties decay with higher frequencies. This reduces the impact to the chassis when the vehicle is driven on a surface like cobblestones where higher frequencies are dominating the excitation. The frequency axis is scaled linearly because of the small frequency range. Due to the sharp bends in the inverse damper characteristic, the force element is not always perfectly in a steady-state condition, which causes small fluctuations in the frequency response, in particular visible in the phase angle ψ. In addition, the response depends on the excitation amplitude A, which is typical for nonlinear force elements. As the damper shows more resistance during its rebound cycle than its compression cycle, the least squares approximation results in a force offset that increases with higher frequencies.

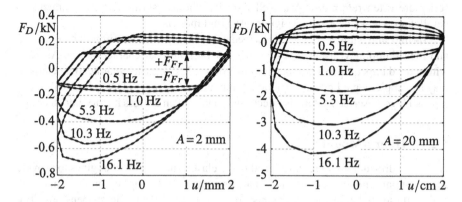

FIGURE 6.18
Force displacement diagrams of a damper topmount combination.

Force displacement diagrams for different amplitudes and frequencies illustrate the complex nonlinear behavior of this force element, Figure 6.18. In particular, at low frequencies ($f = 0.5$ Hz and $f = 1$ Hz) and small amplitudes ($A = 2$ mm) the force displacement curves show typical stick slip cycles with an inclination at the left and right side corresponding with the stiffness $c = 400$ kN/m of the topmount and a magnitude that is nearly entirely determined by the friction force $F_{Fr} = 0.1$ kN. The results for a larger excitation amplitude ($A = 20$ mm) are close to the pure damper response shown in Figure 6.7. However, the compliance of the topmount adds some spring effects that distort the force displacement curves accordingly.

6.2.3   General Dynamic Force Model

To approximate the complex dynamic behavior of bushings and elastic mounts, different spring damper models can be combined. Usually, several dynamic force elements consisting of a spring in series to a damper are arranged in parallel to a single spring that carries the static load and a damper, Figure 6.19. Spring damper elements in

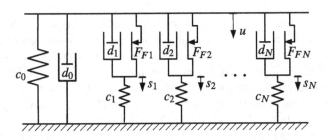

FIGURE 6.19
General dynamic force model.

series are often referred as Maxwell models, whereas spring damper elements in parallel are known as Kelvin-Voigt or just Voigt models.

Springs and dampers may be described quite simply by their stiffness $c_0, c_1 \ldots c_N$ and their damping constants $d_0, d_1 \ldots d_N$, or by nonlinear characteristics. As done within the damper topmount model (*c.f.* Section 6.2.2.3), each dynamic force element may be supplemented by a dry friction component, F_{Fi}, $i = 1(1)N$, which will make it possible to describe hysteresis effects and to take even a stress history into account.

6.2.4   Hydro-Mount

At first, hydro-mounts were introduced in the elastic suspension of engines to provide sufficient damping on one side and to achieve acoustic decoupling on the other side. Today, these dynamic force elements are also employed in axle suspensions. The principle and a dynamic model of a hydro-mount are shown in Figure 6.20. At small deformations, the change in the volume in chamber 1 is compensated by displacements of the membrane. When the membrane reaches the stop, the liquid in chamber 1 is

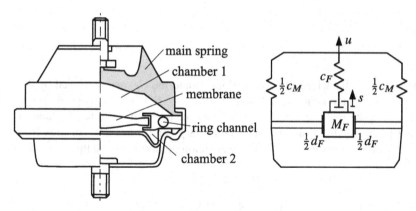

FIGURE 6.20
Hydro-mount: Layout and dynamic model.

pressed through a ring channel into chamber 2. The ratio of the chamber cross section to the ring channel cross section is very large. Thus the fluid is moved through the ring channel at very high speed. This results in remarkable inertia and resistance forces (damping forces). The force effect of a hydro-mount is combined from the elasticity of the main spring and the volume change in chamber 1. The actual mass M_{FR} of the fluid in the ring channel is amplified by the ratio of the cross-section areas A_C and A_R of the chamber and the ring channel and yields the generalized fluid mass

$$M_F = \left(\frac{A_C}{A_R}\right)^2 M_{FR} \tag{6.76}$$

The motions of the fluid mass cause friction losses in the ring channel, which in a first approximation are modeled proportional to the velocity of the generalized fluid mass,

$$F_D = d_F \dot{s} \tag{6.77}$$

where d_F denotes the corresponding damping constant. The force generated by the hydro-mount is given by

$$F = c_M u + F_F(u - s) \tag{6.78}$$

where u and s denote the overall displacement and the displacement of the generalized fluid mass M_F, respectively. The force effect of the main spring $F_M = c_M u$ is simply modeled by a linear spring with the constant c_M and the fluid force $F_F = F_F(u - s)$ combines the compliance of the fluid and the elasticity of the membrane bearing. The fluid compliance is modeled by a linear spring with the stiffness c_F and the membrane bearing is approximated by a rounded clearance. Then, the force F_F will be provided by the function uty_clearance given in Listing 6.7.

Listing 6.7
Function uty_clearance: force characteristic with smoothed clearance

```
1 function f = uty_clearance( x, c, s )
2   if abs(x) >= 2*s
```

```
3       f = sign(x) * c * ( abs(x) - s ) ;
4     else
5       f = sign(x) * c/(4*s) * x^2 ;
6     end
7 end
```

The spring is characterized by its stiffness c and the clearance, defined by the displacement s, is smoothed in the interval $-2s \leq x \leq +2s$ by appropriate parabolas. This is a quite realistic behavior in the case of the membrane bearing and avoids flutter problems that would occur when a sharp bend forms the transition from the clearance to the spring force. Then, the equation of motion for the fluid mass reads as

$$M_F \, \ddot{s} \; = \; -F_F - F_D \tag{6.79}$$

The membrane clearance makes Equation (6.79) nonlinear and affects the overall force in the hydro-mount provided by Equation (6.78) too. The dynamic equations of the hydro-mount are provided in the function fun_06_hydro_mount which is given in Listing 6.8.

Listing 6.8

Function fun_06_hydro_mount: hydro-mount dynamics

```
1 function xdot = fun_06_hydro_mount(t,x,p)
2
3 % get states and define sweep sine excitation
4   s = x(1);   sdot = x(2);
5   arg=2*pi/p.q*log(p.p/(p.p-p.q*t)); u=p.amp*sin(arg);
6 % fluid force including membran clearance
7   ff = uty_clearance((u-s), p.cf, p.sf);
8 % acceleration of generalized fluid mass and state derivatives
9   sddot = ( ff - p.df*sdot ) / p.mf ;
10  xdot = [ sdot; sddot ] ;
11
12 end
```

The MATLAB-Script in Listing 6.6, that computes the frequency response for a damper topmount combination, can easily be adapted to the hydro-mount. The appropriately modified MATLAB-Script is provided in Listing 6.9.

Listing 6.9

Script app_06_hydro_mount.m: frequency response of a hydro-mount

```
1 % define sweep sine excitation
2 f0     =    1;    % lowest frequency [Hz]
3 fe     =  100;    % highest frequency [Hz]
4 n      =   50;    % number of frequency interval
5 p.amp = 0.0005;   % amplitude [m]
6 p.q = (1/n)*log ( fe/f0 );  p.p = p.q / ( f0*( 1 - (f0/fe)^(1/n) ) );
7
8 % model parameter
9 p.mf=25;          % generalized fluid mass [kg]
10 p.cm=125000;     % stiffness of main spring [N/m]
11 p.df=750;        % fluid damping [N/(m/s)]
12 p.cf=100000;     % fluid stiffness [N/m]
13 p.sf=0.0002;     % clearance in mebrane support [m]
14
15 % pre-allocate vars to speed up loop
16 freq=zeros(n+1,1); f0=zeros(n+1,1); fs=zeros(n+1,1); fc=zeros(n+1,1);
17
```

```
18 t0=0; s0=0; sd0=0;  x0 = [ s0 sd0 ]; % initial condition first cycle
19 for ifreq=1:n+1
20
21 % Integrate from t0 to tE results in  freq = 1/(tE-t0)
22   tE=p.p/p.q*(1-exp(-ifreq*p.q));
23   freq(ifreq)=1/(tE-t0); disp(['f=',num2str(freq(ifreq))])
24   [t,xout] = ode45(@(t,x) fun_06_hydro_mount(t,x,p),[t0,tE],x0);
25   t0=tE;  x0=xout(length(t),:).'; % initial conditions for next cycle
26
27 % excitation and element force = force in main spring + fluid force
28   arg=2*pi/p.q*log(p.p./(p.p-p.q*t)); se=sin(arg); ce=cos(arg);
29   u = p.amp*se;  f = p.cm*u;
30   for i=1:length(f)
31     f(i) = f(i) + uty_clearance((u(i)-xout(i,1)),p.cf,p.sf);
32   end
33
34 % curve fit to force (1. harmonic): f = f0 + fs*sin(arg) + fc*cos(arg)
35   A = [ ones(size(arg))  se  ce ] ;  coeff = A \ f ;
36   f0(ifreq) = coeff(1);  % offset
37   fs(ifreq) = coeff(2);  % sine part
38   fc(ifreq) = coeff(3);  % cosine part
39
40 end
41
42 % calculate and plot dynamic stiffness and phase angle
43 cdyn = sqrt(fs.^2+fc.^2)/p.amp; psi  = atan2(fc,fs)*180/pi ;
44 subplot(3,1,1), semilogx(freq,cdyn,'k','Linewidth',1), grid on
45 subplot(3,1,2), semilogx(freq, psi,'k','Linewidth',1), grid on
46 subplot(3,1,3), semilogx(freq,  f0,'k','Linewidth',1), grid on
```

Note that the standard solver ode45 is used for the time integration because the
dynamics of the hydro-mount is smooth and not stiff. The results, plotted in a semi-

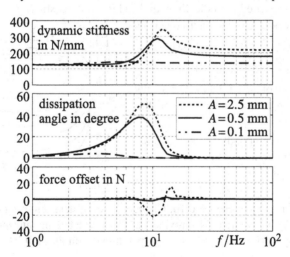

FIGURE 6.21
Frequency response of a hydro-mount at different excitation amplitudes.

logarithmic scaling, are shown in Figure 6.21. The dissipation angle illustrates that
the damping properties of the hydro-mount are limited to a small frequency range and
reaches its maximum at $f \approx 8$ Hz here. In particular, at higher frequencies, damping

is not noticeable at all, which represents a very good compromise between noise isolation at higher frequencies ($f > 20$ Hz) and vibration damping in the range from 5 to 15 Hz. However, the dynamic stiffness hardens significantly at higher frequencies, which will change the soft layout of an engine suspension to hard and transmit the engine shake caused by unbalanced masses to the chassis. The clearance in the membrane bearing bypasses this effect, at least at small amplitudes ($A = 0.5$ mm) but will then provide no damping at all.

Exercises

6.1 Write a MATLAB program that reproduces the force characteristics of an air spring plotted and defined in Figure 6.3. Study the effects of the design volume V_D and the effective cross section A_E on the results.

6.2 According to Figure 6.5, the layout of a U-shaped anti-roll bar is defined by $a = 220$ mm, $b = 700$ mm and $d = 19$ mm. The ratio between the vertical displacements of the wheel center and the attachment points of the anti-roll bar amounts to $i_{arb} = 0.7$ and the shear modulus of steel is given by $G = 85\,000$ N/mm^2.

Calculate the anti-roll bar stiffness with respect to the vertical displacements of the attachment points and the wheel centers.

6.3 A damper is excited periodically with the signal $u = u_0 \sin(2\pi f t)$ on a test rig. The tests are performed with the frequencies

$f_1 = 0.3$ Hz, $f_2 = 0.6$ Hz, $f_3 = 1.0$ Hz,

$f_4 = 1.5$ Hz, $f_5 = 2.5$ Hz, $f_6 = 4.0$ Hz

and an amplitude of $u_0 = 0.04$ m. The results are displayed in a force displacement diagram, $F_D = F_D(u)$.

Generate the corresponding lookup table that describes the nonlinear damper characteristic $F_D = F_D(\dot{u})$.

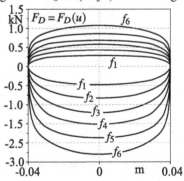

6.4 Verify the results in Figure 6.10. Change the nonlinear damping characteristic, which according to Listing 6.1 is provided as a lookup table, by multiplying the force values with appropriate factors and study its influence on the simulation results.

7

Vertical Dynamics

CONTENTS

7.1 Goals

The aim of vertical dynamics is the tuning of body suspension and damping to guarantee good ride comfort, respectively, a minimal stress of the load at sufficient safety. The stress of the load can be judged fairly well by maximal or integral values of the body accelerations. The wheel load F_z is linked to the longitudinal force F_x and the lateral force F_y by the coefficient of friction. The degressive influence of F_z on F_x and F_y as well as nonstationary processes at the increase of F_x and F_y in the average lead to lower longitudinal and lateral forces at wheel load variations. Maximal driving safety can therefore be achieved with minimal variations in the wheel load. Small variations of the wheel load also reduce road damage. The comfort of a vehicle is subjectively judged by the driver and passengers. The vibration of a vehicle occurs in several directions, contains many frequencies, and changes over time. Via the seat it is transferred to all passengers. The driver is also exposed to the vibration of the steering wheel. Whole-body vibration may cause sensations (e.g., discomfort or annoyance), influence human performance capability, or present a health and safety risk. According to ISO-Directive 2631 [19], root-mean-square (r.m.s.) values of the body accelerations are used to judge the effects of vibration on health and comfort. Because the human response to vibration is a function of frequency, the accelerations are filtered with frequency weighting curves. Different approaches of describing the human sense of vibrations by different metrics can be found in the literature [26].

The road excitation is transferred via the tire to the wheel, via the suspension system to the chassis and via the seat to the passengers. Soft suspension systems reduce the r.m.s. acceleration values of the chassis but will need a large suspension travel. Hence, a good ride comfort will always be a compromise between low acceleration values and limited suspension travel. The chassis acceleration and the suspension travel will thus be used as objective criteria.

7.2   From Complex to Simple Models

For detailed investigations of ride comfort and ride safety, sophisticated road and vehicle models are needed. The three-dimensional vehicle model, shown in Figure 7.1

FIGURE 7.1
Full vehicle model.

and discussed in [61], includes an elastically suspended engine and dynamic seat models. The elasto-kinematics of the wheel suspension is described as fully nonlinear. In addition, dynamic force elements for the damper topmount combination and the hydro-mounts are used. Such sophisticated models not only provide simulation results that are in good conformity to measurements but also make it possible to investigate the vehicle dynamic attitude at an early design stage.

Much simpler models can be used, however, for fundamental studies of ride comfort and ride safety. If the vehicle is mainly driving straight ahead at constant speed, the hub and pitch motion of the chassis as well as the vertical motion of the axles will dominate the overall movement. Then, planar vehicle models can be used. A nonlinear planar model consisting of five rigid bodies with eight degrees of freedom is discussed in [50]. The model, shown in Figure 7.2, considers nonlinear spring

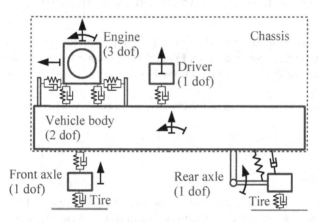

FIGURE 7.2
Sophisticated planar vehicle model with eight degrees of freedom (dof).

characteristics of the vehicle body and the engine suspension, as well as degressive characteristics of the shock absorbers. Even the suspension of the driver's seat is taken into account here. Planar vehicle models suit perfectly with a single track road model.

In a further simplification, the chassis is considered as one rigid body. The corresponding simplified planar model has four degrees of freedom then, which are characterized by the hub and pitch motion of the chassis z_C, θ_C, and the vertical motion of the axles z_{A1} and z_{A2}, Figure 7.3. Assuming small pitch motions ($\theta_C \ll 1$), the equations of motion for this simple planar vehicle model read as

$$M \ddot{z}_C = F_1 + F_2 - M g \tag{7.1}$$

$$\Theta \ddot{\theta}_C = -a_1 F_1 + a_2 F_2 \tag{7.2}$$

$$m_1 \ddot{z}_{A1} = -F_1 + F_{T1} - m_1 g \tag{7.3}$$

$$m_2 \ddot{z}_{A2} = -F_2 + F_{T2} - m_2 g \tag{7.4}$$

where M, m_1, m_2 denote the masses of the chassis, the front, and the rear axle. The inertia of the chassis around an axis located in the chassis center C and pointing into

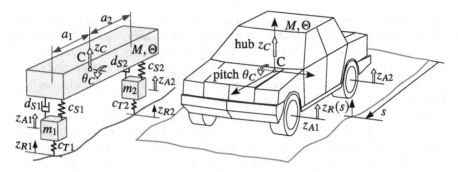

FIGURE 7.3
Simple planar vehicle model for basic comfort and safety analysis.

the lateral direction is described by Θ and a_1, a_2 represent the distances of the chassis center C to the front and rear axle. Finally, F_1, F_2 name the suspension forces and F_{T1}, F_{T2} the tire forces. The restrictions $F_{T1} \geq 0$ and $F_{T2} \geq 0$ will take tire liftoff into account.

The hub and pitch motion of the chassis can be combined to two new coordinates

$$
\begin{array}{c}
z_{C1} = z_C - a_1\,\theta_C \\
z_{C2} = z_C + a_2\,\theta_C
\end{array}
\quad \text{or} \quad
\left[\begin{array}{c} z_{C1} \\ z_{C2} \end{array} \right] =
\left[\begin{array}{cc} 1 & -a_1 \\ 1 & a_2 \end{array} \right]
\underbrace{\left[\begin{array}{c} z_C \\ \theta_C \end{array} \right]}_{}
\qquad (7.5)
$$

$$T_C$$

which describe the vertical motions of the chassis in the front and in the rear, Figure 7.4. Then, the Equations (7.1) and (7.2) arranged in matrix form

$$
\left[\begin{array}{cc} M & 0 \\ 0 & \Theta \end{array} \right]
\left[\begin{array}{c} \ddot{z}_C \\ \ddot{\theta}_C \end{array} \right] =
\left[\begin{array}{c} F_1 + F_2 - M\,g \\ -a_1\,F_1 + a_2\,F_2 \end{array} \right]
\qquad (7.6)
$$

can be written as

$$
\left[\begin{array}{cc} M & 0 \\ 0 & \Theta \end{array} \right]
\underbrace{\frac{1}{a_1+a_2}\left[\begin{array}{cc} a_2 & a_1 \\ -1 & 1 \end{array} \right]}_{T_C^{-1}}
\left[\begin{array}{c} \ddot{z}_{C1} \\ \ddot{z}_{C2} \end{array} \right] =
\underbrace{\left[\begin{array}{cc} 1 & 1 \\ -a_1 & a_2 \end{array} \right]}_{T_C^T}
\left[\begin{array}{c} F_1 \\ F_2 \end{array} \right] +
\left[\begin{array}{c} -M\,g \\ 0 \end{array} \right]
$$

$$(7.7)$$

where the inverse of the transformation matrix T_C defined in Equation (7.5) was used to replace the chassis hub and pitch accelerations by the vertical accelerations of chassis points located above the front and rear axle. It can be seen also that the distribution matrix for the suspension forces F_1 and F_2 is defined by the transposed transformation matrix. Multiplying Equation (7.7) with the inverse of T_C^{-1} finally results in

$$
\frac{1}{(a_1+a_2)^2}\left[\begin{array}{cc} M\,a_2^2 + \Theta & M\,a_1\,a_2 - \Theta \\ M\,a_1\,a_2 - \Theta & M\,a_2^2 + \Theta \end{array} \right]
\left[\begin{array}{c} \ddot{z}_{C1} \\ \ddot{z}_{C2} \end{array} \right] =
\left[\begin{array}{c} F_1 \\ F_2 \end{array} \right] +
\frac{1}{a_1+a_2}\left[\begin{array}{c} -a_2\,M\,g \\ -a_1\,M\,g \end{array} \right]
$$

$$(7.8)$$

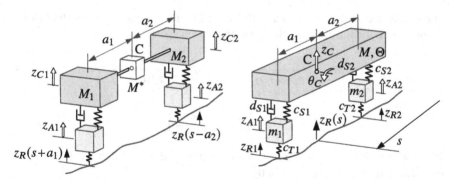

FIGURE 7.4
Chassis split into three point masses.

The off-diagonal elements in the mass matrix $M\,a_1\,a_2-\Theta$ generate a coupling between the chassis acceleration $\ddot{z}_{C1}$ and $\ddot{z}_{C2}$ that are induced by the suspension forces F_1, F_2 and the corresponding parts of the chassis weight $M\,g$. If the inertia of the vehicle happens to satisfy the relation

$$\Theta = M\,a_1\,a_2 \tag{7.9}$$

then, the remaining mass diagonal elements

$$M_1 = \frac{M\,a_2^2 + \Theta}{(a_1+a_2)^2} = \frac{M\,a_2^2 + M\,a_1\,a_2}{(a_1+a_2)^2} = \frac{M\,a_2}{a_1+a_2} \tag{7.10}$$

and

$$M_2 = \frac{M\,a_1^2 + \Theta}{(a_1+a_2)^2} = \frac{M\,a_1^2 + M\,a_1\,a_2}{(a_1+a_2)^2} = \frac{M\,a_1}{a_1+a_2} \tag{7.11}$$

spread the chassis mass to the front and rear according to the distribution of the chassis weight. In this particular case, the equations of motion for the front and rear chassis parts are decoupled and simply read as

$$M_1\,\ddot{z}_{C1} = F_1 - \frac{M\,g\,a_2}{a_1+a_2} \quad \text{and} \quad M_2\,\ddot{z}_{C2} = F_2 - \frac{M\,g\,a_1}{a_1+a_2}. \tag{7.12}$$

These equations, supplemented by the corresponding differential equations for the axles provided by the Equations (7.3) and (7.3), represent two separate models with two degrees of freedom that describe the vertical motions of the axle and the corresponding chassis mass on top of each axle.

The mass and inertia properties of the chassis may also be judged by three point masses M^*, M_1, M_2, which are located in the chassis center C and on top of the front and the rear axle, left image in Figure 7.4. The point masses must satisfy the relations

$$M_1 + M^* + M_2 = M\,, \quad a_1^2 M_1 + a_2^2 M_2 = \Theta\,, \quad a_1 M_1 = a_2 M_2 \tag{7.13}$$

that ensure the same chassis mass, the same inertia, and the same location of the center of gravity. Resolved for the point masses, one gets

$$M_1 = \frac{\Theta}{a_1(a_1+a_2)} \quad \text{and} \quad M_2 = \frac{\Theta}{a_2(a_1+a_2)} \tag{7.14}$$

as well as

$$M^* = M\left(1 - \frac{\Theta}{M\,a_1\,a_2}\right). \tag{7.15}$$

It can be seen that the coupling mass vanishes if $\Theta = M\,a_1\,a_2$ will hold. This relation coincides with Equation (7.9) exactly.

Hence, a vanishing $(M^* = 0)$ or at least a neglectable coupling mass $(M^* \ll M_1, M_2)$ indicate a specific chassis mass distribution that makes it possible to split the planar model with four degrees of freedom into two separate models with two degrees of freedom describing the vertical motions of the axle and the corresponding chassis mass on top of each axle. By using half the chassis and half the axle mass, we finally end up in quarter car models.

The parameters in Table 7.1 show that for a wide range of passenger cars the coupling mass is smaller than the corresponding chassis masses, $M^* < M_1$ and $M^* < M_2$. In these cases, the two mass models or the quarter car model represent quite

TABLE 7.1

Mass and inertia properties of standard vehicles and corresponding model parameter[2]

Vehicles and Model Properties		Mid-Size Car	Full-Size Car	Sports Utility Vehicle	Com-mercial Vehicle	Heavy Truck
Front axle mass	m_1/kg	80	100	125	120	600
Rear axle mass	m_2/kg	80	100	125	180	1100
Center of gravity	a_1/m	1.10	1.40	1.45	1.90	2.90
	a_2/m	1.40	1.40	1.38	1.40	1.90
Chassis mass	M/kg	1100	1400	1950	3200	14300
Chassis inertia	Θ/(kgm²)	1500	2350	3750	5800	50000
Discrete mass model	M_1/kg	545	600	914	925	3592
	M^*/kg	126	200	76	1020	5225
	M_2/kg	429	600	960	1255	5483

a good approximation of the simple planar model. For commercial vehicles and trucks, however, where the coupling mass has the same magnitude as the corresponding chassis masses, the quarter car model will serve for very basic studies only.

---

[2]The commercial vehicle and the heavy truck is assumed to be fully laden, whereas the cars are only partly loaded by two passengers at the front seats.

Finally, the function $z_R(s)$ provides road irregularities in the space domain, where s denotes the distance covered by the vehicle and measured at the chassis center of gravity. Then, the irregularities at the front and the rear axle are given by $z_R(s + a_1)$ and $z_R(s - a_2)$, respectively, where a_1 and a_2 locate the position of the chassis center of gravity C in the longitudinal direction. A quarter car model with a trailing arm suspension was presented in Section 1.6, further quarter car models are provided in this chapter.

For most vehicles, *c.f.* Table 7.1, the axle mass is much smaller than the corresponding chassis mass, $m_i \ll M_i$, $i = 1, 2$. Hence, for a first basic study, axle and chassis motions can be investigated independently. Now, the quarter car model is further simplified to two single mass models, Figure 7.5. The chassis model neglects

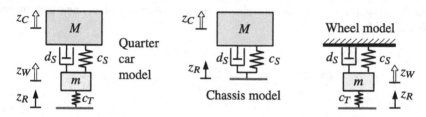

FIGURE 7.5
Simple vertical vehicle models.

the tire deflection and the inertia forces of the wheel. For the high frequent wheel motions, the chassis can be considered fixed to the inertia frame. The equations of motion for the chassis and the wheel model read as

$$M \ddot{z}_C + d_S \dot{z}_C + c_S z_C = d_S \dot{z}_R + c_S z_R \tag{7.16}$$

$$m \ddot{z}_W + d_S \dot{z}_W + (c_S + c_T) z_W = c_T z_R \tag{7.17}$$

where z_W and z_C define the vertical motions of the wheel mass and the corresponding chassis mass with respect to the steady-state position. The constants c_S, d_S describe the suspension stiffness and damping. The dynamic wheel load is calculated by

$$F_T^D = c_T (z_R - z_W) \tag{7.18}$$

where c_T is the vertical or radial stiffness of the tire and z_R denotes the road irregularities. In this simple approach the damping effects in the tire are not taken into account.

7.3 Basic Tuning

7.3.1 Natural Frequency and Damping Ratio

At an ideally even track, the right side of the equations of motion provided in Equations (7.16) and (7.17) vanish because $z_R = 0$ and $\dot{z}_R = 0$ will hold. The remaining homogeneous second-order differential equations can be written in a more general form as

$$\ddot{z} + 2\zeta\omega_0\dot{z} + \omega_0^2 z = 0 \tag{7.19}$$

where ω_0 represents the natural frequency and ζ is a dimensionless parameter called the viscous damping ratio. For the chassis and the wheel model, the new parameters are defined by

$$\left.\begin{array}{l} z \;\rightarrow\; z_C \\ \zeta \;\rightarrow\; \zeta_C \\ \omega_0^2 \rightarrow \omega_{0C}^2 \end{array}\right\} \text{Chassis:} \quad \zeta_C = \frac{d_S}{2\sqrt{c_S M}}, \qquad \omega_{0C}^2 = \frac{c_S}{M} \tag{7.20}$$

$$\left.\begin{array}{l} z \rightarrow z_W \\ \zeta \rightarrow \zeta_W \\ \omega_0^2 \rightarrow \omega_{0W}^2 \end{array}\right\} \text{Wheel:} \quad \zeta_W = \frac{d_S}{2\sqrt{(c_S+c_T)m}}, \qquad \omega_{0W}^2 = \frac{c_S+c_T}{m}$$

The solution of Equation (7.19) is of the type

$$z(t) = z_0\, e^{\lambda t} \tag{7.21}$$

where t denotes the time and z_0, λ are constants yet to determine. Inserting Equation (7.21) into Equation (7.19) results in

$$(\lambda^2 + 2\zeta\omega_0\lambda + \omega_0^2)\, z_0\, e^{\lambda t} = 0 \tag{7.22}$$

Nontrivial solutions $z_0 \neq 0$ are possible if

$$\lambda^2 + 2\zeta\omega_0\lambda + \omega_0^2 = 0 \tag{7.23}$$

will hold. The roots of the characteristic equation (7.23) depend on the magnitude of the viscous damping ratio

$$\begin{array}{ll} \zeta < 1 : & \lambda_{1,2} = -\zeta\omega_0 \pm i\,\omega_0\sqrt{1-\zeta^2} \\[2mm] \zeta \geq 1 : & \lambda_{1,2} = -\omega_0\left(\zeta \mp \sqrt{\zeta^2-1}\right) \end{array} \tag{7.24}$$

Figure 7.6 shows the root locus of the eigenvalues for different values of the viscous damping ratio ζ. For damping ratios $\zeta \geq 1$, the eigenvalues $\lambda_{1,2}$ are both real and negative. Hence, Eq. (7.21) will produce an exponentially decaying solution. If $\zeta < 1$

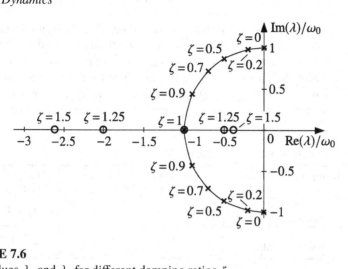

FIGURE 7.6
Eigenvalues λ_1 and λ_2 for different damping ratios ζ.

holds, the eigenvalues $\lambda_{1,2}$ will become complex, where λ_2 is the complex conjugate of λ_1. Now, the solution can be written as

$$z(t) = A\,e^{-\zeta\omega_0 t} \sin\left(\omega_0\sqrt{1-\zeta^2}\,t - \Psi\right) \tag{7.25}$$

where A and Ψ are constants that must be adjusted to given initial conditions $z(0) = z_0$ and $\dot{z}(0) = \dot{z}_0$. The real part $Re\,(\lambda_{1,2}) = -\zeta\omega_0$ is always negative and determines the decay of the solution over time. The imaginary part $Im\,(\lambda_{1,2}) = \omega_0\sqrt{1-\zeta^2}$ defines the actual frequency of the vibration. The actual frequency

$$\omega = \omega_0\sqrt{1-\zeta^2} \tag{7.26}$$

tends to zero, $\omega \to 0$, if the viscous damping ratio will approach the critical damping value $\zeta = 1$. In a more general way, the relative damping may also be judged by the ratio

$$D_\lambda = \frac{-Re(\lambda_{1,2})}{|\lambda_{1,2}|} \tag{7.27}$$

For complex eigenvalues that characterize vibrations with viscous damping ratios $\zeta < 1$, the relative damping ratio equals the viscous damping ratio because the absolute value of the complex eigenvalues is given by

$$|\lambda_{1,2}| = \sqrt{Re(\lambda_{1,2})^2 + Im(\lambda_{1,2})^2} = \sqrt{\left(-\zeta\,\omega_0\right)^2 + \left(\pm\omega_0\sqrt{1-\zeta^2}\right)^2} = \omega_0 \tag{7.28}$$

Then, Equation (7.27) simply results in

$$D_\lambda^{\zeta<1} = \frac{+\zeta\,\omega_0}{\omega_0} = \zeta. \tag{7.29}$$

For $\zeta \geq 1$, the eigenvalues become real but are still negative. Then, its absolute value equals the negative real part and Equation (7.27) will always produce the relative damping ratio $D_\lambda^{\zeta \geq 1} = 1$. In this case, the viscous damping ratio is more sensitive because, according to Equation (7.20), ζ is proportional to the damping value d.

The dimensionless ratios ζ and D_λ can be used to classify the type of the motion, Table 7.2, but they will not serve as proper judging criteria. Even in this simple

TABLE 7.2

Type of motion for single mass models

Damping Ratio / Type of Motion	Undamped Oscillation	Damped Oscillation	Damped Motion
Relative damping ratio	$D_\lambda = 0$	$0 < D_\lambda < 1$	$D_\lambda = 1$
Viscous damping ratio	$\zeta = 0$	$0 < \zeta < 1$	$\zeta \geq 1$

single mass model, the dimensionless damping ratios are a combination of several parameters. Hence, in order to evaluate the effect of the suspension damping d_S on the chassis motions and on the wheel load, the influence of the suspension stiffness c_S, the tire stiffness c_T, and the model masses M, m must also be taken into account.

7.3.2   Minimum Spring Rate

The suspension spring is loaded with the corresponding vehicle weight. For linear spring characteristics, the steady-state spring deflection is calculated from

$$u_0 = \frac{F_0}{c_S} = F_0 \frac{M\,g}{c_S} \tag{7.30}$$

where M denotes the corresponding chassis mass and $F_0 = M\,g$ the preload of the suspension spring, which is supposed to have a linear characteristic defined by its stiffness c_S. For a conventional suspension without level regulation, a load variation $M \to M + \Delta M$ changes the spring deflections from u_0 to $u_0 + \Delta u$. Analogous to Equation (7.30), the additional spring deflection follows from

$$\Delta u = \frac{\Delta M\,g}{c_S} \tag{7.31}$$

However, the suspension travel and hence the additional spring deflection will be limited, $\Delta u \leq \Delta u_{max}$. Then, the suspension spring rate can be estimated by a lower bound

$$c_S \geq \frac{\Delta M_{max}\,g}{\Delta u_{max}} \tag{7.32}$$

where ΔM_{max} denotes the maximum permissible load. In the standard design of a passenger car, the engine is located in the front and the trunk in the rear part of the vehicle. Hence, most of the load is supported by the rear axle suspension.

7.3.3 Example

As an example, we assume that 150 kg of the permissible load of 500 kg are going to the front axle. Then, each front wheel is loaded by $\Delta M_{FW} = 150/2$ kg $= 75$ kg and each rear wheel by $\Delta M_{RW} = (500 - 150)/2$ kg $= 175$ kg. For standard passenger cars, the maximum wheel travel on compression is in the range of $u_{max} \approx 0.08$ m to $u_{max} \approx 0.10$ m. By setting $\Delta u_{max} = u_{max}/2$, we demand that the spring deflection caused by the load should not exceed half of the maximum value. Then, according to Equation (7.32), a lower bound of the spring rate at the front axle can be estimated by

$$c_{SF}^{min} = (75 \text{ kg} * 9.81 \text{ m/s}^2)/(0.08/2 \text{ m}) = 18400 \text{ N/m} \qquad (7.33)$$

The maximum load over one rear wheel amounts here to $\Delta M_{RW} = 175$ kg. Assuming that the suspension travel at the rear axle is slightly larger, $u_{max} \approx 0.10$ m, the minimum spring rate at the rear axle can be estimated by

$$c_{SR}^{min} = (175 \text{ kg} * 9.81 \text{ m/s}^2)/(0.10/2) \text{ m} = 34300 \text{ N/m} \qquad (7.34)$$

which is nearly two times the minimum value of the spring rate at the front axle. To reduce this difference, a spring rate of $c_{SF} = 20\,000$ N/m will be chosen at the front axle as a compromise.

7.3.4 Natural Eigenfrequencies

In Table 7.1 the discrete mass chassis model of a full-size passenger car is described by $M_1 = M_2 = 600$ kg and $M^* = 200$ kg. To separate the model into two decoupled two mass models, we have to neglect the coupling mass or, in order to achieve the same chassis mass, to distribute M^* equally to the front and the rear. Then, the corresponding chassis mass over one front wheel is obtained as

$$M_{FW} = (M_1 + M^*/2)/2 = (600 \text{ kg} + 200/2 \text{ kg})/2 = 350 \text{ kg} \qquad (7.35)$$

According to Equation (7.20), the natural eigenfrequency of the simple chassis model is given by $\omega_{0C}^2 = c_S/M$. Hence, for the chosen spring rate of $c_{SF} = 20000$ N/m, the natural frequency of the unloaded car amounts to

$$f_{0C}^0 = \frac{1}{2\pi} \sqrt{\frac{20000 \text{ N/m}}{350 \text{ kg}}} = 1.2 \text{ Hz} \qquad (7.36)$$

which is a typical value for most passenger cars. Due to the small amount of load, the natural frequency for the loaded chassis front model does not change very much,

$$f_{0C}^L = \frac{1}{2\pi} \sqrt{\frac{20000 \text{ N/m}}{(350 + 75) \text{ kg}}} = 1.1 \text{ Hz} \qquad (7.37)$$

The corresponding chassis mass over the rear axle is given here by

$$M_{RW} = (M_2 + M^*/2)/2 = (600 \text{ kg} + 200/2 \text{ kg})/2 = 350 \text{ kg} \qquad (7.38)$$

The natural frequencies for the quarter car chassis model at the rear axle result in

$$f_{0C}^0 = \frac{1}{2\pi} \sqrt{\frac{34\,300 \text{ N/m}}{350 \text{ kg}}} = 1.6 \text{ Hz} \tag{7.39}$$

$$f_{0C}^L = \frac{1}{2\pi} \sqrt{\frac{34\,300 \text{ N/m}}{(350 + 175) \text{ kg}}} = 1.3 \text{ Hz} \tag{7.40}$$

Now, the frequencies for the loaded and unloaded rear chassis model differ more and are larger than the ones at the front.

Finally, the natural eigenfrequencies of the front and rear wheel model are provided by

$$f_{0W}^F = \frac{1}{2\pi} \sqrt{\frac{c_T + c_S}{m}} = \frac{1}{2\pi} \sqrt{\frac{220000 \text{ N/m} + 20000 \text{ N/m}}{50 \text{ kg}}} = 11.0 \text{ Hz} \tag{7.41}$$

$$f_{0W}^R = \frac{1}{2\pi} \sqrt{\frac{c_T + c_S}{m}} = \frac{1}{2\pi} \sqrt{\frac{220000 \text{ N/m} + 34300 \text{ N/m}}{50 \text{ kg}}} = 11.4 \text{ Hz} \tag{7.42}$$

which are common values for standard passenger cars. For the full-size car under consideration Table 7.1 delivers the wheel mass $m = 50$ kg as half of the axle mass and the same vertical tire stiffness of $c_T = 220000$ N/m was assumed at the front and rear wheel.

7.3.5   Influence of Damping

To investigate the influence of suspension damping on the chassis and wheel motion, the simple vehicle models are exposed to initial disturbances, Figure 7.7. Both time simulations start at $t = 0$ with vanishing velocities, $\dot{z}_C(t = 0) = 0$, $\dot{z}_W(t = 0) = 0$ and displacements, which by $z_C(t = 0) = M\,g/c_S$ and $z_W(t = 0) = (M + m)\,g/c_T$, are adjusted to the system parameter such that the suspension spring and the tire, respectively, are unloaded at the very beginning. The time response of the chassis displacement $z_C(t)$ and the wheel displacement $z_W(t)$ as well as the chassis acceleration $\ddot{z}_C$ and the wheel load or the vertical tire force F_T are shown in Figure 7.7 for different damping rates ζ_C and ζ_W.

The wheel load $F_T = F_T^0 + F_T^D$ is composed of the static wheel load F_T^0 and the dynamic wheel load F_T^D which is defined by Equation (7.18). The wheel model is part of the quarter car model, hence the static wheel load is provided by $F_T^0 = (M + m)\,g = $ (350 kg + 50 kg) $* 9.81$ m/s² $= 3.9$ kN in this particular case. Because the dynamic wheel load F_T^D decays over time, $F_T(t \rightarrow \infty) = F_T^0$ will hold as a consequence.

The natural eigenfrequency of the unloaded chassis model is provided by Equation (7.36) as $f_{0C} = f_{0C}^0 = 1.2$ Hz and Equation (7.41) delivers the natural eigenfrequency of the front wheel as $f_{0W} = f_{0C}^F = 11.0$ Hz. That is why, the time scales in the plots on the left and right of Figure 7.7, which represent the chassis and the wheel dynamics,

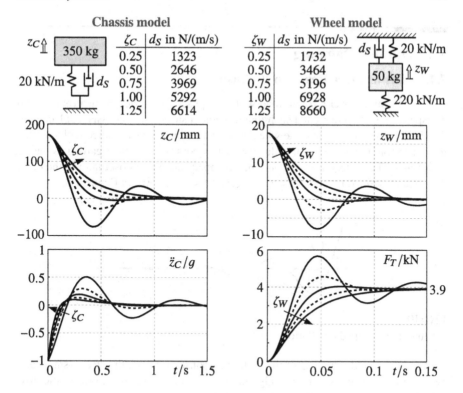

FIGURE 7.7
Time response of simple vehicle models to initial disturbances.

differ in an order of magnitude. This fact also justifies the decoupling of quarter car models into two single mass oscillators.

To achieve the same damping rates for the chassis and the wheel model, different values for the damping parameter d_S were needed. With increased damping, the overshoot effect in the time history of the chassis displacement and the wheel load becomes smaller and smaller until it vanishes completely at $\zeta_C = 1$ and $\zeta_W = 1$.

Usually, as it is here, the corresponding damping values will be different. Hence, a simple linear damper can either avoid overshoots in the chassis motions or in the wheel loads. However, the overshoot in the time history of the chassis accelerations $\ddot{z}_C(t)$ will only vanish for $\zeta_C \to \infty$, which surely is not a desirable configuration, because then, it will take a very long time for the initial chassis displacement to fully disappear.

7.4   Optimal Damping

7.4.1   Disturbance Reaction Problem

A vehicle running over a rough road will be exposed to a series of disturbances. So, a suspension system designed such that the time history of the chassis displacement, the chassis acceleration, and the wheel load to an arbitrary disturbance will approach the corresponding steady-state values as fast as possible may be regarded as perfect or optimal. The typical time response of a damped single-mass oscillator to the initial disturbance defined by $z(t=0) = z_0$ and $\dot{z}(t=0) = 0$ is shown in Figure 7.8.

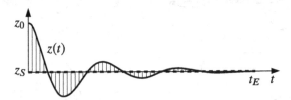

FIGURE 7.8
Evaluating a damped oscillation.

Counting the differences of the system response $z(t)$ from the steady-state value $z_S = 0$ as errors allows one to judge the attenuation. If the overall quadratic error calculated in an appropriate time interval $0 \le t \le t_E$ becomes a minimum,

$$\epsilon^2 = \int_{t=0}^{t=t_E} z(t)^2 \, dt \;\rightarrow\; Min \tag{7.43}$$

then the system will surely approach the steady-state position as fast as possible. In theory, $t_E \rightarrow \infty$ will hold; for practical applications, a finite t_E must be chosen appropriately. For nonlinear systems, Equation (7.43) must be solved numerically, which may be quite time consuming. However, there exists an algebraic solution for linear systems.

In general, linear dynamic systems are described by the state equation

$$\dot{x} = A\,x + B\,u \quad \text{with} \quad x(t = 0) = x_0 \tag{7.44}$$

where x denotes the state vector that collects generalized coordinates and their derivatives, the state matrix A represents the dynamics of the system, the term $B\,u$ describes the excitation, and x_0 describes the initial state. The steady-state solution $x(t \rightarrow \infty) = x_S$ to a step input at $t = 0$ with $u = u_0$ results in

$$0 = A\,x_S + B\,u_0 \quad \text{or} \quad x_S + A^{-1} B\,u_0 \tag{7.45}$$

Applying the state transformation $x = \bar{x} - x_S$ to Equation (7.44) results in

$$\frac{d}{dt}(\bar{x} - x_S) = A\,(\bar{x} - x_S) + B\,u \quad \text{with} \quad x(t = 0) = x_0 = \bar{x}(t_0) - x_S \tag{7.46}$$

which, using Equation (7.45), simplifies to

$$\dot{\bar{x}} = A\bar{x} \quad \text{with} \quad \bar{x}(t_0) = \bar{x}_0 = x_0 + x_S \tag{7.47}$$

So, the response of a linear dynamic system to a step input $u = u_0$ at $t = 0$ is equivalent to an initial disturbance $\bar{x}_0 = x_0 + x_S$ applied to the transformed system. As an extension to Equation (7.43), the transition of the state $\bar{x}(t)$ to its steady-state value $\bar{x}(t \to \infty) = 0$ may be judged by the more general disturbance-reaction problem

$$\epsilon^2 = \int_{t=0}^{t=t_E} \bar{x}^T Q\,\bar{x}\,dt \;\to\; \text{Min.} \tag{7.48}$$

where $Q = Q^T$ is a symmetric matrix of appropriate size that allows one to weight the components of the state vector x individually [9]. For $t_E \to \infty$, the integral in Equation (7.48) is solved by

$$\int_{t=0}^{t\to\infty} \bar{x}^T Q\,\bar{x}\,dt = \bar{x}_0^T R\,\bar{x}_0 \tag{7.49}$$

where $\bar{x}(t \to \infty) = 0$ was taken for granted. The symmetric matrix $R = R^T$ is defined by the Lyapunov equation,

$$A^T R + R A + Q = 0 \tag{7.50}$$

In general, the Lyapunov equation will be solved numerically by appropriate algorithms. For the single-mass oscillator described by Equation (7.19), the state equation (7.47) reads as

$$\underbrace{\begin{bmatrix} \dot{z} \\ \ddot{z} \end{bmatrix}}_{\dot{\bar{x}}} = \underbrace{\begin{bmatrix} 0 & 1 \\ -\omega_0^2 & -2\zeta\omega_0 \end{bmatrix}}_{A} \underbrace{\begin{bmatrix} z \\ \dot{z} \end{bmatrix}}_{\bar{x}} \tag{7.51}$$

Then, the Lyapunov equation

$$\begin{bmatrix} 0 & -\omega_0^2 \\ 1 & -2\zeta\omega_0 \end{bmatrix} \begin{bmatrix} R_{11} & R_{12} \\ R_{12} & R_{22} \end{bmatrix} + \begin{bmatrix} R_{11} & R_{12} \\ R_{12} & R_{22} \end{bmatrix} \begin{bmatrix} 0 & 1 \\ -\omega_0^2 & -2\zeta\omega_0 \end{bmatrix} + \begin{bmatrix} Q_{11} & Q_{12} \\ Q_{12} & Q_{22} \end{bmatrix} \tag{7.52}$$

delivers three linear equations,

$$\begin{aligned} -\omega_0^2 R_{12} - \omega_0^2 R_{12} + Q_{11} &= 0 \\ -\omega_0^2 R_{22} + R_{11} - 2\zeta\omega_0 R_{12} + Q_{12} &= 0 \\ R_{12} - 2\zeta\omega_0 R_{22} + R_{12} - 2\zeta\omega_0 R_{22} + Q_{22} &= 0 \end{aligned} \tag{7.53}$$

which can easily be solved for the elements of R. The first and third equations deliver

$$R_{12} = \frac{Q_{11}}{2\omega_0^2}, \quad R_{22} = \frac{1}{4\zeta\omega_0}\left(\frac{Q_{11}}{\omega_0^2} + Q_{22}\right) \tag{7.54}$$

Finally, the second equation provides

$$R_{11} = \left(\zeta + \frac{1}{4\zeta}\right)\frac{Q_{11}}{\omega_0} - Q_{12} + \frac{\omega_0}{4\zeta}Q_{22} \qquad (7.55)$$

If the initial state is characterized by an initial displacement $z(t = 0) = z_0$ and a vanishing initial velocity $\dot{z}(t = 0) = 0$, the disturbance reaction problem defined in Equation (7.49) will simplify to

$$\int\limits_{t=0}^{t\rightarrow\infty} x^T(t)\,Q\,x(t)\,dt = [z_0\ 0]^T \begin{bmatrix} R_{11} & R_{12} \\ R_{12} & R_{22} \end{bmatrix} \begin{bmatrix} z_0 \\ 0 \end{bmatrix} = z_0^2 R_{11}$$

$$= z_0^2 \left[\left(\zeta + \frac{1}{4\zeta}\right)\frac{Q_{11}}{\omega_0} - Q_{12} + \frac{\omega_0}{4\zeta}Q_{22}\right] \qquad (7.56)$$

where ζ and ω_0 denote the viscous damping and the natural eigenfrequency of the single-mass oscillator.

7.4.2 Optimal Safety

Ride safety may be judged by dynamic wheel load variations. In the absence of road irregularities $z_R = 0$, the dynamic wheel load computed in Equation (7.18) simplifies to $F_T^D = -c_T z_W$. Then, optimal ride safety is achieved by

$$\epsilon_S^2 = \int\limits_{t=0}^{t\rightarrow\infty} \left(F_T^D\right)^2 dt = \int\limits_{t=0}^{t\rightarrow\infty} \left(-c_T z_W\right)^2 dt = \int\limits_{t=0}^{t\rightarrow\infty} c_T^2 z_W^2\,dt \rightarrow \text{Min.} \qquad (7.57)$$

This demand can easily be transformed to the corresponding general disturbance reaction problem,

$$\epsilon_S^2 = \int\limits_{t=0}^{t\rightarrow\infty} c_T^2 z_W^2\,dt = \int\limits_{t=0}^{t\rightarrow\infty} [z_W\ \dot{z}_W] \begin{bmatrix} c_T^2 & 0 \\ 0 & 0 \end{bmatrix} \begin{bmatrix} z_W \\ \dot{z}_W \end{bmatrix} \rightarrow \text{Min.} \qquad (7.58)$$

where $x_W^T = [z_W\ \dot{z}_W]$ denotes the state vector of the wheel model and the weighting matrix Q is defined by the coefficients

$$Q_{11} = c_T^2, \quad Q_{12} = 0, \quad Q_{22} = 0 \qquad (7.59)$$

According to Equation (7.56), the integral in Equation (7.58) is solved by

$$\epsilon_S^2 = \int\limits_{t=0}^{t\rightarrow\infty} c_T^2 z_W^2\,dt = z_{0W}^2 \left(\zeta_W + \frac{1}{4\zeta_W}\right)\frac{c_T^2}{\omega_{0W}} \qquad (7.60)$$

where the specific coefficients of the weighting matrix Q provided by Equation (7.59) were already taken into account.

A soft tire ($c_T \to 0$) makes the safety criteria in Equation (7.60) small ($\epsilon_S^2 \to 0$) and thus reduces the dynamic wheel load variations. However, the tire spring stiffness cannot be reduced to arbitrary low values because this would cause too large tire deformations and finally damage the tire.

Small wheel masses ($m \to 0$) will increase the natural eigenfrequency ($\omega_{0W} = \sqrt{(c_S + c_T)/m}$) and thus reduce the safety criteria in Equation (7.60) too. So, the use of light metal rims will indeed improve the ride safety of a car because of the wheel weight reduction.

In principle, large values for the natural eigenfrequency ω_{0W} could be achieved by hardening the suspension spring $c_S \to \infty$ but this will contradict good driving comfort.

With fixed values for c_T and ω_{0W}, the merit function in Equation (7.60) will become a minimum if

$$\frac{\partial \epsilon_S^2}{\partial \zeta_W} = \frac{z_{0W}^2}{\omega_{0W}} \left(1 + \frac{-1}{4\zeta_W^2} \right) c_T^2 = 0 \tag{7.61}$$

will hold. Hence, a viscous damping rate of

$$\zeta_W^{opt}\big|_{Safety} = \frac{1}{2} \tag{7.62}$$

will guarantee optimal ride safety by minimizing the merit function in Equation (7.60). According to Equation (7.20), this corresponds to the damping parameter

$$d_S^{opt}\big|_{Safety} = \sqrt{(c_S + c_T)m} \tag{7.63}$$

7.4.3 Optimal Comfort

To judge the ride comfort, the hub motion of the chassis z_C and its acceleration $\ddot{z}_C$ can be used as objective criteria. Hence, the demand

$$\epsilon_C^2 = \int_{t=0}^{t=t_E} \left[(\alpha \ddot{z}_C)^2 + (\theta z_C)^2 \right] dt \to \text{Min.} \tag{7.64}$$

will guarantee optimal ride safety. By the factors α and β the acceleration and the hub motion can be weighted differently.

The equation of motion for the chassis model provided by Equation (7.16) can be resolved for the acceleration

$$\ddot{z}_C = - \left(\omega_{0C}^2 z_C + 2\zeta_C \omega_{0C} \dot{z}_C \right) \tag{7.65}$$

where the absence of road irregularities $z_R = 0$, $\dot{z}_R = 0$ was taken into account and, as done in Equation (7.20), the system parameter M, c_S, and d_S are summarized in the viscous damping ratio ζ_C and in the natural eigenfrequency ω_{0C}. Then, the problem in Equation (7.64) can be written as

$$
\epsilon_C^2 = \int_{t=0}^{t=t_E} \left[\alpha^2 \left(\omega_{0C}^2 z_C + 2\zeta_C \omega_{0C} \dot{z}_C \right)^2 + \beta^2 z_C^2 \right] dt
$$

$$
= \int_{t=0}^{t=t_E} [z_C \;\; \dot{z}_C] \begin{bmatrix} \alpha^2 \left(\omega_{0C}^2 \right)^2 + \beta^2 & \alpha^2 \omega_{0C}^2 2\zeta_C \omega_{0C} \\ \alpha^2 \omega_{0C}^2 2\zeta_C \omega_{0C} & \alpha^2 (2\zeta_C \omega_{0C})^2 \end{bmatrix} \begin{bmatrix} z_C \\ \dot{z}_C \end{bmatrix} \rightarrow \text{Min.}
$$

(7.66)

where $x_C^T = [\; z_C \;\; \dot{z}_C \;]$ is the state vector of the chassis model and the weighting matrix Q is defined by the coefficients

$$
Q_{11} = \alpha^2 \left(\omega_{0C}^2 \right)^2 + \beta^2, \quad Q_{12} = 2\alpha^2 \zeta_C \omega_{0C}^3, \quad Q_{22} = 4\alpha^2 \zeta_C^2 \omega_{0C}^2
$$

(7.67)

According to Equation (7.56), the integral in Equation (7.66) evaluating the ride comfort is solved by

$$
\epsilon_C^2 = z_{C0}^2 \left[\left(\zeta_C + \frac{1}{4\zeta_C} \right) \frac{Q_{11}}{\omega_{0C}} - Q_{12} + \frac{\omega_{0C}}{4\zeta_C} Q_{22} \right]
$$

$$
= z_{0C}^2 \left[\left(\zeta_C + \frac{1}{4\zeta_C} \right) \frac{\alpha^2 \left(\omega_{0C}^2 \right)^2 + \beta^2}{\omega_{0C}} - 2\alpha^2 \zeta_C \omega_{0C}^3 + \frac{\omega_{0C}}{4\zeta_C} 4\alpha^2 \zeta_C^2 \omega_{0C}^2 \right]
$$

(7.68)

$$
= z_{0C}^2 \left[\alpha^2 \frac{\omega_{0C}^3}{4\zeta_C} + \beta^2 \left(\zeta_C + \frac{1}{4\zeta_C} \right) \frac{1}{\omega_{0C}} \right]
$$

By setting $\alpha = 1$ and $\beta = 0$, the time history of the chassis acceleration $\ddot{z}_C$ is weighted only. Equation (7.68) then simplifies to

$$
\epsilon_C^2 \big|_{\ddot{z}_C} = z_{0C}^2 \frac{\omega_{0C}^3}{4\zeta_C}
$$

(7.69)

which will become a minimum if either the viscous damping ratio tends to infinity or the natural frequency to zero. As mentioned before, $\zeta_C \rightarrow \infty$ surely is not a desirable configuration. A low natural frequency $\omega_{0C} \rightarrow 0$ is achieved by a soft suspension spring $c_S \rightarrow 0$ or a large chassis mass $M \rightarrow \infty$. However, a large chassis mass is not economical and the suspension stiffness is limited by the loading conditions. Hence, weighting the chassis accelerations only does not lead to a specific result for the system parameter here.

Practical results can be achieved if the attenuation of the chassis acceleration $\ddot{z}_C$ and the chassis displacement z_C are evaluated simultaneously. To do so, appropriate weighting factors must be chosen. In the equation of motion for the chassis mass (7.16), the terms $M \ddot{z}_C$ and $c_S z_C$ are added. Hence, $\alpha = M$ and $\beta = c_S$ or

$$
\alpha = 1 \quad \text{and} \quad \beta = \frac{c_S}{M} = \omega_{0C}^2
$$

(7.70)

will provide weighting factors that are automatically adjusted to the relevant system parameter. Now, Equation (7.68) reads as

$$\epsilon_C^2 = z_{0C}^2 \left[\frac{\omega_{0C}^3}{4\zeta_C} + \left(\omega_{0C}^2 \right)^2 \left(\zeta_C + \frac{1}{4\zeta_C} \right) \frac{1}{\omega_{0C}} \right] = z_{0C}^2 \omega_{0C}^2 \left[\frac{\omega_{0C}}{2\zeta_C} + \zeta_C\,\omega_{0C} \right] \quad (7.71)$$

Again, good ride comfort will be achieved by $\omega_{0C} \to 0$. For finite natural frequencies, Equation (7.71) becomes a minimum if the viscous damping rate ζ_C satisfies

$$\frac{d\,\epsilon_C^2|_{z_C}}{d\,\zeta_C} = z_{0C}^2\,\omega_{0C}^2 \left[\frac{-\omega_{0C}}{2\zeta_C^2} + \omega_{0C} \right] = 0 \quad (7.72)$$

Hence, a viscous damping rate of

$$\zeta_C^{Comfort} = \frac{1}{2}\sqrt{2} \quad (7.73)$$

or a damping parameter of

$$d_S^{opt}\Big|_{Comfort}^{\zeta_C = \frac{1}{2}\sqrt{2}} = \sqrt{2\,c_S M} \quad (7.74)$$

will provide optimal comfort by minimizing the merit function in Equation (7.71).

7.4.4 Example

For the passenger car approximated by quarter car models with sprung and unsprung masses of $M = 350$ kg and $m = 50$ kg and spring rates for the suspension and the tire of $c_S = 20\,000$ N/m and $c_T = 220\,000$ N/m the damping parameter for optimal ride safety will now amount to

$$d_S^{opt}\Big|_{Safety}^{\zeta_W = \frac{1}{2}} = \sqrt{(20\,000\ \text{N/m} + 220\,000\ \text{N/m}) * 50\ \text{kg}} = 3464\ \text{N/(m/s)} \quad (7.75)$$

and the one for optimal ride comfort to

$$d_S^{opt}\Big|_{Comfort}^{\zeta_C = \frac{1}{2}\sqrt{2}} = \sqrt{2 * 20\,000\ \text{N/m} * 350\ \text{kg}} = 3742\ \text{N/(m/s)} \quad (7.76)$$

As it is here, the values will not coincide in general. Hence, a vehicle suspension with a simple linear damper can either provide optimal ride comfort or optimal ride safety, but not both. Improvements may be achieved by nonlinear, dynamic, and active or semi-active force elements. But then, more complex vehicle models must be used.

7.5   Practical Aspects

7.5.1   General Remarks

The disturbance reaction problem in combination with two single-mass oscillators for the chassis and the wheel just provides a first approach to determine optimal values for the damping parameter. In practice, the performance of vehicles driving on real roads is of greater importance. The merit function defined in Equation (7.57) represents an objective criteria to judge the ride safety of a vehicle. In general, road irregularities are of a stochastic nature. According to Section 2.3.1, the standard deviation, which is the square root of the variance, characterizes the statistical properties of a stationary Gaussian random process. Then, optimal ride safety may be achieved by minimizing the variance of the dynamic wheel load,

$$\sigma_S^2 = \frac{1}{t_E - t_0} \int_{t_0}^{t_E} \left(\frac{F_T^D(t)}{F_T^S} \right)^2 dt \;\rightarrow\; \text{Min.} \tag{7.77}$$

where the time history of the dynamic wheel load $F_T^D = F_T^D(t)$ is normalized to the static wheel load F_T^S and evaluated in an appropriate time interval from $t = t_0$ to $t = t_E$.

The demand in Equation (7.64) that provides optimal ride comfort in combination with the weighting factors defined in Equation (7.70) will hold for the simple chassis model only. In practice, the focus in evaluating the ride comfort is usually placed on the chassis accelerations at the seat mountings. During the optimization process, at least the suspension travel have to be monitored too, because it is limited by the suspension design. As a consequence, optimal driving comfort may be achieved now by minimizing an appropriate combination of the acceleration variance and the suspension travel,

$$\sigma_C^2 = \frac{1}{t_E - t_0} \int_{t_0}^{t_E} \left[\left(\frac{\ddot{z}(t)}{g} \right)^2 + \left(\frac{s(t)}{s_R} \right)^2 \right] dt \;\rightarrow\; \text{Min.} \tag{7.78}$$

where instead of using weighting factors, the acceleration $a = a(t)$ is normalized to the constant of gravity g and the suspension travel $s = s(t)$ to a reference value s_R.

For linear systems, the covariance analysis can be used to determine the corresponding variances. As shown in [36], an algebraic solution for a single wheel mass oscillator is possible again. It turns out, that a viscous damping ratio of $\zeta_W = 0.5$ minimizes the wheel load variance and hence will provide optimal ride safety in the case of stochastic excitation. This value exactly coincides with the value found by solving the corresponding disturbance reaction problem.

7.5.2   Quarter Car Model on Rough Road

In standard vehicle suspension systems, springs and dampers are mounted between the wheel and the chassis. As a consequence, body and wheel/axle motions will be affected simultaneously. That is why a simple quarter car model will be used later instead of the separate mass models for the chassis and the wheel. The equations of motion for the quarter car model, shown in Figure 7.5, are given by

$$\begin{aligned} M\,\ddot{z}_C &= F_S + F_D - M\,g \\ m\,\ddot{z}_W &= F_T - F_S - F_D - m\,g \end{aligned} \tag{7.79}$$

where the displacements of the chassis z_C and the wheel z_W are measured from the equilibrium position. The terms Mg and mg represent the weights of the chassis and the wheel. Assuming linear characteristics, the suspension forces and the wheel load are provided by

$$F_S = F_S^0 + c_S\,(z_W - z_C), \quad F_D = d_S\,(\dot{z}_W - \dot{z}_C), \quad F_T = F_T^0 + c_T\,(z_R - z_W) \tag{7.80}$$

where z_W describes the irregularities of the road, the constants c_S, d_S characterize the stiffness and damping properties of the suspension, and c_T denotes the tire stiffness. The preloads of the suspension spring and the tire are simply given by

$$F_S^0 = M g, \quad F_T^0 = (M + m)\,g \tag{7.81}$$

and the demand $F_T \geq 0$ will take wheel liftoff into account.

The MATLAB-Script app_07_qcm_simple provided in Listing 7.1 performs simulations of a simple quarter car model driving at constant velocity on a rough road with different values of the suspension damping.

Listing 7.1

Script app_07_qcm_simple.m: evaluating ride comfort and safety on random road

```
1 p.grav=9.81;              % constant of gravity in m/s^2
2 p.M=350;     p.m=50;      % chassis & wheel mass (quarter car) in kg
3 p.cS=20000; p.cT=220000;  % suspension & tire stiffness in N/m
4 p.v=100/3.6;              % vehicle velocity in km/h->m/s
5 p.FS0=p.M*p.grav;         % spring preload in N
6 p.FT0=(p.m+p.M)*p.grav;   % tire preload in N
7
8 % calculate amplitudes and random phases for pseudo random road
9 Phi0=10e-6; w=2; n=1000; Omin=2*pi/200; Omax=2*pi/0.2;
10 dOm = (Omax-Omin)/(n-1); p.Om  = Omin:dOm:Omax;
11 Om0 = 1; Phi = Phi0.*(p.Om./Om0).^(-w);  p.Amp = sqrt(2*Phi*dOm);
12 rng('default'); p.Psi = 2*pi*rand(size(p.Om));
13
14 dSvar=1000:200:5000;  % define range of suspension damping in N/(m/s)
15 sR = 0.05;            % define reference value for susp. travel in m
16
17 % simulation time and initial values adjusted to road height
18 tE=10;   zr0=sum(p.Amp.*sin(p.Psi)); x0 = [ zr0; zr0; 0; 0 ];
19
20 % calculate merit functions for different damping values
21 eps_saf=zeros(size(dSvar));   eps_com=zeros(size(dSvar));
22 for k=1:length(dSvar)
23
24     p.dS=dSvar(k);  disp(['simulation for dS=',num2str(p.dS)])
25     [tout,xout] = ode45(@(t,x) fun_07_qcm_simple(t,x,p),[0,tE],x0);
26
27 % normalized suspension travel
28     st = ( xout(:,1)-xout(:,2) ) / sR ;
29
30 % get normalized acceleration and normalized dynamic wheel load
31     zcdd = zeros(size(tout));  FDn = zeros(size(tout));
32     for i=1:length(tout)
33        [xdot,FT] = fun_07_qcm_simple(tout(i),xout(i,:).',p);
34        zcdd(i) = xdot(3)/p.grav; FDn(i)= FT/p.FT0-1;
35     end
36
37 % merit functions (variation of dyn. wheel load, accel. & susp travel)
```

```
38  eps_saf(k) = var(FDn);  eps_com(k) = var(zcdd) + var(st);
39
40 end
41
42 % plot criteria versus damping values
43 plot(dSvar,eps_saf,'r'), hold on, grid on
44 plot(dSvar,eps_com,'--k'), legend('safety','comfort')
```

As in Section 2.3.4, the random road, defined by its power spectral density, is approximated by a series of sine functions. In line 14, the seed for the random number generator is explicitly set to the MATLAB default. This will force the MATLAB-Function rand to produce exactly the same series of random numbers not only when a new MATLAB session is started, but also each time when running the script app_07_qcm_simple. The data correspond with the single-mass models for the wheel and the chassis. According to Equation (7.78), the ride comfort is judged by the sum of the effective values of the normalized chassis acceleration and the suspension travel. The effective value of the dynamic wheel load normalized to the static load evaluates the ride safety. The MATLAB-Function var computes the variances. The state equation of the simple quarter car model as well as the computation of pseudo-random road irregularities are provided in the function fun_07_qcm_simple, which is given in Listing 7.2.

Listing 7.2

Function fun_07_qcm_simple: simple quarter car model on random road

```
1 function [xp,FT] = fun_07_qcm_simple(t,x,p)
2
3 % local state variables and pseudo random road
4 zc=x(1); zw=x(2); zcd=x(3); zwd=x(4);
5 s=p.v*t; zr=sum(p.Amp.*sin(p.Om*s+p.Psi));
6
7 FS = p.FS0 + p.cS*(zw-zc);      % linear spring force
8 FD = p.dS*(zwd-zcd);            % linear damper force
9 FT = max([0, p.FT0+p.cT*(zr-zw)]); % wheel load (including lift off)
10
11 zcdd = (     FS + FD - p.M*p.grav ) / p.M ;  % chassis acceleration
12 zwdd = (FT - FS - FD - p.m*p.grav ) / p.m ;  % wheel acceleration
13
14 xp = [ zcd; zwd; zcdd; zwdd ]; % state derivatives
15
16 end
```

A simple inspection of the resulting plot in Figure 7.9 provides the damping parameter d_S^S = 2600 Ns/m and d_S^C = 3000 Ns/m, which correspond to the minimum of the variances $\sigma_S^2(d_S)$ and $\sigma_C^2(d_S)$ and thus will provide optimal ride safety or optimal comfort respectively. Again, either optimal safety or optimal comfort can be achieved here. Compared to the results of the single-mass models computed in Equations (7.75) and (7.76), the damping parameter that grants optimal safety is 25%, and the one that provides optimal comfort is 20% less than the corresponding values.

Optimization is always a delicate task. The results strongly depend on the merit function and on the complexity of the vehicle model. As shown in [61], the engine suspension for instance will have a significant influence on the chassis accelerations. Results that are in good conformity to measurements demand complex vehicle models, which include friction effects and nonlinearities. In addition, matching objective

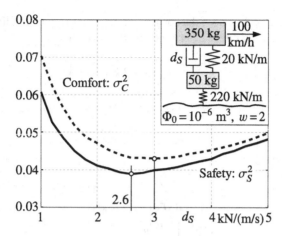

FIGURE 7.9
Variances to judge ride comfort and safety for different damping values.

comfort criteria with the human sense of comfort is still a problem. The corresponding DIN/ISO directive [19] tries to evaluate the human exposure to whole-body vibration via frequency-dependent weighting functions. However, the main focus is placed here on physical fatigue and health hazards.

7.6 Nonlinear Suspension Forces

7.6.1 Progressive Spring

In order to reduce the spring rate and to avoid too large spring deflections when loaded, nonlinear spring characteristics are usually used, Figure 7.10. Adding soft bump stops, the overall spring force in the compression mode $u \geq 0$ can be modeled by the nonlinear function

$$F_S = F_S^0 + c_0 u \left(1 + k \left(\frac{u}{\Delta u}\right)^2\right) \tag{7.82}$$

where F_S^0 denotes the spring preload, c_0 describes the spring rate at $u = 0$, and $k \geq 0$ characterizes the intensity of the nonlinearity. This approach may also be used to design air springs which, as shown in Section 6.1.2, provide a progressive nonlinear force characteristic that can be tuned by varying the air spring volume and its effective cross-section area. A simple linear characteristic generates at $u = \Delta u$ the value $F_S^{lin}(\Delta u) = F_S^0 + c_S \Delta u$. To achieve the same value with the nonlinear spring,

$$F_S^0 + c_0 \Delta u (1 + k) = F_S^0 + c_S \Delta u \quad \text{or} \quad c_0 (1 + k) = c_S \tag{7.83}$$

must hold, where c_S describes the spring rate of the corresponding linear character-istics. The local spring rate is determined by the derivative

$$\frac{dF_S}{du} = c_0 \left(1 + 3k \left(\frac{u}{\Delta u}\right)^2\right) \tag{7.84}$$

Then, the spring rate for the loaded car at $u = \Delta u$ is given by

$$c_L = c_0 \left(1 + 3k\right) \tag{7.85}$$

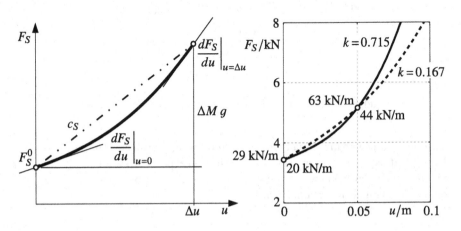

FIGURE 7.10
Principle and realizations of nonlinear spring characteristics.

The intensity of the nonlinearity k can be fixed, for instance, by choosing an appropriate spring rate for the unloaded vehicle. With $c_0 = 20\,000$ N/m $= 20$ kN/m the spring rates on the front and rear axle will here be the same for the unloaded vehicle. According to Equation (7.34), the stiffness value of a corresponding linear spring amounts to $c_S = 34\,300$ N/m $= 34.3$ kN/m. Then, Equation (7.83) delivers the intensity of the nonlinear spring by the value of

$$k = \frac{c_S}{c_0} - 1 = \frac{34.300}{20} - 1 = 0.715 \tag{7.86}$$

The solid line in the left plot of Figure 7.10 shows the resulting nonlinear spring characteristic that is characterized by the spring rates $c_0 = 20$ kN/m and $c_L = c_0 \left(1 + 3k\right) = 20 * (1 + 3 * 0.715) = 62.9$ kN/m for the unloaded and the loaded vehicle. Again, the natural frequencies

$$f_{0C}^0 = \frac{1}{2\pi} \sqrt{\frac{20\,kN/m}{350\,kg}} = 1.20\,\text{Hz}, \quad f_{0C}^L = \frac{1}{2\pi} \sqrt{\frac{62.9\,kN/m}{(350+175)\,kg}} = 1.74\,\text{Hz} \tag{7.87}$$

for the unloaded and the loaded vehicle will differ quite a lot.

The unloaded and the loaded vehicle have the same natural frequencies if

$$\frac{c_0}{M} = \frac{c_L}{M + \Delta M} \quad \text{or} \quad \frac{c_L}{c_0} = \frac{M + \Delta M}{M} \tag{7.88}$$

holds. Combining this relationship with Equation (7.85) yields

$$1 + 3\,k = \frac{M}{M + \Delta M} \quad \text{or} \quad k = \frac{1}{3}\left(\frac{M + \Delta M}{M} - 1\right) = \frac{1}{3}\frac{\Delta M}{M} \tag{7.89}$$

Hence, for the quarter car model with $M = 350\ kg$ and $\Delta M = 175\ kg$ the intensity of the nonlinear spring reduces to $k = 1/3 * 175/350 = 0.167$. This value, combined with the corresponding linear spring stiffness $c_S = 34.3$ kN/m, will produce the dotted line in Figure 7.10. The spring rates $c_0 = c_S/(1 + k) = 34.3$ N/m / $(1 + 0.1667) =$ 29.4 kN/m and $c_L = c_0(1 + 3k) = 29.400$ kN/m $* (1 + 3 * 0.1667) = 44.1$ kN/m, which apply for the unloaded and the loaded vehicle result from Equations (7.84) and (7.85). Now, the natural frequency for the unloaded $f_{0C}^0 = \sqrt{c_0/M} = 1.46\ Hz$ and the loaded vehicle $f_{0C}^0 = \sqrt{c_L/(M + \Delta M)} = 1.46\ Hz$ are indeed the same.

7.6.2 Nonlinear Spring and Nonlinear Damper

The equations of motion for a simple quarter car model are provided in Equation (7.79). The spring travel is defined by

$$u = z_W - z_C \tag{7.90}$$

where z_W and z_C describe the vertical displacements of the wheel and the chassis mass measured from the equilibrium position, Figure 7.11. Similar to Equation (7.82),

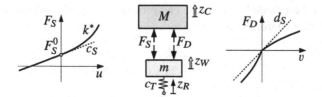

FIGURE 7.11
Quarter car model with nonlinear spring and damper elements.

the spring force is modeled by

$$F_S = F_S^0 + \begin{cases} c_S\,u & u < 0 \\ c_S\,u\left(1 + k^*\,u^2\right) & u \geq 0 \end{cases} \tag{7.91}$$

where c_S defines the stiffness at $u = 0$ and $F_S^0 = M\,g$ denotes the preload. The parameter $k^* \geq 0$ characterizes the nonlinearity in the compression cycle, $u > 0$, where $k^* = 0$ simply results in a linear spring with stiffness c_S. The nonlinear spring design in Section 7.6.1 provides at the rear axle the data

$$c_S = 29\,400 \text{ N/m} \quad \text{and} \quad k^* = \frac{k}{(\Delta u)^2} = \frac{0.1667}{(0.05 \text{ m})^2} = 66.7 \text{ m}^{-2} \qquad (7.92)$$

The tire stiffness is given further on by $c_T = 220\,000$ N/m and the chassis and wheel mass are determined by $M = 350$ kg and $m = 50$ kg. The function in Listing 7.2 computes linear suspension forces. Replacing the line

```
FS = p.FS0 + p.cS*(zw-zc);  % linear spring force
```

by the code lines

```
u = zw - zc;  % spring displacement (u>0: compression)
if u<0
  FS = p.FS0 + p.cS*u;  % rebound (linear)
else
  FS = p.FS0 + p.cS*u*(1+p.kS*u^2);  % compression (progressive)
end
```

will provide a progressive spring characteristic according to Equation (7.91). In addition, the element p.kS must be added to the parameter structure and assigned to the value of 66.7 in Listing 7.1. In addition, the spring constant must be changed to the value $c_S = 29\,400$ N/m, which now holds for the rear axle.

The damper element is arranged parallel to the spring, which simply acts between the chassis and the wheel mass. Then, the time derivative of the suspension travel

$$\dot{u} = v = \dot{z}_W - \dot{z}_C \qquad (7.93)$$

defines the damper velocity, where the sign convention is consistent with the spring travel. As a consequence, $u < 0$, $v < 0$ characterize tension or rebound and $u \geq 0$, $v \geq 0$ compression. As shown in Section 6.1.4, nonlinear damper characteristics are usually modeled via lookup tables. Typically, suspension dampers will have more resistance in the rebound than in the compression cycle. The piecewise linear but overall nonlinear characteristic

$$F_D(v) = \begin{cases} d_{Rb}\, v & v < 0 \\ d_{Cp}\, v & v \geq 0 \end{cases} \qquad (7.94)$$

defined by the constants d_{Rb} and d_{Cp} takes this effect into account. According to Equation (7.93), $v < 0$ characterizes rebound and $v \geq 0$ the compression cycle. As a consequence, d_{Rb} describes the damping properties during rebound and d_{Cp} in the compression mode. A slight modification in Listing 7.2 will make the nonlinear damper available in the simple quarter car model. At first, the line

```
FD = p.dS*(zwd-zcd);  % linear damper force
```

must be replaced with the code lines

```
v = zwd-zcd;  % velocity (v>0: compression)
if v<0
  FD = p.dRb*v;  % rebound
else
  FD = p.dCp*v;  % compression
end
```

Then, the element p.dS must be replaced or simply supplemented by the elements p.dRb and p.dCp which have to be assigned with appropriate values in Listing 7.1. By setting $d_{Rb} = d_{Cp} = d_S$, the linear damper characteristic is still available.

As done in Equation (7.80), the tire is modeled by a linear spring characterized by the constant c_T and the demand $F_T \geq 0$ takes wheel liftoff into account.

7.6.3   Some Results

At first, the influence of the nonlinear spring on ride comfort and safety is studied. As done in Section 7.5.2, the vehicle is driven with constant velocity on a rough road. The corresponding simulation results, computed for different values of a linear damper ($d_{Rb} = d_{Cp} = d_S$) are plotted in Figure 7.12. A comparison with the results

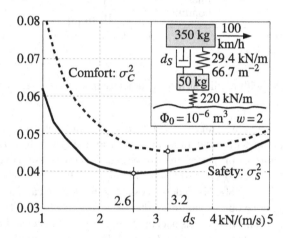

FIGURE 7.12
Comfort and safety for a progressive nonlinear spring and a linear damper.

in Figure 7.9 shows that the damping value $d_S^S = 2600$ Ns/m, which provides optimal ride safety, is still valid. However, the damping value that grants optimal ride comfort has slightly increased to $d_S^S = 3200$ Ns/m. The single-mass approach ends up in the algebraic equations (7.63) and (7.74), which factor in the mass and stiffness properties of the quarter car model. Hence, increasing the suspension stiffness from $c_S = 20$ kN/m in Figure 7.9 to $c_S = 29.4$ kN/m in Figure 7.12 affects ride comfort and ride safety. However, the influence of the increased suspension stiffness on ride safety will hardly be noticeable because the suspension stiffness is much smaller than the tire stiffness, which was not changed here.

In general, a nonlinear damper layout with $d_{Cp} < d_{Rb}$ will generate less impact on the chassis when driving over single bumps because in the first part of the compression cycle, spring and damper force have the same sign and induce large chassis accelerations. Progressive spring characteristics, which are employed at the rear suspension in particular, will intensify this effect. Corresponding simulation results are plotted in Figure 7.13. The quarter car model, equipped with a linear and a nonlinear

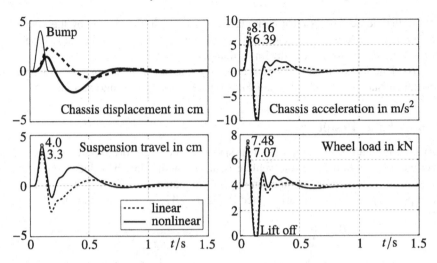

FIGURE 7.13
Quarter car model with nonlinear force characteristics driving over a bump.

damper, is driven hereby with the velocity v = 80 km/h over a single obstacle. A cosine-shaped bump with a height of H = 0.04 m and a length of L = 3.0 m was used here. The damping properties in the rebound and compression cycle, are characterized by the constants d_{Rb} = 4200 Ns/m and d_{Cp} = 1400 Ns/m. The average value $d_S = (d_{Rb} + d_{Cp})/2$ = 2800 Ns/m describes the equivalent linear damper. Compared to the linear damper layout, the nonlinear damper characteristic results in significantly reduced peak values for the chassis acceleration (6.39 m/s² instead of 8.16 m/s²) and for the wheel load (7.07 kN instead of 7.48 kN). However, the tire liftoff at $t \approx 0.16\,s$ could not be avoided here. As a quid pro quo, the nonlinear damper requires a larger suspension travel (4.0 cm instead of 3.3 cm). Both layouts provide similar overall damping.

The nonlinear spring force is described by Equation (7.91) and according to Equation (7.92), defined by the stiffness c_S = 29 400 N/m that holds in the design position and the parameter k^* = 66.7 m^{-2} that characterizes the progressive nonlinearity. The linear spring characteristics are simple realized by k^* = 0.

While crossing a bump, large damper velocities occur in general. A low but constant damping rate in the compression mode will generate large damping forces that induce unnecessary chassis accelerations. Damper layouts that generate a degressive characteristic may reduce this impact on the chassis when crossing a bump but they in turn will increase the suspension travel.

7.7 Sky Hook Damper

7.7.1 Modeling Aspects

In a standard layout of a wheel/axle suspension system, the damper acts between the wheel or axle and the chassis, Figure 7.14a. For a linear characteristic, the damper force is given by

$$F_D = d_S (\dot{z}_W - \dot{z}_C) \qquad (7.95)$$

where d_S denotes the damping constant and $\dot{z}_C$, $\dot{z}_W$ are the time derivatives of the absolute vertical body and wheel displacements. A standard damper is mounted in between the wheel body and the chassis and it just generates a force that depends on the relative velocity $\dot{z}_W - \dot{z}_C$.

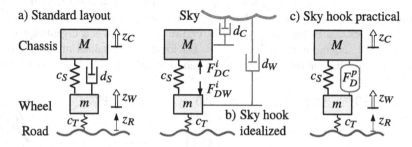

FIGURE 7.14
Quarter car model with a standard and a sky hook damper.

The sky hook damping concept starts with two independent dampers mounted between the chassis and the sky and the wheel or axle and the sky, Figure 7.14b. In this idealized configuration the forces $F_{DW}^i = d_W \dot{z}_W$ and $F_{DC}^i = -d_C \dot{z}_C$ are applied to the wheel and to the chassis. In this basic investigation, a linear dependency on the velocities was assumed for the sake of simplicity. However, a practical realization is only possible if the damper or a controlled, or even active, force element is mounted in between the wheel/axle and the chassis, Figure 7.14c. As a best possible solution, the damping force

$$F_D = d_W \dot{z}_W - d_C \dot{z}_C \qquad (7.96)$$

can then be applied to the chassis and its opposite to the wheel/axle. However, the single damping constant d_S, which according to Equation (7.95) characterizes the standard damper, is extended to two design parameters d_W and d_C now. The special case $d_C = d_W = d_S$ represents the standard layout with a linear damper.

7.7.2 Eigenfrequencies and Damping Ratios

The equations of motion for a simple quarter car model are provided in Equation (7.79). The linear suspension spring force and the vertical tire force are defined in

Equation (7.80). Using the force definitions in Equations (7.96) and (7.80), the equations of motion provided in Equation (7.79) can be transformed to the state equation

$$
\underbrace{\begin{bmatrix} \dot{z}_C \\ \dot{z}_W \\ \ddot{z}_C \\ \ddot{z}_W \end{bmatrix}}_{\dot{x}} = \underbrace{\begin{bmatrix} 0 & 0 & 1 & 0 \\ 0 & 0 & 0 & 1 \\ -\frac{c_S}{M} & \frac{c_S}{M} & -\frac{d_C}{M} & \frac{d_W}{M} \\ \frac{c_S}{m} & -\frac{c_S+c_T}{m} & \frac{d_C}{m} & -\frac{d_W}{m} \end{bmatrix}}_{A} \underbrace{\begin{bmatrix} z_C \\ z_W \\ \dot{z}_C \\ \dot{z}_W \end{bmatrix}}_{x} + \underbrace{\begin{bmatrix} 0 \\ 0 \\ 0 \\ \frac{c_T}{m} \end{bmatrix}}_{B} \begin{bmatrix} z_R \end{bmatrix} u
\tag{7.97}
$$

where the weight forces Mg and mg were compensated by the preloads F_S^0 and F_T^0. The term Bu describes the excitation, x denotes the state vector, and A is the state matrix. In this linear approach the tire liftoff is not taken into consideration.

The eigenvalues λ of the state matrix A characterize the eigendynamics[3] of the quarter car model. In case of complex eigenvalues, the damped eigenfrequencies are given by the imaginary parts, $\omega = Im(\lambda)$, and according to Equation (7.27), $\zeta = D_\lambda = -Re(\lambda)/|\lambda|$ evaluates the damping ratio. Figure 7.15 shows the eigenfrequencies $f = \omega/(2\pi)$ and the damping ratios $\zeta = D_\lambda$ for different values of the damping parameter d_S. Optimal ride comfort with a damping ratio of $\zeta_C = \frac{1}{2}\sqrt{2} \approx 0.7$ for the

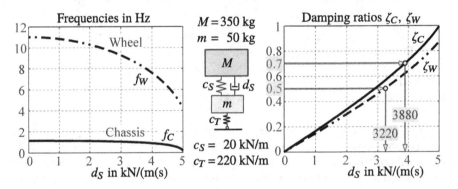

FIGURE 7.15
Eigenfrequencies and damping ratios of a quarter car model with standard damper.

chassis motion can be achieved with the damping parameter $d_S = 3.88$ kN/(m/s) = 3880 N/(m/s), and the damping parameter $d_S = 3.22$ kN/(m/s) = 3220 N/(m/s) would provide for the wheel motion a damping ratio of $\zeta_W = 0.5$, which corresponds to minimal wheel load variations. These damping parameters are very close to the values 3742 N/(m/s) and 3464 N/(m/s) that were calculated in Equations (7.76) and (7.75) with the single mass-models. Hence, the very simple single-mass models can be used for a first damper layout. Usually, as it is here, optimal ride comfort and optimal ride safety cannot both be achieved with a standard linear damper.

---

[3]The MATLAB command [EV,EW]=eig(A) computes the eigenvectors (columns of EV) and eigenvalues (main diagonal elements of EW) of the matrix A.

The practical realization of a sky hook damper, modeled by Equation (7.96), provides with d_C and d_W two design parameters. Their influence on the eigenfrequencies f and the damping ratios ζ is shown in Figure 7.16. The sky hook

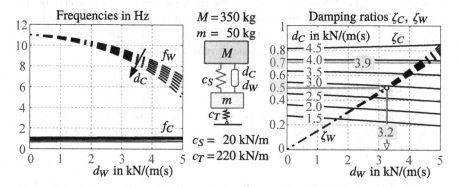

FIGURE 7.16
Quarter car model with sky hook damper.

damping parameters d_C and d_W have a nearly independent influence on the damping ratios. The chassis damping ratio ζ_C mainly depends on d_C, and the wheel damping ratio ζ_W mainly depends on d_W. Hence, the damping of the chassis and the wheel motion can be adjusted to nearly each design goal. Here, a sky hook damper with $d_C = 3.9$ kN/(m/s) = 3900 N/(m/s) and $d_W = 3.2$ kN/(m/s) = 3200 N/(m/s) would generate the damping ratios $\zeta_C = 0.7$ and $\zeta_W = 0.5$ and thus combine ride comfort and ride safety within one damper layout.

7.7.3 Technical Realization

By modifying the damper law in Equation (7.96) to

$$F_D = d_W \dot{z}_W - d_C \dot{z}_C = \underbrace{\frac{d_W \dot{z}_W - d_C \dot{z}_C}{\dot{z}_W - \dot{z}_C}}_{d_{sky}} (\dot{z}_W - \dot{z}_C) \qquad (7.98)$$

the practical approach to a sky hook damper can be realized by a standard damper in the form of Equation (7.95). The new damping parameter d_{sky} now nonlinearly depends on the absolute vertical velocities of the chassis and the wheel $d_{sky} = d_{sky}(\dot{z}_C, \dot{z}_W)$. As a standard damper operates in a dissipative mode only, the damping parameter will be restricted to positive values, $d_{sky} \geq 0$. Hence, the passive realization of a sky hook damper will only match with some properties of the ideal damper law in Equation (7.96). But, compared with the standard damper, it still can provide better ride comfort combined with sufficient ride safety.

Some continuous damping control (CDC) systems are based on the sky hook control strategy. Twin-tube dampers equipped with electromagnetically activated additional external valves make it possible to regulate the flow of the hydraulic fluid between the inner and outer damper tubes and thus vary the resistance of the damper. Vehicle sensors provide the accelerations of the chassis and the wheels, which are processed in a control unit to the desired damping parameter within milliseconds. Most commercial CDC-systems take the actual driving situation into account too. For example, a temporarily increased damping rate will reduce the pitch and roll reaction of the vehicle when cornering fast or braking hard. However, a lower damping parameter will grant a rather smooth ride on rough country roads.

7.7.4   Simulation Results

The function fun_07_qcm_skyhook given in Listing 7.3 provides the state equation for a simple quarter car model with a nonlinear suspension spring and a sky hook damper modeled dynamically.

Listing 7.3
Function fun_07_qcm_skyhook: dynamics of a quarter car model with a sky hook damper

```
1  function [xdot,out] = fun_07_qcm_skyhook(t,x,p)
2
3  % quarter car states & dynamic damping parameter
4  zc=x(1); zw=x(2);   zcd=x(3); zwd=x(4);   d=x(5);
5
6  % obstacle @ sx = o_x0, sy = o_y0
7  sx = p.v_vel*t; sy=0; % actual position of vehicle
8  out.zr = uty_obstacle(sx-p.ox0,sy-p.oy0,p.otp,p.ol,p.ow,p.oh);
9
10 % tire force
11 out.ft = max(0,p.FT0+p.cT*(out.zr-zw));
12
13 % nonlinear spring
14 u  = zw - zc; % displacement (u>0: compression)
15 if u<0
16   fs = p.FS0 + p.cS*u; % rebound
17 else
18   fs = p.FS0 + p.cS*u*(1+p.kS*u^2); % compression
19 end
20
21 % damper velocity (vD>0: compr.) and damping force
22 vD = zwd - zcd;   fd = d*vD;
23
24 % compute skyhook damping parameter and its time derivative
25 if abs(vD) > 0.001
26   d_sky = max( 0, (p.dW*zwd-p.dC*zcd)/vD );
27 else
28   d_sky = d; % keep old value
29 end
30 ddot = (d_sky - d) / p.TdS;   % first order dynamics
31
32 % chassis and wheel acceleration
33 zcdd = (            (fs+fd) ) / p.M   - p.grav ;
34 zwdd = ( out.ft - (fs+fd) ) / p.m   - p.grav ;
35
36 % state derivatives
37 xdot = [ zcd; zwd; zcdd; zwdd;   ddot ];
38
39 end
```

The function `uty_obstacle` defined in Listing 2.1 provides the actual road height. Again, the tire is modeled by a linear spring characterized by the constant `p.cT` and the code line 11 takes the wheel liftoff into account too. As done in Section 7.6.2, the parameter `p.cS` and `p.kS` model a nonlinear spring force.

In reality it will be impossible to change the resistance of a damper in an instant. That is why, the controlled damper is characterized by a dynamic damping parameter d, which is simply defined here by a first-order differential equation

$$T_{dS}\,\dot{d} = d_{sky} - d \tag{7.99}$$

The time constant T_{dS} characterizes the damper dynamics and Equation (7.98) provides the sky hook damping parameter d_{sky}, which is restricted to positive values. As the computation of the sky hook damping parameter d_{sky}, will become critical at small damper velocities, a new sky hook damping parameter is calculated only if $|v_D| > 0.001$ m/s holds.

The MATLAB-Script `app_07_qcm_skyhook` provided in Listing 7.4 sets the data for a simple quarter car model equipped with a nonlinear suspension spring and a sky hook damper, performs simulations where the vehicle drives with constant speed over an obstacle, does some post-processing, and finally plots the results.

Listing 7.4
Script `app_07_qcm_skyhook_main.m`: simulation of a quarter car model with nonlinear spring and sky hook damper crossing an obstacle

```
1  p.grav=9.81;                  % constant of gravity in m/s^2
2  p.M=350; p.m=50;              % chassis and wheel mass in kg
3  p.cS=29400; p.kS=66.7;        % spring rate in N/m and nonlinearity in 1/m^2
4  p.dC=4850; p.dW=3200;         % sky hook damping parameter in N/(m(s)
5  p.TdS=0.005;                  % time constant in s
6  p.cT=220000;                  % tire stiffness in N/m
7
8  p.v_vel=80/3.6;               % vehicle velocity, km/h --> m/s
9  p.ox0=2.0; p.oy0=0; p.otp=2;  % obstacle pos in m and type
10 p.ow=10; p.oh=0.04; p.ol=3;   % obstacle width, height and length in m
11
12 % preloads
13 p.FS0 = p.M*p.grav;  p.FT0 = p.FS0 + p.m*p.grav;
14
15 % time interval & initial states (steady state & average damping)
16 t0=0; te=1;  x0 =[ 0; 0; 0; 0; (p.dC+p.dW)/2];
17
18 % perform simulation
19 [t,x] = ode23(@(t,x) fun_07_qcm_skyhook(t,x,p),[t0,te],x0);
20
21 % chassis acceleration, wheel load, road height, and suspension travel
22 ac=zeros(size(t)); FT=zeros(size(t)); zr=zeros(size(t));
23 for i=1:length(t)
24    [xdot,out]=fun_07_qcm_skyhook(t(i),x(i,:).',p);
25    ac(i)=xdot(3); FT(i)=out.ft; zr(i)=out.zr;
26 end
27 st = x(:,2) - x(:,1);
28
29 % plot results
30 subplot(2,2,1), plot(t,x(:,5),'k','Linewidth',1),
31    title('damping constant'), grid on
32 subplot(2,2,2), plot(t,ac,'k','Linewidth',1)
33    title('chassis acceleration'), grid on
```

```
34 subplot(2,2,3), plot(t,st,'k','Linewidth',1)
35    hold on, plot(t,zr,'--b','Linewidth',1)
36    title('suspension travel & bump'), grid on
37 subplot(2,2,4), plot(t,FT,'k','Linewidth',1)
38    title('wheel load'), grid on
```

As done in Section 7.7.2, the eigendynamics of the system were analyzed first. The stiffer spring (c_S = 29.4 kN/m instead of c_S = 20 kN/m) results in the sky hook damping parameter d_C = 4850 N/(m/s) and d_W = 3200 N/(m/s), which correspond with damping ratios of ζ_C = 0.7 and ζ_W = 0.5 and thus will provide optimal ride comfort and optimal ride safety.

The solid lines plotted in Figure 7.17 represent the simulation results. For compar-

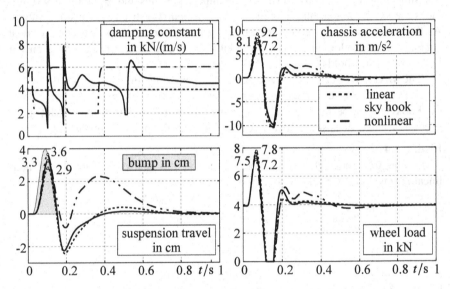

FIGURE 7.17
Quarter car model with different damper types crossing a bump.

ison, the corresponding results of simulations with a linear and a nonlinear damper are shown also. The linear damper is simply realized by setting both sky hook parameters to the average damping parameter, $d_C = d_W = (4850 + 3200)/2 = 4025$ N/(m/s). As done in Section 7.6.2, the nonlinear damper was modeled by piecewise linear functions characterized by the parameters d_{Cp} = 2000 N/(m/s) and d_{Rb} = 6000 N/(m/s), which describe the damper resistance in the compression and rebound mode. In both cases, the dynamic approach was applied. The small time constant of $TdS = 0.005$ s makes rather quick changes of the damping constant possible. Whereas the damping parameter d simply jumps from $d = d_{Cp}$ to $d = d_{Rb}$ when the nonlinear damper is employed, it varies over a wide range when the sky hook damper is used instead. Compared to the linear damper, the sky hook damper generates significantly smaller peak values in the chassis acceleration ($9.2 \rightarrow 8.1$ m/s²) and the wheel load ($7.8 \rightarrow 7.5$ kN) but it will require an increased suspension travel ($2.9 \rightarrow 3.3$ kN) instead. The nonlinear damper reduces the peak values in the chassis acceleration and the wheel load further on ($8.1 \rightarrow 7.2$ m/s^2 and $7.5 \rightarrow 7.2$ kN) but needs still more suspension travel ($3.3 \rightarrow 3.6$ kN) and shows a significant second overshoot at $t \approx 3.7$ s in the time history of the suspension travel. So, the performance of the sky hook damper regarding comfort and safety may be judged clearly superior to the linear and slightly better than the nonlinear damper layout. By applying a more sophisticated control strategy, further enhancements will be possible.

Exercises

7.1 Write a MATLAB function that provides the equations of motion for the simple planar vehicle model discussed in Section 7.2 in the form of $\dot{x} = f(t, x)$, where $x = [\, z_C;\ \theta_C;\ z_{A1};\ z_{A2};\ \dot{z}_C;\ \dot{\theta}_C;\ \dot{z}_{A1};\ \dot{z}_{A2}\,]$ defines the state vector and the data are supplied by global variables. The generalized coordinates z_C, θ_C z_{A1}, and z_{A2} are measured from the steady-state position.

Then, assuming linear characteristics, the suspension forces are defined by

$$
F_1 = \underbrace{Mg \frac{a_2}{a_1+a_2}}_{F_1^0} + c_{S1} \underbrace{\left[z_{A1} - (z_C - a_1\theta_C) \right]}_{s_1} + d_{S1} \underbrace{\left[\dot{z}_{A1} - (\dot{z}_C - a_1\dot{\theta}_C) \right]}_{\dot{s}_1},
$$

$$
F_2 = \underbrace{Mg \frac{a_1}{a_1+a_2}}_{F_2^0} + c_{S2} \underbrace{\left[z_{A2} - (z_C + a_2\theta_C) \right]}_{s_2} + d_{S2} \underbrace{\left[\dot{z}_{A2} - (\dot{z}_C + a_2\dot{\theta}_C) \right]}_{\dot{s}_2}
$$

where F_1^0, F_2^0 name the preloads, c_{S1}, c_{S2}, d_{S1}, d_{S2} characterize the stiffness and damping properties of the suspension, and the displacements s_1, s_2 and their time derivatives $\dot{s}_1$, $\dot{s}_2$ are defined such that positive values represent compression.

Assuming that the tires at each axle can be modeled by a linear spring with the constant c_T, the tire forces are defined by

$$
F_{T1} = F_1^0 + m_1g + c_{T1}(z_{R1} - z_{A1}),
$$

$$
F_{T2} = F_2^0 + m_2g + c_{T2}(z_{R2} - z_{A2}),
$$

where the road irregularities at the front and rear axle are described by $z_{R1} = z(s + a_1)$ and $z_{R2} = z(s - a_2)$ finally. Note, that the tire liftoff is not taken into account in this linear approach.

Derive the linear state equation in the form of $\dot{x} = A x + B u$, where the vector $u = [\, z_{R1};\ z_{R2}\,]$ contains the road irregularities. Combine the two quarter car models defined in Figure 7.9 and Figure 7.12 by neglecting the spring nonlinearity and by setting $a_1 = a_2 = 1.3$ m. Then, calculate the eigenvalues of the state matrix A and derive the eigenfrequencies and damping ratios.

7.2 Estimate appropriate values for the suspension stiffness for the vehicles characterized by the data in Table 7.1 by assuming realistic loading scenarios. Compute suitable values for the suspension damping that will provide optimal comfort and safety. Check the results by analyzing the performance of corresponding simple quarter car and planar models on rough road and when crossing different obstacles.

7.3 The chassis mass of a four-wheeler or quad is given with $m_C = 180$ kg, the driver has a mass of $m_D = 80$ kg and the masses of the front and rear axles amount to $m_1 = 30$ kg and $m_2 = 35$ kg. The horizontal position of the center of gravity for the chassis and the driver is defined by $a_1 = 0.72$ m and $a_2 = 0.68$ mm. The four-wheeler may be loaded by a mass not more than $m_L = 70$ kg.

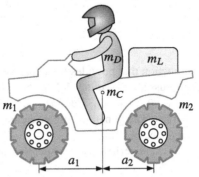

Model the chassis and driver by two point masses M_1 and M_2 located directly above the front and rear axle. Determine the coupling mass M^* and, if present, neglect its influence by adding appropriate parts of it to M_1 and M_2.

Compute the steady-state axle loads for the unloaded four-wheeler ($m_L = 0$).

Estimate the minimal possible overall spring rate for the rear axle suspension, by assuming that a potential load will be placed directly above the rear axle.

Determine a reasonable value for the overall spring rate at the front axle suspension, by taking into account that the rebound reaction at the front axle should be limited to an acceptable amount when the driver leaves the four-wheeler.

Compute an appropriate overall damping parameter for the front and rear axle suspensions, when $c_{TF} = 70$ kN/m and $c_{TR} = 80$ kN/m determine the vertical stiffness of one single front or rear tire respectively.

8

Longitudinal Dynamics

CONTENTS

8.1 Dynamic Wheel Loads

8.1.1 Simple Vehicle Model

The vehicle is considered as one rigid body that moves along an ideally even and horizontal road, Figure 8.1. At each axle, the forces in the wheel contact points are combined into one normal and one longitudinal force. If aerodynamic forces (drag,

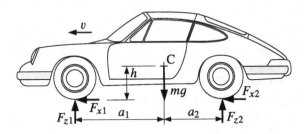

FIGURE 8.1
Simple vehicle model.

positive and negative lift) are neglected at first, the equations of motions in the x-, z-plane will read as

$$m\,\dot{v} = F_{x1} + F_{x2} \tag{8.1}$$

$$0 = F_{z1} + F_{z2} - m\,g \tag{8.2}$$

$$0 = -F_{z1}\,a_1 + F_{z2}\,a_2 - (F_{x1} + F_{x2})\,h \tag{8.3}$$

where $\dot{v} = a_x$ indicates the vehicle's acceleration, m is the mass of the vehicle, $a_1 + a_2$ is the wheel base, and h is the height of the center of gravity.

These are only three equations for the four unknown forces F_{x1}, F_{x2}, F_{z1}, F_{z2}. But, if we insert Equation (8.1) in Equation (8.3), we can eliminate two unknowns at a stroke

$$0 = -F_{z1}\,a_1 + F_{z2}\,a_2 - m\,\dot{v}\,h \tag{8.4}$$

Equations (8.2) and (8.4) can be resolved for the axle loads now

$$F_{z1} = m\,g\,\frac{a_2}{a_1 + a_2} - \frac{h}{a_1 + a_2}\,m\,\dot{v} \tag{8.5}$$

$$F_{z2} = m\,g\,\frac{a_1}{a_1 + a_2} + \frac{h}{a_1 + a_2}\,m\,\dot{v} \tag{8.6}$$

The static parts

$$F_{z1}^{st} = m\,g\,\frac{a_2}{a_1 + a_2} \quad \text{and} \quad F_{z2}^{st} = m\,g\,\frac{a_1}{a_1 + a_2} \tag{8.7}$$

describe the weight distribution according to the horizontal position of the center of gravity. The height h of the center of gravity only influences the dynamic part of the axle loads,

$$F_{z1}^{dyn} = -m g \frac{h}{a_1 + a_2} \frac{\dot{v}}{g} \quad \text{and} \quad F_{z2}^{dyn} = +m g \frac{h}{a_1 + a_2} \frac{\dot{v}}{g} \tag{8.8}$$

When accelerating $\dot{v} > 0$, the front axle is relieved as the rear is when decelerating $\dot{v} < 0$.

8.1.2 Influence of Grade

For a vehicle on a grade, the equations of motion, defined by Equations (8.1) to (8.3), can easily be extended to

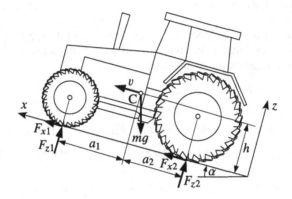

FIGURE 8.2
Vehicle on grade.

$$m \dot{v} = F_{x1} + F_{x2} - m g \sin \alpha$$
$$0 = F_{z1} + F_{z2} - m g \cos \alpha \tag{8.9}$$
$$0 = -F_{z1} a_1 + F_{z2} a_2 - (F_{x1} + F_{x2}) h$$

where α denotes the grade angle, Figure 8.2. Now, the axle loads are given by

$$F_{z1} = \overbrace{m g \cos \alpha \frac{a_2 - h \tan \alpha}{a_1 + a_2}}^{F_{z1}^{st}} \overbrace{- \frac{h}{a_1 + a_2} m \dot{v}}^{F_{z1}^{dyn}} \tag{8.10}$$

$$F_{z2} = \underbrace{m g \cos \alpha \frac{a_1 + h \tan \alpha}{a_1 + a_2}}_{F_{z2}^{st}} \underbrace{+ \frac{h}{a_1 + a_2} m \dot{v}}_{F_{z2}^{dyn}} \tag{8.11}$$

where the dynamic parts F_{z1}^{dyn}, F_{z2}^{dyn} remain unchanged and the static parts F_{z1}^{st}, F_{z2}^{st} also depend on the grade angle α and the height of the center of gravity h.

8.1.3   Aerodynamic Forces

The shape of most vehicles or specific wings mounted on the vehicle produce aerodynamic forces and torques. The effect of these aerodynamic forces and torques can

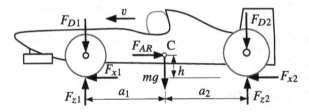

FIGURE 8.3
Vehicle with aerodynamic forces.

be represented by a resistant force applied at the center of gravity and "down forces" acting at the front and rear axle, Figure 8.3. If we assume a positive driving speed, $v > 0$, the equations of motion will read as

$$m\,\dot{v} = F_{x1} + F_{x2} - F_{AR}$$
$$0 = F_{z1} - F_{D1} + F_{z2} - F_{D2} - m\,g \qquad (8.12)$$
$$0 = -(F_{z1} - F_{D1})\,a_1 + (F_{z2} - F_{D2})\,a_2 - (F_{x1} + F_{x2})\,h$$

where F_{AR} and F_{D1}, F_{D2} describe the air resistance and the down forces. For the axle loads, we get

$$F_{z1} = \overbrace{F_{D1} + m\,g\,\frac{a_2}{a_1 + a_2}}^{F_{z1}^{st}} \overbrace{- \frac{h}{a_1 + a_2}\,(m\,\dot{v} + F_{AR})}^{F_{z1}^{dyn}} \qquad (8.13)$$

$$F_{z2} = \underbrace{F_{D2} + m\,g\,\frac{a_1}{a_1 + a_2}}_{F_{z2}^{st}} \underbrace{+ \frac{h}{a_1 + a_2}\,(m\,\dot{v} + F_{AR})}_{F_{z2}^{dyn}} \qquad (8.14)$$

The down forces F_{D1}, F_{D2} increase the static axle loads, and the air resistance F_{AR} generates an additional dynamic term.

8.2   Maximum Acceleration

8.2.1   Tilting Limits

Ordinary automotive vehicles can only apply pressure forces to the road. If we take the demands $F_{z1} \geq 0$ and $F_{z2} \geq 0$ into account, Equations (8.10) and (8.11) will result in

$$\frac{\dot{v}}{g} \le \frac{a_2}{h} \cos\alpha - \sin\alpha \quad \text{and} \quad \frac{\dot{v}}{g} \ge -\frac{a_1}{h} \cos\alpha - \sin\alpha \tag{8.15}$$

These two conditions can be combined into one

$$-\frac{a_1}{h} \cos\alpha \le \frac{\dot{v}}{g} + \sin\alpha \le \frac{a_2}{h} \cos\alpha \tag{8.16}$$

Here, aerodynamic forces are not taken into account. Then, the maximum achievable accelerations ($\dot{v} > 0$) and decelerations ($\dot{v} < 0$) are limited by the grade angle α and the position a_1, a_2, h of the center of gravity. For $\dot{v} \to 0$, the tilting condition Equation (8.16) results in

$$-\frac{a_1}{h} \le \tan\alpha \le \frac{a_2}{h} \tag{8.17}$$

which characterizes the climbing and downhill capacity of a vehicle.

The presence of aerodynamic forces complicates the tilting condition. However, aerodynamic forces become important only at high speeds. But then, the vehicle acceleration is normally limited by the engine power.

8.2.2   Friction Limits

The maximum acceleration is also restricted by the friction conditions

$$|F_{x1}| \le \mu F_{z1} \quad \text{and} \quad |F_{x2}| \le \mu F_{z2} \tag{8.18}$$

where the same friction coefficient μ has been assumed at the front and the rear axle. In the limit case

$$F_{x1} = \pm \mu F_{z1} \quad \text{and} \quad F_{x2} = \pm \mu F_{z2} \tag{8.19}$$

the linear momentum in Equation (8.9) can be written as

$$m\,\dot{v}_{max} = \pm\mu\,(F_{z1} + F_{z2}) - m\,g\,\sin\alpha \tag{8.20}$$

Using Equations (8.10) and (8.11), one obtains

$$\left(\frac{\dot{v}}{g}\right)_{max} = \pm\mu\cos\alpha - \sin\alpha \tag{8.21}$$

That means climbing ($\dot{v} > 0$, $\alpha > 0$) or downhill stopping ($\dot{v} < 0$, $\alpha < 0$) requires at least a friction coefficient $\mu \ge \tan|\alpha|$. On a horizontal road ($\alpha = 0$), the maximum longitudinal acceleration is simply determined by the coefficient of friction

$$\left(\frac{\dot{v}}{g}\right)_{max} = \pm\mu \quad \text{or} \quad a_x^{max} = \pm\mu\,g \tag{8.22}$$

According to the vehicle dimensions a_1, a_2, h and the magnitude of the friction coefficient μ, the maximal acceleration or deceleration is restricted either by Equation (8.16) or by Equation (8.21).

If we take aerodynamic forces into account, the maximum acceleration and deceleration on a horizontal road ($\alpha = 0$) will be limited by

$$-\mu \left(1 + \frac{F_{D1}}{mg} + \frac{F_{D2}}{mg}\right) - \frac{F_{AR}}{mg} \le \frac{\dot{v}}{g} \le \mu \left(1 + \frac{F_{D1}}{mg} + \frac{F_{D2}}{mg}\right) - \frac{F_{AR}}{mg} \tag{8.23}$$

In particular the aerodynamic forces enhance the braking performance of the vehicle.

8.3 Driving and Braking

8.3.1 Single Axle Drive

With the rear axle driven in limit situations, $F_{x1} = 0$ and $F_{x2} = \mu F_{z2}$ hold. Then, using Equation (8.6) the linear momentum Equation (8.1) results in

$$m\,\dot{v}_{RWD} = \mu m g \left[\frac{a_1}{a_1 + a_2} + \frac{h}{a_1 + a_2}\frac{\dot{v}_{RWD}}{g}\right] \tag{8.24}$$

where the subscript *RWD* indicates the rear-wheel drive. Hence, the maximum acceleration for a rear-wheel driven vehicle is given by

$$\frac{\dot{v}_{RWD}}{g} = \frac{\mu}{1 - \mu \dfrac{h}{a_1 + a_2}} \frac{a_1}{a_1 + a_2} \tag{8.25}$$

By setting $F_{x1} = \mu F_{z1}$ and $F_{x2} = 0$, the maximum acceleration for a front-wheel driven vehicle can be calculated in a similar way. One gets

$$\frac{\dot{v}_{FWD}}{g} = \frac{\mu}{1 + \mu \dfrac{h}{a_1 + a_2}} \frac{a_2}{a_1 + a_2} \tag{8.26}$$

where the subscript *FWD* denotes front-wheel drive. Depending on the parameter μ, a_1, a_2, and h, the accelerations may be limited by the tilting condition $\frac{\dot{v}}{g} \le \frac{a_2}{h}$. The maximum accelerations of a single-axle driven vehicle are plotted in Figure 8.4. Besides the friction limits, the hazard of tilting must be taken into account. On a horizontal road ($\alpha = 0$), Equation (8.16) yields

$$\frac{\dot{v}}{g} \le \frac{a_2}{h} \quad \text{or} \quad \frac{a_2}{a_1 + a_2} \ge \frac{h}{a_1 + a_2}\frac{\dot{v}}{g} \tag{8.27}$$

where the inequality was rearranged and divided by the wheel base $a_1 + a_2$. According to Equation (8.21), the maximum acceleration is limited to the friction coefficient on a horizontal road

$$\left(\frac{\dot{v}}{g}\right)_{max} = \mu \tag{8.28}$$

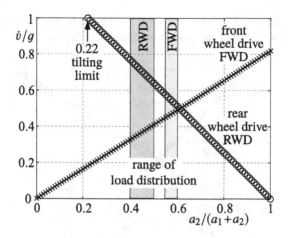

FIGURE 8.4
Single-axle driven passenger car: $\mu = 1$, $h = 0.55\,m$, $a_1 + a_2 = 2.5\,m$.

In this case, the tilting condition Equation (8.27) just delivers

$$\frac{a_2}{a_1 + a_2} \geq \frac{h}{a_1 + a_2}\mu = \frac{0.55}{2.5}1 = 0.22 \tag{8.29}$$

For rear-wheel driven passenger cars, the parameter $a_2/(a_1+a_2)$ describing the static axle load distribution is usually in the range of $0.4 \leq a_2/(a_1+a_2) \leq 0.5$. For $\mu = 1$ and $h = 0.55$, this results in maximum accelerations in the range of $0.77 \geq \dot{v}/g \geq 0.64$. Front-wheel driven passenger cars usually cover the range $0.55 \leq a_2/(a_1+a_2) \leq 0.60$, which produces accelerations in the range of $0.45 \leq \dot{v}/g \geq 0.49$. Hence, rear-wheel driven vehicles can accelerate much faster than front-wheel driven vehicles.

8.3.2 Braking at Single Axle

If only the front axle is braked, $F_{x1} = -\mu F_{z1}$ and $F_{x2} = 0$ will hold in the limit case. With Equation (8.5) one gets from Equation (8.1),

$$m\,\dot{v}_{FWB} = -\mu m g \left[\frac{a_2}{a_1 + a_2} - \frac{h}{a_1 + a_2}\frac{\dot{v}_{FWB}}{g} \right] \tag{8.30}$$

where the subscript FWB indicates front-wheel braking. Then, the maximum deceleration is given by

$$\frac{\dot{v}_{FWB}}{g} = -\frac{\mu}{1 - \mu \dfrac{h}{a_1 + a_2}}\frac{a_2}{a_1 + a_2} \tag{8.31}$$

If only the rear axle is braked ($F_{x1} = 0$, $F_{x2} = -\mu F_{z2}$), one will obtain the maximum deceleration

$$\frac{\dot{v}_{RWB}}{g} = -\frac{\mu}{1 + \mu \dfrac{h}{a_1 + a_2}}\frac{a_1}{a_1 + a_2} \tag{8.32}$$

where the subscript *RWB* denotes rear-wheel braking. The maximum decelerations of a single-axle braked vehicle are plotted in Figure 8.5. Depending on the parameter μ,

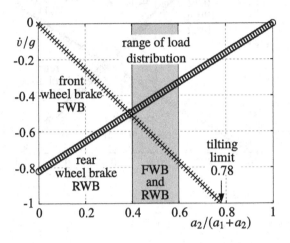

FIGURE 8.5
Single-axle braked passenger car: $\mu = 1$, $h = 0.55\ m$, $a_1 + a_2 = 2.5\ m$.

a_1, a_2, and h, the decelerations may be limited by the tilting condition $\dot{v}/g \geq -a_1/h$, which in the limit case $\dot{v}/g = -\mu$ yields

$$-\mu \geq -\frac{a_1}{h} \quad \text{or} \quad -a_1 \leq -\mu h \quad \text{or} \quad (a_1+a_2) - a_1 \leq (a_1+a_2) - \mu h \quad (8.33)$$

and finally results in

$$\frac{a_2}{a_1+a_2} \leq 1 - \mu\,\frac{h}{a_1+a_2} = 1 - 1\,\frac{0.55}{2.5} = 0.78 \quad (8.34)$$

For passenger cars, the load distribution parameter $a_2/(a_1+a_2)$ usually covers the range of 0.4 to 0.6. If only the front axle is braked, decelerations from $\dot{v}/g = -0.51$ to $\dot{v}/g = -0.77$ will be achieved. This is quite a large value compared to the deceleration range of a braked rear axle, which is in the range of $\dot{v}/g = -0.49$ to $\dot{v}/g = -0.33$. Therefore, the braking system at the front axle will have a redundant design in general.

8.3.3   Braking Stability

A small yaw disturbance of the vehicle, indicated by the side slip angle β, will cause slip angles at the wheels. Two extreme braking scenarios are shown in in Figures 8.6 and 8.7, where the profile pattern of the tire is fully visible at locked wheels and slurred to gray at rotating ones.

If the front wheels are locked, the tire friction forces F_1 and F_2 will point into the opposite direction of sliding velocity v, which just equals the driving velocity of the vehicle, Figure 8.6. The forces F_1 and F_2, which are approximately equal in

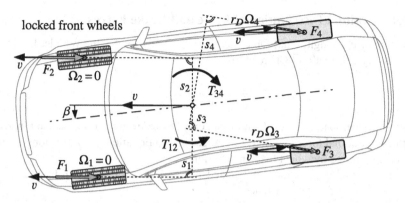

FIGURE 8.6
Braking with locked front wheels.

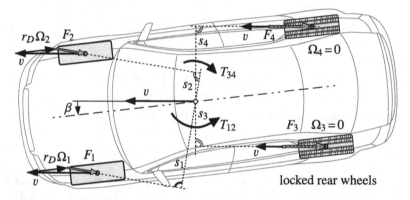

FIGURE 8.7
Braking with locked rear wheels.

magnitude, generate the torque T_{12} with respect to the vehicle center. As the lever arm s_1 of the force F_1 is slightly larger than the lever arm s_2 of the force F_2, the torque T_{12} will increase the side slip angle β and produce a destabilizing effect to the vehicle. The rear wheels are also braked but will still rotate. Here, each sliding velocity in the contact point is the result of the driving velocity v and the corresponding transport velocity $r_D\Omega_3$ or $r_D\Omega_4$, respectively. As a consequence, the sliding velocities and hence the corresponding tire forces F_3 and F_4 point in a direction that is inclined to the driving direction. Now, the lever arm s_3 of the force F_3 is significantly smaller than the lever arm s_4 of the force F_4. The resulting torque $T_{34} = s_4 F_4 - s_3 F_3$ is aligned opposite to the yaw disturbance. In magnitude, it is larger than T_{12} and will thus produce an overall stabilizing effect on the vehicle.

However, if the rear wheels are locked and the front wheels are still rotating, $s_1 \gg s_2$ and $s_3 < s_4$ will hold, Figure 8.7. Then, the destabilizing torque T_{12}, is in magnitude, larger than the stabilizing torque T_{34}, which as a result will increase the yaw disturbance β, thus representing an unstable braking situation.

8.3.4   Optimal Distribution of Drive and Brake Forces

The sum of the longitudinal forces accelerates or decelerates the vehicle. In dimensionless style, Equation (8.1) reads as

$$\frac{\dot{v}}{g} = \frac{F_{x1}}{mg} + \frac{F_{x2}}{mg} \tag{8.35}$$

A certain acceleration or deceleration can only be achieved by different combinations of the longitudinal forces F_{x1} and F_{x2}. According to Equation (8.19), the longitudinal forces are limited by wheel load and friction.

The optimal combination of F_{x1} and F_{x2} will be achieved when front and rear axle have the same skid resistance:

$$F_{x1} = \pm v\,\mu\,F_{z1} \quad \text{and} \quad F_{x2} = \pm v\,\mu\,F_{z2} \tag{8.36}$$

Using Equations (8.5) and (8.6), one obtains

$$\frac{F_{x1}}{mg} = \pm v\,\mu \left(\frac{a_2}{h} - \frac{\dot{v}}{g} \right) \frac{h}{a_1 + a_2} \tag{8.37}$$

and

$$\frac{F_{x2}}{mg} = \pm v\,\mu \left(\frac{a_1}{h} + \frac{\dot{v}}{g} \right) \frac{h}{a_1 + a_2} \tag{8.38}$$

Combining Equation (8.35) with Equations (8.37) and (8.38) results in

$$\frac{\dot{v}}{g} = \pm v\,\mu \tag{8.39}$$

Here, F_{x1} and F_{x2} are assumed to have the same sign. This means that either both axles are driven or braked. Finally, inserting Equation (8.39) in Equations (8.37) and (8.38) yields

$$\frac{F_{x1}}{mg} = \frac{\dot{v}}{g} \left(\frac{a_2}{h} - \frac{\dot{v}}{g} \right) \frac{h}{a_1 + a_2} \tag{8.40}$$

and

$$\frac{F_{x2}}{mg} = \frac{\dot{v}}{g} \left(\frac{a_1}{h} + \frac{\dot{v}}{g} \right) \frac{h}{a_1 + a_2} \tag{8.41}$$

Depending on the desired acceleration $\dot{v} > 0$ or deceleration $\dot{v} < 0$, the longitudinal forces that grant the same skid resistance at both axles can be calculated now.

The curve of optimal drive and brake forces for a typical passenger car is plotted in Figure 8.8. The abscissa represents the longitudinal force at the front axle and the ordinate the one at the rear axle. Braking forces are indicated by negative longitudinal forces. In the diagram they are pointing to the left and upward. Rearranging Equation (8.35) generates the lines of constant acceleration and constant deceleration

$$\frac{F_{x2}}{mg} = \frac{\dot{v}}{g} - \frac{F_{x1}}{mg} \tag{8.42}$$

which are also plotted in Figure 8.8 in the range of $-1.0 \le \dot{v}/g \le 1.0$.

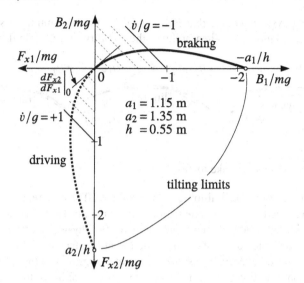

FIGURE 8.8
Optimal distribution of driving and braking forces.

The initial gradient of the curve of optimal drive and brake forces only depends on the steady-state distribution of the wheel loads. From Equations (8.40) and (8.41) it follows:

$$\frac{d\,\dfrac{F_{x1}}{mg}}{d\,\dfrac{\dot{v}}{g}} = \left(\frac{a_2}{h} - 2\frac{\dot{v}}{g}\right)\frac{h}{a_1 + a_2} \tag{8.43}$$

and

$$\frac{d\,\dfrac{F_{x2}}{mg}}{d\,\dfrac{\dot{v}}{g}} = \left(\frac{a_1}{h} + 2\frac{\dot{v}}{g}\right)\frac{h}{a_1 + a_2} \tag{8.44}$$

For $\dot{v}/g = 0$, the initial gradient just remains as

$$\left.\frac{d\,F_{x2}}{d\,F_{x1}}\right|_0 = \frac{a_1}{a_2} \tag{8.45}$$

At the tilting limits $\dot{v}/g = -a_1/h$ and $\dot{v}/g = +a_2/h$, no longitudinal forces can be applied at the lifting axle.

8.3.5 Different Distributions of Brake Forces

Practical applications aim at approximating the optimal distribution of brake forces by a linear, a limited, or a reduced distribution of brake forces as good as possible in a range of physical interest that is bounded by the maximum possible friction coefficient μ_M, Figure 8.9. When braking, the stability of a vehicle depends on the

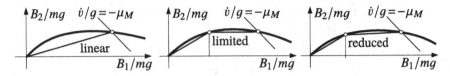

FIGURE 8.9
Different distributions of brake forces.

potential of generating a lateral force at the rear axle. Thus, a greater skid (locking) resistance is realized at the rear axle than at the front axle. Therefore, the brake force distributions are all below the optimal curve in the physically relevant area. This restricts the achievable deceleration, in particular at low friction values.

Because the optimal curve depends on the center of gravity of the vehicle, an additional safety margin must be installed when designing real brake force distributions. The distribution of brake forces is often fitted to the axle loads. There, the influence of the height of the center of gravity, which may also vary significantly on trucks, is not taken into account and must be compensated by a safety margin from the optimal curve. Only the control of brake pressure in anti-lock systems provides an optimal distribution of brake forces independently of loading conditions.

8.3.6   Braking in a Turn

The braking stability becomes apparent when a vehicle is braked while cornering. Different braking scenarios, including no braking at all, are shown in Figure 8.10. At the beginning, the vehicle, a standard passenger car, is cornering with a driving velocity of $v = v_0 = 80$ km/h on a radius of $R \approx 100$ m, which results in a lateral acceleration of $a_y \approx (80/3.6)^2/100 = 4.94$ m/s^2. All braking scenarios start at $t = 3$ s. In the standard case, the braking torques at the front wheels are raised within 0.1 s to 900 Nm and at the rear wheels to 270 Nm, which stops the vehicle in barely 4 seconds. The inner front wheel strongly decelerates at first, upper left plot in Figure 8.11. This brings the wheel close to a locking situation and reduces the transmittable lateral force. As a consequence, the lateral acceleration and the yaw angular velocity collapse for a short time, lower plots in Figure 8.11. The beginning deceleration of the vehicle increases the wheel load at the front axle, which enables the front tires to transmit larger lateral forces and the lateral acceleration and the yaw angular velocity again. As the vehicle has reduced its velocity in the meantime, the peak value of the lateral acceleration is below the initial value. However, the resulting peak in the yaw angular velocity exceeds the initial value and causes the vehicle to turn into the corner slightly. When the vehicle comes to a standstill at $t \approx 7$ s, the compliance of the tire causes oscillations of the longitudinal tire forces, upper left plot in Figure 8.11.

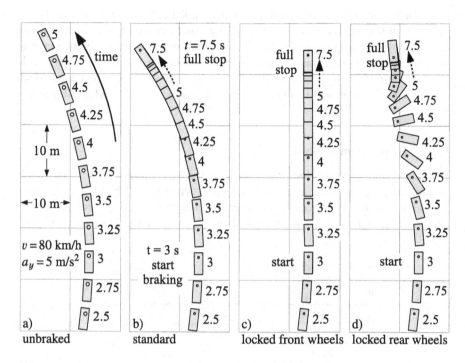

FIGURE 8.10
Braking in a turn with different scenarios.

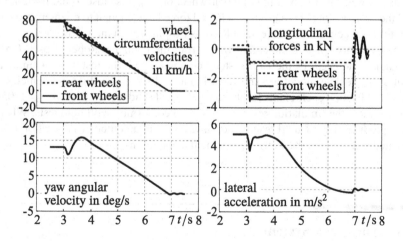

FIGURE 8.11
Baking in a turn with standard braking force distribution.

If large braking torques of 1500 Nm are applied only at the front wheels, the vehicle will stop in nearly the same time. But, the front wheels will lock now and cause the vehicle to go straight ahead instead of further cornering, third graph in Figure 8.10. If the same braking torques are put on the rear wheels only, the vehicle becomes unstable, rotates around, is then stabilized by the locked rear axle, which has come to the front, and finally comes to a standstill, as in the rightmost graph in Figure 8.11.

8.3.7 Braking on μ-Split

If a vehicle without an anti-lock system is braked on a μ-split surface, then the wheels running on μ_{low} will lock in an instant, thus providing small braking forces only. The wheels on the side of μ_{high}, however, generate large braking forces, $F_1 \gg F_2$ and $F_3 > F_4$, Figure 8.12. The rear wheel on μ_{low} is locked and provides no lateral

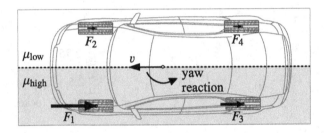

FIGURE 8.12
Yaw reaction when braking on μ-split.

guidance at all. At full braking, the rear wheel on μ_{high} is close to the friction limit and therefore is not able to produce a lateral force large enough to counteract the yaw impact. As a consequence, the vehicle starts to spin around the vertical axis.

Screen shots of a commercial trailer from the company Robert Bosch GmbH, explaining the need for the EPS system, compared with the results of a simulation with a full vehicle model are shown in Figure 8.13. Despite different vehicles and estimated friction coefficients for the dry ($\mu_{high} = 1$) and the icy part ($\mu_{low} = 0.05$) of the test track, the simulation results are in good conformity with field tests. Whereas the reproducibility of field tests is not always given, a computer simulation can be repeated exactly with the same environmental conditions.

8.4 Anti-Lock System

8.4.1 Basic Principle

On hard braking maneuvers, large longitudinal slip values occur. Then, the stability and/or steerability is no longer given because nearly no lateral forces can be generated.

$t = 0$ $\longrightarrow$

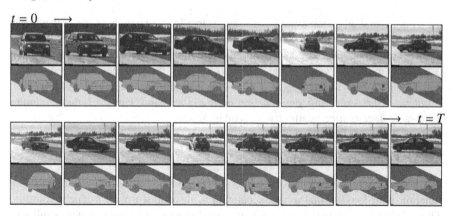

$\longrightarrow$ $t = T$

FIGURE 8.13
Braking on μ-split: Field test and simulation [47].

By controlling the brake torque or brake pressure, respectively, the longitudinal slip can be restricted to values that allow considerable lateral forces.

Here, the angular wheel acceleration $\dot{\Omega}$ is used as a control variable. Angular accelerations of the wheel are derived from the measured angular speeds of the wheel by differentiation. The rolling condition is fulfilled with a longitudinal slip of $s_x = 0$. Then

$$r_D \dot{\Omega} = \ddot{x} \tag{8.46}$$

holds, where r_D is the dynamic tire radius and $\ddot{x}$ describes the longitudinal acceleration of the vehicle. According to Equation (8.21), the maximum acceleration/deceleration of a vehicle depends on the friction coefficient, $|\ddot{x}| = \mu g$. For a given friction coefficient μ, a simple control law can be realized for each wheel,

$$|\dot{\Omega}| \le \frac{1}{r_D} |\ddot{x}| = \frac{1}{r_D} |\mu g| \tag{8.47}$$

Because no reliable possibility to determine the local friction coefficient μ between tire and road has yet been found, useful information can only be gained from Equation (8.47) at optimal conditions on dry road. Therefore, the longitudinal slip s_x is used as a second control variable. In order to calculate longitudinal slips, a reference speed

$$v_{ref} = \frac{1}{N} \sum_{i=1}^{N} r_i \Omega_i \tag{8.48}$$

is calculated by averaging all measured wheel speeds and is then used for the calculation of slip at all wheels. However, this method is too imprecise at low speeds. Therefore, no control is applied below a limit velocity. Problems also arise when all wheels lock simultaneously, which may happen on icy roads, for example.

The control of the brake torque is done via the brake pressure, which can be "increased," "held," or "decreased" by three-way valves. To prevent vibrations, the decrease is usually made slower than the increase.

Commercial anti-lock systems use the "select low" principle at the rear wheels. When braking on μ-split, the braking pressure at both rear wheels is controlled by the wheel running on μ_{low}. As a result, the longitudinal force of the wheel on μ_{high} is bounded to the low value of the one on μ_{low}. Thus, the yaw reaction is reduced on one hand and the wheel on μ_{high} can provide a significantly larger lateral force that counteracts the remaining yaw reaction on the other hand. However, the maximum achievable deceleration is slightly reduced by this.

8.4.2 Demonstration Model

The simple planar vehicle model in Section 8.1.1 just delivers the wheel loads as a function of the longitudinal vehicle acceleration, but cannot provide a general solution. That is why, the planar vehicle model is extended by a suspension system similar to the simple chassis model illustrated in Figure 7.5. As a consequence, the longitudinal motion of the vehicle x must be supplemented by the vertical displacement z and the pitch motion θ to a full planar vehicle motion, Figure 8.14. In a simple linear

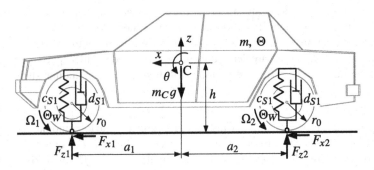

FIGURE 8.14
Planar vehicle brake model.

approach, the wheel loads on a flat and horizontal road are now provided by

$$F_{z1} = m\,g\,\frac{a2}{a_1 + a_2} - c_{S1}(z - a_1\theta) - d_{S1}(\dot{z} - a_1\dot{\theta})$$

$$F_{z2} = mg\,\frac{a1}{a_1 + a_2} - c_{S2}(z + a_2\theta) - d_{S2}(\dot{z} + a_2\dot{\theta})$$

(8.49)

where $c_{S1}, c_{S1}, d_{S2}, d_{S2}$ characterize the stiffness and damping properties of the front and the rear suspension system. Small pitch motions ($|\theta| \ll 1$) as well as small vertical displacements ($|z| \ll h$) are taken for granted in this linear model approach. As done in Section 4.3, the dynamics of the wheel rotations (Ω_1, Ω_2) and an appropriate tire model for the longitudinal forces (F_{x1}, F_{x2}) must be taken into account, too. Then, the equations of motion for the simple planar vehicle model, provided in Section 8.1.1, extend to

$$m\,\ddot{x} = F_{x1} + F_{x2}$$
$$m\,\ddot{z} = F_{z1} + F_{z2} - m\,g \qquad (8.50)$$
$$\Theta\,\ddot{\theta} = -F_{z1}\,a_1 + F_{z2}\,a_2 - (F_{x1} + F_{x2})\,h$$

where the velocity v is substituted via $\dot{x} = v$ by the longitudinal motion x of the vehicle. Such kind of planar models may consider either the full vehicle with two front and two rear wheels or just half of the vehicle with one front and one rear wheel. In the latter case, m and Θ represent the mass and the inertia of half of the vehicle and F_{x1}, F_{z1} as well as F_{x2}, F_{z2} describe the longitudinal and vertical tire forces applied at one front and one rear wheel.

In practice, the functionality of an anti-lock control unit is very complex. Here, a simple three-point controller is applied for demonstration purposes. As distinct from reality, the simplified controller is based on the features of the lateral tire force characteristics. The TMeasy tire model provides the global derivative fos of the longitudinal tire force F_x versus the longitudinal slip s_x. At the slip value $s_x = s_x^M$, where the maximum longitudinal force $F_x = F_x^M$ occurs, fos $= F_x^M / s_x^M$ will hold, Figure 8.15. Global derivatives with fos $>$ fos^M and fos $<$ fos^M indicate that the

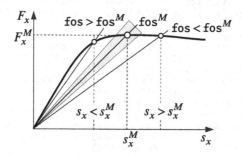

FIGURE 8.15
Global derivatives at the longitudinal force characteristics.

longitudinal slip s_x is smaller or larger than the slip s_x^M which defines the location of the maximum longitudinal force. To realize a simple three-point controller that increases, holds, and decreases the braking torques appropriately, a margin around the critical global derivative, defined by increasing and decreasing the value of fos^M by $\pm 20\,\%$, is used in addition. The control strategy is explained in Table 8.1. The controller of a commercial anti-lock system processes several signals and operates with more sophistication, however.

TABLE 8.1
Three-point control strategy for an ABS-like control

s_x too small	fos $> 1.20 *$ fos^M	increase braking torque
s_x appropriate	$0.80 *$ fos$^M <$ fos $< 1.20 *$ fos^M	hold braking torque
s_x too large	fos $< 0.80 *$ fos^M	decrease braking torque

The function given in Listing 8.1 provides the state equation for the simple planar vehicle model. It includes the longitudinal tire deflections and dynamic braking torques generated by a three-point controller.

Listing 8.1

Function fun_08_abs_brake_dyn: planar vehicle model dynamics with ABS-like control

```
1  function [xdot,out] = fun_08_abs_brake_dyn(t,x,p)
2
3  % get states
4  posx = x( 1); posz = x( 2); pitch = x( 3); % vehicle pos. & orient.
5  velx = x( 4); velz = x( 5); pdot  = x( 6); % vehicle velocities
6  o    = x( 7: 8);     % angular velocities of front and rear wheels
7  xt   = x( 9:10);     % longitudinal tire deflections
8  tqbctrl = x(11:12);  % controlled braking torques
9
10 % wheel loads (linear suspension force characteristics)
11 out.fz = [ p.fz0(1)-p.cS1*(posz-p.a1*pitch)-p.dS1*(velz-p.a1*pdot);
12            p.fz0(2)-p.cS2*(posz+p.a2*pitch)-p.dS2*(velz+p.a2*pdot) ];
13
14 % get maximum braking torque via linear interpolation and default
15 tqb_mx = interp1(p.tqb_t,p.tqb_v,t);   out.TB=[0;0];
16
17 % tire and wheel dynamics: front wheel (i=1), rear wheel (i=2)
18 xtdot=[0;0]; odot=[0;0]; fos=[0;0]; wf=[0;0]; out.fx=[0;0]; fosm=[0;0];
19 for i=1:2
20   % modified transport velocities and long. slips
21   vc = p.r0*o(i); vt = abs(vc); vtn = vt+p.vn;
22   vs = (velx-vc); out.sx(i)=-vs/vtn; sc = abs(out.sx(i));
23   % simple linear wheel load influence to tire parameter
24   wf(i) = out.fz(i)/p.fzn;   fosm(i) = p.fxm/p.sxm*wf(i);
25   pt = [ p.dfx0*wf(i), p.fxm*wf(i), p.sxm, p.fxs*wf(i), p.sxs];
26   % generalized tire characteristics (sx only)
27   [~,fos(i)] = tmy_fcombined( sc, pt(1),pt(2),pt(3),pt(4),pt(5) );
28   % time deriv. of long. tire deflection and long. dynamic tire force
29   xtdot(i) = -(p.cx*xt(i)*vtn+fos(i)*vs)/(p.dx*vtn+fos(i));
30   out.fx(i) = p.cx*xt(i) + p.dx*xtdot(i);
31   % applied braking torque (enhanced dry friction model)
32   out.TB(i) = -p.r0*out.fx(i) + p.dn*o(i);
33   out.TB(i) = sign(out.TB(i))*min(abs(out.TB(i)),max(0,tqbctrl(i)));
34   % angular momentum wheel (no driving torque)
35   odot(i) = ( -out.TB(i) - p.r0*out.fx(i) ) / p.ThetaW;
36 end
37
38 % vehicle equations of motion: long., vert., and pitch
39 velxdot = ( out.fx(1) + out.fx(2) ) / p.mass;
40 velzdot = ( out.fz(1) + out.fz(2) ) / p.mass   - p.grav;
41 pdotdot = (-p.a1*out.fz(1) + p.a2*out.fz(2) ...
42            - p.h * ( out.fx(1) + out.fx(2) ) ) / p.Theta;
43
44 % ABS-like control (simple 3-point brake torque control)
45 tqbctrldot=[0;0]; odx=1.2*p.grav/p.r0;  % "hold" & max. wheel decel.
46 for i=1:2
47   % "increase torque"
48   if fos(i)>1.20*fosm(i) && tqbctrl(i)<tqb_mx && abs(odot(i))<odx
49     tqbctrldot(i) = p.dtqb_inc*abs(velx)/(abs(velx)+1);
50   end
51   % "decrease torque"
52   if fos(i)<0.80*fosm(i) && tqbctrl(i)>0.0
53     tqbctrldot(i) = -p.dtqb_dec*abs(velx)/(abs(velx)+p.vn);
54   end
```

```
55 end
56
57 % state derivatives (vehicle, wheels, tires, braking torque)
58 xdot = [ velx;velz;pdot; velxdot;velzdot;pdotdot; ...
59           odot; xtdot; tqbctrldot ];
60
61 end
```

The linear suspension model realized in Equation (8.49) delivers the wheel loads o.fz, where the elements p.fz0(1) and p.fz0(2) of the parameter structure provide their steady-state values. The loop from line 20 to 41 computes at first the global derivatives fos(i) of the longitudinal tire characteristics via the function tmy_fcombined which is provided by Listing 3.3. Note, that the initial inclination, the maximum, and the sliding force provided by the elements p.dfx0, p.fxm, and p.fxs of the parameter structure, are adjusted by the weighting factor wf(i)=o.fz(i)/p.fzn to the actual wheel loads. Then, the dynamic tire forces o.fx(i) are calculated and the enhanced braking torque model delivers the required braking torques o.tqba(i) at the wheels. Finally, the wheel accelerations owhdot(i) are computed, where the state variables tqbctrl(i) represent the controlled braking torques at the front (i=1) and the rear wheel (i=2).

The equations of motion (8.50) deliver the longitudinal, the vertical, and the pitch acceleration.

The simple three-point control just distinguishes between increase, hold, and decrease. The increase and decrease conditions, shown in Table 8.1 are supplemented by practical restrictions, that the controlled braking torque is limited to the provided maximum braking torque and must be positive according to the definition used here. In addition, the condition for torque increase takes Equation (8.47) into account, whereby the maximum possible tire/road friction coefficient was roughly estimated by $\mu = 1.2$.

By multiplying the braking torque changes tqbctrldot(i) at the front (i=1) and the rear wheel (i=2) with the term abs(velx)/(abs(velx)+1.0), the changes in the braking torques become less intensive, as done in practice, when the vehicle comes close to a standstill velx → 0.

The MATLAB-Script in Listing 8.2 provides the parameter structure p that characterizes the planar vehicle model, performs a simulation, and plots some results. Because the three-point control is anything but smooth, Heun's method with a step size of 1 ms (hstep=0.001) is used here instead of a standard MATLAB ode-solver.

Listing 8.2
Script app_08_abs_brake_sim.m: ABS-like brake control

```
 1 v0 = 80/3.6;          % initial vehicle velocity, km/h -> m/s
 2 p.grav = 9.81;        % constant of gravity in m/s^2
 3 p.mass  = 1400/2;     % corresp. vehicle mass in kg
 4 p.Theta = 2000/2;     % corresp. vehicle inertia in kgm^2
 5 p.a1 = 1.2;           % distance vehicle center front axle in m
 6 p.a2 = 1.3;           % distance vehicle center rear axle in m
 7 p.h  = 0.55;          % height of vehicle center in m
 8 p.cS1 = 20000;        % suspension stiffness front in N/m
 9 p.dS1 =  3000;        % suspension damping front in N/(m/s)
10 p.cS2 = 25000;        % suspension stiffness rear in N/m
11 p.dS2 =  3200;        % suspension damping rear in N/(m/s)
```

```
12 p.ThetaW = 1.2;       % wheel inertia in kgm^2
13 p.r0 = 0.3;           % unloaded wheel radius in m
14
15 % long. tire characteristic for payload fz=fzn only
16 muf     = 1.0;        % friction adjustment factor
17 p.fzn   = 3000;       % payload in N
18 p.dfx0  = 100000;     % initial incl. long. force char. in N/-
19 p.fxm   = 3200*muf;   % maximum long. force in N
20 p.sxm   = 0.1*muf;    % sx where fx=fxm
21 p.fxs   = 3000*muf;   % long. sliding force in N
22 p.sxs   = 0.8*muf;    % sx where fx=fxs
23 p.cx    = 160000;     % longitudinal tire stiffness in N/m
24 p.dx    = 500;        % longitudinal tire damping in N/(m/s)
25 p.vn    = 0.01;       % fictitious velocity in m/s
26
27 % steady state wheel loads
28 p.fz0 = p.mass*p.grav/(p.a1+p.a2)*[ p.a2   p.a1 ];
29
30 % appropriate "damping" constant in enhanced braking torque model
31 p.dn = p.r0*sqrt(p.cx*p.ThetaW);
32
33 % brake control via simple three-point ABS-like control
34 p.dtqb_inc = p.r0*p.fxm/0.1;  % braking torque increase in Nm/s
35 p.dtqb_dec = p.dtqb_inc/2;    % braking torque decrease in Nm/s
36
37 % set braking torque via lookup table: time t in s, torque value in Nm
38 p.tqb_t = [ 0.0 0.05 0.10 0.80 0.85 1.00 ]*round(1.5+v0/(muf*p.grav));
39 p.tqb_v = [ 0.0 0.00 1.00 1.00 0.00 0.00 ]*2*p.fxm*p.r0;
40
41 % simulation interval and initial states
42 hstep=0.001; iout=25; t=0; tE=max(p.tqb_t); N=round(tE/(iout*hstep))+1;
43 x = [ 0; 0; 0;  v0; 0; 0;  v0/p.r0; v0/p.r0; 0; 0;  0; 0 ];
44
45 % pre-allocate vars to speed up loop
46 tout=zeros(N,1); xout=zeros(N,length(x));
47 sxi=zeros(N,2);fxi=sxi;fzi=sxi;TBi=sxi;xpi=zeros(size(xout));
48
49 disp(['time simulation: t=0 to t=',num2str(tE)])
50 for io = 1 : N
51   [xdot,out] = fun_08_abs_brake_dyn(t,x,p);  % generate output
52   tout(io)=t; xout(io,:)=x.';
53   sxi(io,:)=out.sx.'; fxi(io,:)=out.fx.'; fzi(io,:)=out.fz.';
54   TBi(io,:)=out.TB.'; xpi(io,:)=xdot.';
55   if io < N    % simple Heun-solver (second order constant step size)
56     for i=1:iout
57       xdot1 = fun_08_abs_brake_dyn(t,x,p);  t = t + hstep;
58       xdot2 = fun_08_abs_brake_dyn(t,x+hstep*xdot1,p);
59       xdot  = 0.5*(xdot1+xdot2);        x = x + hstep*xdot;
60     end
61   end
62 end
63
64 % plot some results
65 subplot(2,3,1)
66 plot(tout,TBi(:,1:2)/1000), grid on, ylim([-0.8,2])
67 title('applied braking torques in kNm'), legend('front','rear')
68 subplot(2,3,2)
69 plot(tout,xpi(:,4)/p.grav), grid on, ylim([-1.2,0.4])
70 legend('normalized vehicle acceleration [-]')
71 subplot(2,3,3);
72 plot(tout,[p.r0*xout(:,7:8),xout(:,4)]), grid on, ylim([-5,25])
73 title('velocities in m/s'), legend('r \Omega_1','r \Omega_2','v')
74 subplot(2,3,4)
```

```
75 plot(tout,fxi(:,1:2)/(p.mass*p.grav)),  grid on, ylim([-0.9,0.4])
76 title('normalized longitudinal forces'), legend('front','rear')
77 subplot(2,3,5)
78 plot(tout,fzi(:,1:2)/(p.mass*p.grav)),  grid on, ylim([0,1])
79 title('normalized vertical forces'), legend('front','rear')
80 subplot(2,3,6)
81 plot(tout,sxi(:,1:2)), grid on, ylim([-0.18,0.02])
82 title('longitudinal slips'), legend('front','rear')
```

The numerical damping of the enhanced braking torque model as well as the amount of the braking torque increase and decrease are adjusted to the wheel and tire parameter. As in practice, the braking torque decrease is smaller than (here just the half of) the braking torque increase. The friction adjustment factor muf makes it possible to perform simulations with different tire/road friction conditions.

Some results of simulations with muf=1 (dry road) and muf=0.5 (wet road) are plotted in Figure 8.16 and Figure 8.17. The simulations are performed with the initial

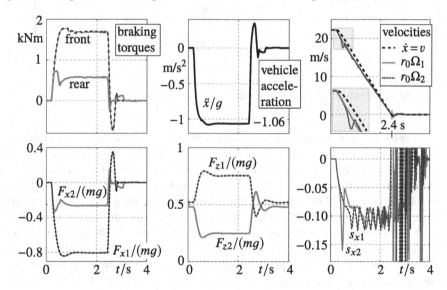

FIGURE 8.16
Braking on dry road ($\mu = 1.0$) with ABS-like three-point control.

velocity of 80 km/h or 22.22 m/s, respectively. The time interval and the time history of the braking torque input are automatically adapted to the friction adjustment factor in order to come to a full stop regardless of the tire/road friction properties.

On dry road (muf=1), the vehicle decelerates in a steady state with $\dot{v}/g \approx -1.06$ and comes to a full stop at $t \approx 2.4$ s, Figure 8.16. The wet surface reduces the deceleration to $\dot{v}/g \approx -0.53$, which stops the vehicle in double the time, Figure 8.17. The steady-state values of the deceleration correspond very well with the tire characteristic, which defines the maximum force $F_x^M = 3200$ N at the payload $F_z^N = 3000$ N, thus modeling a tire/road friction coefficient of $\mu = \mu_R = 3200/3000 = 1.07$. On the wet road, the friction adjustment factor muf=0.5 reduces this value to $\mu = 0.5 * 1.07 = 0.535$. As defined in the lookup table, the braking is initiated at $t = 0.05$ s, where the preset

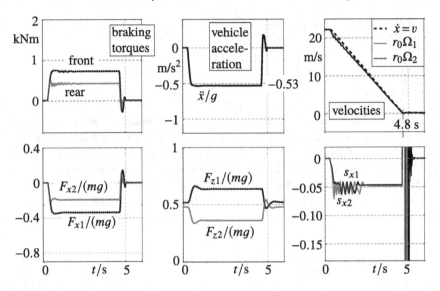

FIGURE 8.17
Braking on wet road ($\mu = 0.5$) with ABS-like three-point control.

braking torque is increased emergency-like within 0.05 s to its maximum value. As a consequence, the three-point controller generates rapidly rising braking torques, which decelerate the wheels much faster than the vehicle itself. The deceleration of the vehicle ($\dot{v} < 0$) results in a dynamic wheel load shift from the rear to the front, which makes it possible to transmit larger longitudinal forces $|F_{x1}| > |F_{x2}|$ at the front than on the rear. That is why the rear wheel decelerates slightly faster than the front wheel, which is highlighted by a zoomed window in the upper right plot of Figure 8.16. In order to prevent the wheels from locking, the controller starts to reduce the braking torque applied to the wheels. As they come close to a rolling condition, the braking torques are held and then increased again. This rather crude control input causes the longitudinal slips s_{x1} and s_{x2} to oscillate during the braking maneuver rather rapidly about the slip values (-0.10 on dry and -0.05 on wet road), which will grant maximum longitudinal forces, lower right plots in Figures 8.16 and 8.17. According to the definition in Equation (3.80) the longitudinal slips will tend to $\pm 1/\text{p.vn} = 1/0.01 = 100$ at standstill, where $v \to 0$ as well as $\Omega_1 \to 0$ and $\Omega_2 \to 0$ hold. That is why, the slips exceed the plotting limits in this time period.

In a steady state, the normalized vertical wheel loads correspond with the results obtained in Section 8.1.1. Inserting the data $a_1 = 1.2$ m, $a_2 = 1.3$ m, and $h = 0.55$ m into Equations (8.5) and (8.6) results in

$$
\begin{array}{lllll}
\dot{v}/g = -1.0 & | & F_{z1}/(mg) & = & 0.70 & F_{z2}/(mg) & = & 0.30 \\
\dot{v}/g = -0.5 & | & F_{z1}/(mg) & = & 0.59 & F_{z2}/(mg) & = & 0.41
\end{array}
\tag{8.51}
$$

In a steady state, the normalized longitudinal forces are close to the optimal ones, which are defined by Equations (8.40) and (8.41) and will produce

$$
\begin{array}{llll}
\dot{v}/g = -1.0 & | \quad F_{x1}/(m\,g) = -0.740 & F_{x2}/(m\,g) = -0.260 \\
\dot{v}/g = -0.4 & | \quad F_{x1}/(m\,g) = -0.315 & F_{x2}/(m\,g) = -0.185
\end{array}
\tag{8.52}
$$

Regardless of the vehicle properties, an anti-lock system provides braking forces or braking torques, respectively, which are close to the optimal one. Besides avoiding completely locked wheels, this is a mayor benefit of an anti-lock system.

The slightly increased complexity of the vehicle model takes the hub and pitch motion of the chassis into account too. On hard braking maneuvers, a strong pitch reaction will occur hereby, left plot in Figure 8.18. As can be seen in the rightmost

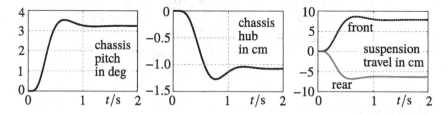

FIGURE 8.18
Brake pitch reaction of simple planar vehicle model at $\dot{v}/g = -1$.

plot of Figure 8.18, the movements of the chassis result in large suspension travel. In a steady state, the compression at the front axle amounts to 7 cm and the rebound at the rear axle to slightly more than 6 cm. The values differ due to different suspension spring rates at the front ($c_{SF} = 20\,000$ N/m) and at the rear axle ($c_{SR} = 25\,000$ N/m) and cause the chassis to perform a slight hub-motion in addition.

Most drivers do not like their vehicle to take a deep "bow" when braking. The straightforward solution to harden the suspension is not practicable, because this will worsen the comfort significantly. A possible solution is discussed in the next section.

8.5 Drive and Brake Pitch

8.5.1 Enhanced Planar Vehicle Model

The planar vehicle model shown in Figure 8.19 is more sophisticated. It consists of five rigid bodies: the left half of the chassis and the left wheel as well as the left knuckle of each axle. The coordinates x_C, z_C, and θ_C characterize the longitudinal, the vertical, and the pitch motion of the chassis. The centers of each wheel and knuckle are supposed to coincide and z_1, z_2 represent their vertical motion relative to the chassis. Finally, the rotation angles φ_1 and φ_2 characterize the wheel rotations relative to the knuckles. The height of the chassis center is defined by $h = r + h_R$,

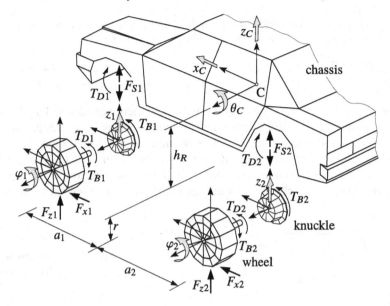

FIGURE 8.19
Sophisticated planar vehicle model.

where r denotes the wheel radius and h_R the height of the chassis center with respect to the wheel center. The distances a_1 and a_2 define the location of the chassis center with respect to the front and rear axle.

The suspension forces acting between the knuckle and the chassis are labeled F_{S1} and F_{S2}. At the wheels, drive torques T_{D1}, T_{D2}; brake torques T_{B1}, T_{B2}; longitudinal forces F_{x1}, F_{x2}; and the wheel loads F_{z1}, F_{z2} apply. In general, the brake torques are directly supported by the wheel bodies, whereas the drive torques are transmitted by the drive shafts to the chassis.

The velocity and the angular velocity of the chassis are given by

$$
v_{0C,0} = \begin{bmatrix} \dot{x}_C \\ 0 \\ 0 \end{bmatrix} + \begin{bmatrix} 0 \\ 0 \\ \dot{z}_C \end{bmatrix} \quad \text{and} \quad \omega_{0C,0} = \begin{bmatrix} 0 \\ \dot{\theta}_C \\ 0 \end{bmatrix} \tag{8.53}
$$

The kinematic analysis of a double wishbone suspension system, performed in Section 5.4.6, shows that knuckle and wheel will not move simply up and down in practice. For example, the movements of the wheel center at a simple trailing arm suspension, discussed in Section 1.6, are determined by the rotation of the trailing arm with the angle β about the revolute joint in B, Figure 8.20. The design position of the trailing arm is defined by the angle β_0. Then, the longitudinal and vertical motion of the wheel center W with respect to the chassis are just given by

$$
x = a \sin(\beta_0 + \beta) \quad \text{and} \quad z = a \cos(\beta_0 + \beta) \tag{8.54}
$$

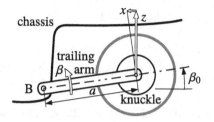

FIGURE 8.20
Trailing arm suspension, where the knuckle is fixed to the trailing arm.

where a denotes the distance of the wheel center W to the joint in B. Within this planar model, the vertical movements of the front and rear wheel center (z_1, z_1) are used as generalized coordinates. Then, the longitudinal (x_1, x_2) and rotational motions (β_1, β_2) of each knuckle are described via functions of the corresponding vertical motions

$$x_1 = x_1(z_1), \quad \beta_1 = \beta_1(z_1) \quad \text{and} \quad x_2 = x_2(z_2), \quad \beta_2 = \beta_2(z_2) \tag{8.55}$$

Under normal driving situations, small vertical displacements $(|z_C| \ll r + h_R)$ and a small pitch motion $(|\theta_C| \ll 1)$ may be taken for granted. Then, the velocities of the knuckles and wheels are obtained by

$$v_{0K_1,0} = v_{0W_1,0} = \begin{bmatrix} \dot{x}_C \\ 0 \\ 0 \end{bmatrix} + \begin{bmatrix} 0 \\ 0 \\ \dot{z}_C \end{bmatrix} + \begin{bmatrix} -h_R\,\dot{\theta}_C \\ 0 \\ -a_1\,\dot{\theta}_C \end{bmatrix} + \begin{bmatrix} \frac{\partial x_1}{\partial z_1}\dot{z}_1 \\ 0 \\ \dot{z}_1 \end{bmatrix} \tag{8.56}$$

$$v_{0K_2,0} = v_{0W_2,0} = \begin{bmatrix} \dot{x}_C \\ 0 \\ 0 \end{bmatrix} + \begin{bmatrix} 0 \\ 0 \\ \dot{z}_C \end{bmatrix} + \begin{bmatrix} -h_R\,\dot{\theta}_C \\ 0 \\ +a_2\,\dot{\theta}_C \end{bmatrix} + \begin{bmatrix} \frac{\partial x_2}{\partial z_2}\dot{z}_2 \\ 0 \\ \dot{z}_2 \end{bmatrix} \tag{8.57}$$

where higher-order terms were neglected. In general, the constraint motions, here defined by Equation (8.55), may depend on more than one generalized coordinate. That is why, partial derivatives are used in this approach. Finally, the angular velocities of the knuckles and wheels are defined by

$$\omega_{0K_1,0} = \begin{bmatrix} 0 \\ \dot{\theta}_C \\ 0 \end{bmatrix} + \begin{bmatrix} 0 \\ \frac{\partial \beta_1}{\partial z_1}\dot{z}_1 \\ 0 \end{bmatrix}, \quad \omega_{0W_1,0} = \begin{bmatrix} 0 \\ \dot{\theta}_C \\ 0 \end{bmatrix} + \begin{bmatrix} 0 \\ \frac{\partial \beta_1}{\partial z_1}\dot{z}_1 \\ 0 \end{bmatrix} + \begin{bmatrix} 0 \\ \dot{\varphi}_1 \\ 0 \end{bmatrix} \tag{8.58}$$

$$\omega_{0K_2,0} = \begin{bmatrix} 0 \\ \dot{\theta}_C \\ 0 \end{bmatrix} + \begin{bmatrix} 0 \\ \frac{\partial \beta_2}{\partial z_2}\dot{z}_2 \\ 0 \end{bmatrix}, \quad \omega_{0W_2,0} = \begin{bmatrix} 0 \\ \dot{\theta}_C \\ 0 \end{bmatrix} + \begin{bmatrix} 0 \\ \frac{\partial \beta_2}{\partial z_2}\dot{z}_2 \\ 0 \end{bmatrix} + \begin{bmatrix} 0 \\ \dot{\varphi}_2 \\ 0 \end{bmatrix} \tag{8.59}$$

Collecting the time derivatives of the generalized coordinates in the vector of generalized velocities

$$z = \begin{bmatrix} \dot{x}_C & \dot{z}_C & \dot{\theta}_C & \dot{z}_1 & \dot{\varphi}_1 & \dot{z}_2 & \dot{\varphi}_2 \end{bmatrix}^T \tag{8.60}$$

the velocities and angular velocities given by Equations (8.53), (8.56), (8.57), (8.58), and (8.59) can be written as

$$v_{0i} = \sum_{j=1}^{7} \frac{\partial v_{0i}}{\partial z_j} z_j \quad \text{and} \quad \omega_{0i} = \sum_{j=1}^{7} \frac{\partial \omega_{0i}}{\partial z_j} z_j \tag{8.61}$$

The partial velocities $\partial v_{0i}/\partial z_j$ and partial angular velocities $\partial \omega_{0i}/\partial z_j$ for the planar vehicle model are arranged in Tables 8.2 and 8.3. The last column lists the forces and torques that apply to the corresponding model body, where g denotes the constant of gravity and m_C, m_{K1}, m_{W1}, m_{K2}, m_{W2}, Θ_C, Θ_{K1}, Θ_{W1}, Θ_{K2}, Θ_{W1} denote the mass and inertia properties.

TABLE 8.2
Partial velocities and applied forces

| Bodies | \multicolumn{7}{c}{Partial Velocities $\partial v_{0i}/\partial z_j$} | Applied Forces |
	$\dot{x}_C$	$\dot{z}_C$	$\dot{\theta}_C$	$\dot{z}_1$	$\dot{\varphi}_1$	$\dot{z}_2$	$\dot{\varphi}_2$	F_i^a
Chassis	1	0	0	0	0	0	0	0
m_C	0	0	0	0	0	0	0	0
	0	1	0	0	0	0	0	$F_{S1}+F_{S2}-m_Cg$
Knuckle	1	0	$-h_R$	$\frac{\partial x_1}{\partial z_1}$	0	0	0	0
front	0	0	0	0	0	0	0	0
m_{K1}	0	1	$-a_1$	1	0	0	0	$-F_{S1}-m_{K1}g$
Wheel	1	0	$-h_R$	$\frac{\partial x_1}{\partial z_1}$	0	0	0	F_{x1}
front	0	0	0	0	0	0	0	0
m_{W1}	0	1	$-a_1$	1	0	0	0	$F_{z1}-m_{W1}g$
Knuckle	1	0	$-h_R$	0	0	$\frac{\partial x_2}{\partial z_2}$	0	0
rear	0	0	0	0	0	0	0	0
m_{K2}	0	1	a_2	0	0	1	0	$-F_{S2}-m_{K2}g$
Wheel	1	0	$-h_R$	0	0	$\frac{\partial x_2}{\partial z_2}$	0	F_{x2}
rear	0	0	0	0	0	0	0	0
m_{W2}	0	1	a_2	0	0	1	0	$F_{z2}-m_{W2}g$

8.5.2 Equations of Motion

As shown in Section 1.5.3 the accelerations and angular accelerations split into two parts. Similar to Equation (1.26), one gets

$$a_{0i} = \sum_{j=1}^{7} \left(\frac{\partial v_{0i}}{\partial z_j} \dot{z}_j + a_{0i}^R \right) \quad \text{and} \quad \alpha_{0i} = \sum_{j=1}^{7} \left(\frac{\partial \omega_{0i}}{\partial z_j} \dot{z}_j + \alpha_{0i}^R \right) \tag{8.62}$$

The remaining terms in the accelerations,

$$a_{0i}^R = \sum_{j=1}^{7} \frac{d}{dt}\left(\frac{\partial v_{0i}}{\partial z_j} \right) \dot{z}_j \quad \text{and} \quad \alpha_{0i}^R = \sum_{j=1}^{7} \frac{d}{dt}\left(\frac{\partial \omega_{0i}}{\partial z_j} \right) \dot{z}_j \tag{8.63}$$

TABLE 8.3

Partial angular velocities and applied torques

Bodies	Partial Angular Velocities $\partial\omega_{0i}/\partial z_j$							Applied Torques
	$\dot{x}_C$	$\dot{z}_C$	$\dot{\theta}_C$	$\dot{z}_1$	$\dot{\varphi}_1$	$\dot{z}_2$	$\dot{\varphi}_2$	T_i^a
Chassis Θ_C	0	0	0	0	0	0	0	0
	0	0	1	0	0	0	0	$-T_{D1}-T_{D2}-a_1 F_{S1}+a_2 F_{S2}$
	0	0	0	0	0	0	0	0
Knuckle front Θ_{K1}	0	0	0	0	0	0	0	0
	0	0	1	$\frac{\partial\beta_1}{\partial z_1}$	0	0	0	T_{B1}
	0	0	0	0	0	0	0	0
Wheel front Θ_{W1}	0	0	0	0	0	0	0	0
	0	0	1	$\frac{\partial\beta_1}{\partial z_1}$	1	0	0	$T_{D1}-T_{B1}-r\,F_{x1}$
	0	0	0	0	0	0	0	0
Knuckle rear Θ_{K2}	0	0	0	0	0	0	0	0
	0	0	1	0	0	$\frac{\partial\beta_2}{\partial z_2}$	0	T_{B2}
	0	0	0	0	0	0	0	0
Wheel rear Θ_{W2}	0	0	0	0	0	0	0	0
	0	0	1	0	0	$\frac{\partial\beta_2}{\partial z_2}$	1	$T_{D2}-T_{B2}-r\,F_{x2}$
	0	0	0	0	0	0	0	0

will contain only higher-order terms here and may be neglected for basic studies. Jourdain's principle of virtual power delivers the equations of motion in the form of

$$M\,\dot{z} = Q \tag{8.64}$$

where

$$M(i,j) = \sum_{k=1}^{5}\left(\frac{\partial v_{0k}}{\partial z_i}\right)^T m_k \frac{\partial v_{0k}}{\partial z_j} + \sum_{k=1}^{5}\left(\frac{\partial \omega_{0k}}{\partial z_i}\right)^T \Theta_k \frac{\partial \omega_{0k}}{\partial z_j} \tag{8.65}$$

defines the elements of the 7×7 mass matrix and the 7×1 generalized force vector

$$Q(i) = \sum_{k=1}^{5}\left(\frac{\partial v_{0k}}{\partial z_i}\right)^T F_k^e + \sum_{k=1}^{5}\left(\frac{\partial \omega_{0k}}{\partial z_i}\right)^T M_k^e \tag{8.66}$$

only processes applied forces and torques in this particular case.

8.5.3 Equilibrium

Introducing the abbreviations

$$m_1 = m_{K1}+m_{W1}, \quad m_2 = m_{K2}+m_{W2}, \quad m_V = m_C+m_1+m_2, \quad h = h_R+r \tag{8.67}$$

the components of the vector of generalized forces and torques will read as

$$
\begin{aligned}
Q(1) &= F_{x1} + F_{x2}, \\
Q(2) &= F_{z1} + F_{z2} - m_V\, g \\
Q(3) &= -a_1(F_{z1} - m_1 g) + a_2(F_{z2} - m_2 g) - h(F_{x1} + F_{x2}) \\
Q(4) &= F_{z1} - m_1 g - F_{S1} + \tfrac{\partial x_1}{\partial z_1} F_{x1} + \tfrac{\partial \beta_1}{\partial z_1}(T_{D1} - r\, F_{x1}) \\
Q(5) &= T_{D1} - T_{B1} - r\, F_{x1}, \\
Q(6) &= F_{z2} - m_2 g - F_{S2} + \tfrac{\partial x_2}{\partial z_2} F_{x2} + \tfrac{\partial \beta_2}{\partial z_2}(T_{D2} - r\, F_{x2}) \\
Q(7) &= T_{D2} - T_{B2} - r\, F_{x2}
\end{aligned}
\tag{8.68}
$$

Without any driving and braking torques,

$$
T_{D1} = 0, \quad T_{D2} = 0, \quad T_{B1} = 0, \quad T_{B2} = 0
\tag{8.69}
$$

one gets the steady-state longitudinal forces, the suspension preloads, and the wheel loads as

$$
\begin{aligned}
F_{x1}^{st} &= 0 & F_{x2}^{st} &= 0 \\
F_{S1}^{st} &= \frac{a_2}{a_1 + a_2}\, m_C g & F_{S2}^{st} &= \frac{a_1}{a_1 + a_2}\, m_C g \\
F_{z1}^{st} &= \frac{a_2}{a_1 + a_2}\, m_C g + m_1 g & F_{z2}^{st} &= \frac{a_1}{a_1 + a_2}\, m_C g + m_2 g
\end{aligned}
\tag{8.70}
$$

As distinct from Equation (8.7), the static wheel loads are now composed of the corresponding chassis and the axle weight.

8.5.4 Driving and Braking

Assuming that on accelerating or decelerating the vehicle, the wheels neither slip nor lock, then the contact point velocities must vanish. By taking the Equations (8.56), (8.58), and (8.57), (8.59) into account one gets

$$
\begin{aligned}
\left(\dot{\theta}_C + \frac{\partial \beta_1}{\partial z_1}\dot{z}_1 + \dot{\varphi}_1\right) r &= \dot{x}_C - h_R\,\dot{\theta}_C + \frac{\partial x_1}{\partial z_1}\dot{z}_1 \\
\left(\dot{\theta}_C + \frac{\partial \beta_2}{\partial z_2}\dot{z}_2 + \dot{\varphi}_2\right) r &= \dot{x}_C - h_R\,\dot{\theta}_C + \frac{\partial x_2}{\partial z_2}\dot{z}_2
\end{aligned}
\tag{8.71}
$$

In a steady state, the pitch motion of the body and the vertical motion of the knuckles and the wheels reach constant values,

$$
\beta_C = \theta_C^{st} = const., \quad z_1 = z_1^{st} = const., \quad z_2 = z_2^{st} = const.
\tag{8.72}
$$

Then Equation (8.71) simplifies to

$$
r\,\dot{\varphi}_1 = \dot{x}_C \quad \text{and} \quad r\,\dot{\varphi}_2 = \dot{x}_C
\tag{8.73}
$$

In addition, the time derivative of the generalized velocities reduces to

$$\dot{z}^{st} = \begin{bmatrix} \ddot{x}_C & 0 & 0 & 0 & \frac{1}{r}\ddot{x}_C & 0 & \frac{1}{r}\ddot{x}_C \end{bmatrix}^T \tag{8.74}$$

in this particular case. Then, the equation of motion (8.64) will result in

$$m_V \ddot{x}_C = F_{x1}^{dyn} + F_{x2}^{dyn} \tag{8.75}$$

$$0 = F_{z1}^{dyn} + F_{z2}^{dyn} \tag{8.76}$$

$$\left[-h_R(m_1 + m_2) + \frac{\Theta_{W1}}{r} + \frac{\Theta_{W2}}{r} \right] \ddot{x}_C = -a_1 F_{z1}^{dyn} + a_2 F_{z2}^{dyn} - h(F_{x1}^{dyn} + F_{x2}^{dyn}) \tag{8.77}$$

$$\left[\frac{\partial x_1}{\partial z_1} m_1 + \frac{\partial \beta_1}{\partial z_1} \frac{\Theta_{W1}}{r} \right] \ddot{x}_C = F_{z1}^{dyn} - F_{S1}^{dyn} + \frac{\partial x_1}{\partial z_1} F_{x1}^{dyn} + \frac{\partial \beta_1}{\partial z_1} (T_{D1} - r F_{x1}^{dyn}) \tag{8.78}$$

$$\frac{\Theta_{W1}}{r} \ddot{x}_C = T_{D1} - T_{B1} - r F_{x1}^{dyn} \tag{8.79}$$

$$\left[\frac{\partial x_2}{\partial z_2} m_2 + \frac{\partial \beta_2}{\partial z_2} \frac{\Theta_{W2}}{r} \right] \ddot{x}_C = F_{z2}^{dyn} - F_{S2}^{dyn} + \frac{\partial x_2}{\partial z_2} F_{x2}^{dyn} + \frac{\partial \beta_2}{\partial z_2} (T_{D2} - r F_{x2}^{dyn}) \tag{8.80}$$

$$\frac{\Theta_{W2}}{r} \ddot{x}_C = T_{D2} - T_{B2} - r F_{x2}^{dyn} \tag{8.81}$$

where Equations (8.72), (8.73), and (8.67) were taken into account and the steady-state spring forces, longitudinal forces, and wheel loads have been separated into steady-state and dynamic terms

$$F_{xi}^{st} = F_{xi}^{st} + F_{xi}^{dyn}, \quad F_{zi}^{st} = F_{zi}^{st} + F_{zi}^{dyn}, \quad F_{Si}^{st} = F_{Si}^{st} + F_{Si}^{dyn}, \quad i = 1, 2 \tag{8.82}$$

Combining Equation (8.75) with Equations (8.79) and (8.81) simply results in

$$\ddot{x}_C = \frac{(T_{D1} + T_{D2} - T_{B1} - T_{B2})/r}{m_V + \Theta_{W1}/r^2 + \Theta_{W2}/r^2} \tag{8.83}$$

The terms in the numerator characterize the overall driving and braking forces and the denominator represents the generalized vehicle mass, which expresses the fact that besides the vehicle mass m_V, the wheels with the inertias Θ_{W1} and Θ_{W2} must be accelerated or decelerated too. The tire forces are obtained as

$$F_{x1}^{dyn} = \frac{T_{D1} - T_{B1}}{r} - \frac{\Theta_{W1}}{r^2} \ddot{x}_C \quad \text{and} \quad F_{x2}^{dyn} = \frac{T_{D2} - T_{B2}}{r} - \frac{\Theta_{W2}}{r^2} \ddot{x}_C \tag{8.84}$$

$$F_{z1}^{dyn} = -\left(\frac{\Theta_{W1} + \Theta_{W2}}{r} + m_1 r + m_2 r + m_C h \right) \frac{\ddot{x}_C}{a_1 + a_2} \tag{8.85}$$

$$F_{z2}^{dyn} = \left(\frac{\Theta_{W1} + \Theta_{W2}}{r} + m_1 r + m_2 r + m_C h \right) \frac{\ddot{x}_C}{a_1 + a_2} \tag{8.86}$$

Neither the vehicle acceleration or deceleration nor the tire forces are affected by the kinematic properties of the suspension.

In general, the inertias of the wheels are small compared to the vehicle mass on normal passenger cars and may be neglected in practice. Then, the dynamic wheel loads simplify to

$$F^{dyn}_{z1,z2} \approx \mp (m_1 r + m_2 r + m_C h) \frac{\ddot{x}_C}{a_1 + a_2} = \mp m_V \frac{h_V}{a_1 + a_2} \ddot{x}_C \qquad (8.87)$$

where $m_V = m_1 + m_2 + m_C$ denotes the overall vehicle mass and h_V defines the height of the vehicle center of gravity. This coincides perfectly with the results obtained in Section 8.1.1.

A simple rearrangement of Equations (8.78) and (8.80) provides the suspension forces as

$$F^{dyn}_{S1} = F^{dyn}_{z1} + \left(\frac{\partial x_1}{\partial z_1} - r \frac{\partial \beta_1}{\partial z_1} \right) F^{dyn}_{x1} + \frac{\partial \beta_1}{\partial z_1} T_{D1} - \left[\frac{\partial x_1}{\partial z_1} m_1 + \frac{\partial \beta_1}{\partial z_1} \frac{\Theta_{W1}}{r} \right] \ddot{x}_C \qquad (8.88)$$

$$F^{dyn}_{S2} = F^{dyn}_{z2} + \left(\frac{\partial x_2}{\partial z_2} - r \frac{\partial \beta_2}{\partial z_2} \right) F^{dyn}_{x2} + \frac{\partial \beta_2}{\partial z_2} T_{D2} - \left[\frac{\partial x_2}{\partial z_2} m_2 + \frac{\partial \beta_2}{\partial z_2} \frac{\Theta_{W2}}{r} \right] \ddot{x}_C \qquad (8.89)$$

A complete algebraic solution is very cumbersome and will result in extremely complicated expressions. The suspension forces F_{S1}, F_{S2} and the wheel loads F_{z1}, F_{z2} support the chassis and the knuckles with the wheels. In a steady state, it holds that

$$\begin{aligned} F^{dyn}_{S1} &= c_{S1} z^{dyn}_1 , \\ F^{dyn}_{S2} &= c_{S2} z^{dyn}_2 , \\ F^{dyn}_{z1} &= -c_{T1}(z^{dyn}_C - a_1 \theta^{dyn}_C + z^{dyn}_1) , \\ F^{dyn}_{z2} &= -c_{T2}(z^{dyn}_C + a_2 \theta^{dyn}_C + z^{dyn}_2) , \end{aligned} \qquad (8.90)$$

where linear spring characteristics are assumed and the tire liftoff is not taken into account.

8.5.5 Drive Pitch

The MATLAB-Script in Listing 8.3 provides the data of the enhanced planar vehicle model, sets the driving and braking torques, calculates for a fast steady-state acceleration the forces, the suspension travel, the pitch and hub motion of the chassis, and finally plots the chassis pitch angle versus different inclinations of the front knuckle/wheel motion.

Listing 8.3

MATLAB-Script app_08_pitch_reaction.m: anti-squat

```
1 grav = 9.81;      % constant of gravity in m/s^2
2 r    = 0.30;      % tire radius in m
3 h    = 0.59;      % height of center of gravity in m
4 a1   = 1.20;      % center of gravity --> front axle in m
5 a2   = 1.30;      % center of gravity --> rear axle in m
6 mC   = 1200/2;    % mass of chassis (half vehicle) in kg
7 m1   = 50;        % mass of front knuckle and wheel in kg
8 m2   = 50;        % mass of rear knuckle and wheel in kg
9 ThW1 = 1.2;       % inertia of front wheel in kgm^2
```

```
10 ThW2  = 1.2;          % inertia of rear wheel in kgm^2
11 cS1   = 22000;        % stiffness of front suspension in N/m
12 cS2   = 28600;        % stiffness of rear suspension in N/m
13 cT1   = 220000;       % stiffness of front wheel in N/m
14 cT2   = 200000;       % stiffness of rear wheel in N/m
15
16 % overall and generalized vehicle mass
17 mV=mC+m1+m2;   mg = mV+ThW1/r^2+ThW2/r^2;
18
19 % set norm. vehicle acceleration (vdg>0) or deceleration (vdg<0)
20 vdg = 1.0;
21
22 % use simple planar model to estimate optimal torque distribution
23 hV   = ( m1*r + m2*r + mC*h ) / mV;   % height of COG vehicle
24 a1V  = ( a1*mC + (a1+a2)*m2 ) / mV;   % vehicle COG --> front axle
25 a2V  = ( a2*mC + (a1+a2)*m1 ) / mV;   % vehicle COG --> rear axle
26 Fx1_opt = mg*vdg*grav*(a2V-hV*vdg)/(a1+a2); % opt. long. force front
27 Fx2_opt = mg*vdg*grav*(a1V+hV*vdg)/(a1+a2); % opt. long. force rear
28 if vdg < 0
29   TD1 = 0;   TD2 = 0;   TB1 = -r*Fx1_opt;   TB2 = -r*Fx2_opt;
30 else
31   TD1 = r*Fx1_opt;   TD2 = r*Fx2_opt;   TB1 = 0;   TB2 = 0;
32 end
33
34 % default axle kinematics
35 dx1 = 0; dx2 = 0; % inclination of front and rear wheel motion
36 db1 = 0; db2 = 0; % y-rotation of front and rear knuckle in rad/m
37
38 % change inclination of front knuckle/wheel motion
39 ivar=25;   dx_min=tan(-12/180*pi);   dx_max=tan(12/180*pi);
40
41 for i=1:ivar
42
43   dx1 = dx_min + (dx_max-dx_min)*(i-1)/(ivar-1);
44
45 % vehicle acceleration
46   xddot = 1/r*(TD1+TD2-TB1-TB2)/mg;
47
48 % dynamic tire forces and required friction coefficients
49   Fx1=(TD1-TB1)/r - ThW1/r^2*xddot;
50   Fx2=(TD2-TB2)/r - ThW2/r^2*xddot;
51   Fz1=-xddot*((ThW1+ThW2)/r+(m1+m2)*r+mC*h)/(a1+a2);
52   Fz2= xddot*((ThW1+ThW2)/(r*h)+(m1+m2)*r/h+mC)*h/(a1+a2);
53
54 % dynamic suspension forces
55   FS1=Fz1+(dx1-r*db1)*Fx1+db1*TD1-(dx1*m1+db1*ThW1/r)*xddot;
56   FS2=Fz2+(dx2-r*db2)*Fx2+db2*TD2-(dx2*m2+db2*ThW2/r)*xddot;
57
58 % suspension travel and chassis motions
59   z1 = FS1/cS1;   z2=FS2/cS2;
60   bC = (Fz1/cT1-Fz2/cT2+z1-z2)/(a1+a2);   zC = a1*bC-z1-Fz1/cT1;
61
62 % plot pitch angle in degree
63   plot(dx1,bC*180/pi,'ok'), hold on, grid on, ylim([-5,0])
64
65 end
```

By specifying a normalized vehicle acceleration, the required driving or braking torques can be computed. Equation (8.83) delivers the overall driving torque $T_D = T_{D1} + T_{D2}$ or the overall braking torque $T_B = T_{B1} + T_{B2}$, respectively, and Equations (8.40) and (8.41) provide the optimal longitudinal forces F_{x1}^{opt} and F_{x2} hereby. The multiplication with the wheel radius r generates the corresponding torques, where the statements TB1 = -r*Fx1_opt and TB2 = -r*Fx2_opt take into account that the braking torques are defined positive whereas negative longitudinal forces represent braking in forward drive.

TABLE 8.4

Vehicle with plain suspension kinematics extremely accelerated ($\ddot{x}_C/g = 1$)

Longitudinal Tire Force	Wheel Load Static+Dynamic	Suspension Forces	Travel	Chassis Motion
$F_{x1}=1991.5$ N	$F_{z1}=2013$ N	$F_{S1}=-1538$ N	$z_1=-70$ mm	$z_C=10.5$ mm
$F_{x2}=4875.5$ N	$F_{z2}=4854$ N	$F_{S2}=+1538$ N	$z_2=54$ mm	$\theta_C=-3.17°$

The complete results, obtained by simply omitting the semicolon at the corresponding code lines, for a plain suspension kinematics (db1=0, dx1=0, db2=0, dx2=0), where both wheels just move straight up and down, are collected in Table 8.4. In this particular case, the required friction coefficients at the front and rear wheels amount to $\mu_1 = |F_{x1}|/F_{z1} = 0.98931$ and $\mu_2 = |F_{x2}|/F_{z2} = 1.0044$, which is very close to an all-wheel drive with perfect driving torque distribution, where $\mu_1 = \mu_2 = |\dot{v}|/g = 1$ would hold. Hence, the simple model shown in Figure 8.1, which was used here to compute or estimate respectively the driving torques, already delivers practical results.

The suspension kinematics are characterized by the parameters db1 = $\partial\beta_1/\partial z_1$ and db2 = $\partial\beta_2/\partial z_2$, which describe the y-rotation of the knuckles caused by their vertical motions z_1 and z_2, as well as by the parameters dx1 = $\partial x_1/\partial z_1$ and dx2 = $\partial x_2/\partial z_2$, which define the inclination of the wheel motions in the x-z-plane. The MATLAB-Script in Listing 8.3 varies the inclination dx1 of the front knuckle/wheel movement only. A simple and straightforward code extension will produce the results plotted in Fig. 8.21. The drive shafts transmit the driving torques directly from the chassis to the wheels in a standard drive train layout. That is why the knuckle rotations caused by the jounce and rebound motion of the knuckles, here characterized by db1 and db2, will have no influence at all on the steady-state chassis pitch angle when the vehicle accelerates. "Anti-squat" suspension kinematics, which are characterized by dx1 > 0 and dx2 < 0, will reduce the pitch angle θ_C when the vehicle is accelerated. However, a longitudinal motion of the knuckle caused by the suspension travel (dx1 ≠ 0 and dx2 ≠ 0) forces the wheel to spin faster or slower when riding on a rough road and may excite drive train vibrations. In addition, a front axle designed with dx1 > 0 will cause severe problems when crossing a bump in forward drive, because knuckle and wheel are moved toward the bump on compression. In practice, only sport cars are able to accelerate really fast. However, these vehicles have a rather low center of gravity, which reduces the pitch reaction anyway. In addition, a strong drive pitch reaction indicates the power of the engine and is therefore more welcome than a brake pitch.

8.5.6   Brake Pitch

Vanishing driving torques TD1 = 0, TD2 = 0 and braking torques of TB1 = 1575 Nm, TB2 = 563.4 Nm will decelerate the vehicle with $\ddot{x}_C/g = -1$. The results for plain

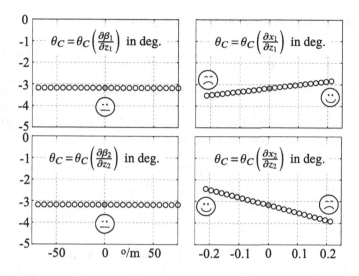

FIGURE 8.21
Influence of suspension kinematics on the chassis pitch when accelerating ($\dot{v}/g = 1$).

suspension kinematics (db1=0, dx1=0, db2=0, dx2=0), where both wheels just move straight up and down, are shown in Table 8.5. The required friction coefficients

TABLE 8.5
Vehicle with plain suspension kinematics fully braked ($\ddot{x}_C/g = -1$)

Longitudinal Tire Force	Wheel Load Static+Dynamic	Suspension Forces	Travel	Chassis Motion
$F_{x1} = -5120$ N	$F_{z1} = 5089$ N	$F_{S1} = +1538$ N	$z_1 = +70$ mm	$z_C = -10.5$ mm
$F_{x2} = -1747$ N	$F_{z2} = 1778$ N	$F_{S2} = -1538$ N	$z_2 = -54$ mm	$\theta_C = +3.17°$

$\mu_1 = |F_{x1}|/F_{z1}$ and $\mu_2 = |F_{x2}|/F_{z2}$ are again close to one, thus representing an optimal braking force distribution. The results correspond quite well with those shown in Figure 8.18 generated with the simple planar model. The simple planar model, presented in Section 8.1.1, neglects the compliance of the tires and the masses of the wheels. Thus, the parameters csf = 20000 N/m and csr = 25000 N/m represent an overall suspension stiffness that describes the effect of a suspension spring in series to the tire spring. The corresponding parameters, given in Listing 8.3, indeed result in

$$c_{sf} = \frac{c_{S1}\, c_{T1}}{c_{S1} + c_{T1}} = \frac{22000 * 220000}{22000 + 220000} = 20000 \text{ N/m} \qquad (8.91)$$

$$c_{sr} = \frac{c_{S2}\, c_{T2}}{c_{S2} + c_{T2}} = \frac{28600 * 200000}{28600 + 200000} = 25022 \text{ N/m} \qquad (8.92)$$

The masses of the chassis $m_C = 600$ kg and the wheels $m_1 = m_2 = 50$ kg amount to the overall vehicle mass of 700 kg, which represents half of a vehicle. Here, the height of the overall vehicle center above the road is given by

$$h_V = \frac{m_C h + m_1 r + m_2 r}{m_c + m_1 + m_2} = \frac{600 * 0.59 + 50 * 0.3 + 50 * 0.3}{600 + 50 + 50} = 0.55 \text{ m} \quad (8.93)$$

which matches the value used in the simple model approach. The suspension travel plotted in Figure 8.18 includes the tire deflection too. That is why the steady-state results of the dynamic simulation $z_1^{st} \approx 73$ mm and $z_2^{st} \approx -59$ mm are slightly larger than the corresponding values given in Table 8.5.

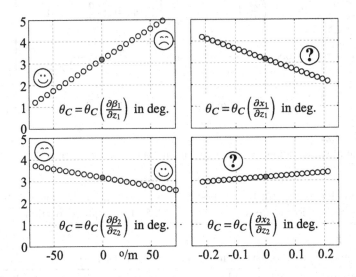

FIGURE 8.22
Influence of suspension kinematics on the chassis pitch when braking ($\dot{v}/g = -1$).

"Anti-dive" suspension kinematics, which are characterized by db1 < 0 and dx1 > 0 as well as db2 > 0 and dx2 < 0, will reduce the pitch angle θ_C significantly when the vehicle is braked, Figure 8.22. As mentioned before, a longitudinal motion of the knuckle caused by the suspension travel (dx1 ≠ 0, dx2 ≠ 0) is not really desirable. That is why most axle layouts will realize the "anti-dive" effect by just rotating the knuckles appropriately (db1 < 0, dx1 = 0 and db2 > 0, dx2 = 0).

8.5.7 Brake Pitch Pole

The pitch of the vehicle caused by braking will be felt as annoying, if too distinct. The brake pitch angle can be reduced by rotating the knuckle appropriately during suspension travel. For real suspension systems, the brake pitch pole can be calculated from the motions of the wheel contact points in the x-, z-plane, Figure 8.23. Increasing the pitch pole height above the track level means a decrease in the brake pitch angle. However, the pitch pole is not set above the height of the center of gravity in practice, because the front of the vehicle would then rise at braking.

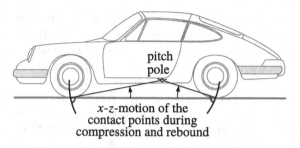

FIGURE 8.23
Brake pitch pole.

Exercises

8.1 A minibus with a wheel base of $a = a_1 + a_2 = 4$ m is characterized by the following parameters:

	unladen	laden
Axle load front	$F_{Z1} = 15$ kN	$F_{Z1} = 20$ kN
Axle load rear	$F_{Z2} = 14$ kN	$F_{Z2} = 23$ kN
COG height	$h = 1.2$ m	$h = 1.4$ m

Determine the mass of the vehicle and the horizontal position of the center of gravity (COG) defined by the parameters a_1 and a_2 for the unladen and laden vehicle.

The wheels have radius $r = 0.372$ m and the vehicle is supposed to decelerate with $\dot{v}/g = -0.6$ now. Calculate the required braking torques at the front and rear axle when an optimal distribution is taken for granted.

8.2 The weight $G = 660$ kN of a heavy-duty dumper distributes to the front and rear axle with the ratio of 1:2 on a horizontal road. The height of the center of gravity is determined by $h = 2500$ mm hereby. Each wheel has radius of $r = 938.5$ mm and the wheel base amounts to $a = 3700$ mm.

Determine the downhill and climbing capacity of the vehicle, $\alpha_D \leq \alpha \leq \alpha_C$.

The dumper moves downhill now and is supposed to decelerate with $\dot{v} = -2$ m/s^2. The inclination of the road is given by $\alpha = -17.72°$. Determine the required coefficient of friction between the tires and the road when the dumper decelerates with an optimal braking force distribution. Specify the required braking torques too.

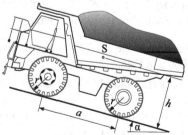

8.3 A pickup is all-wheel driven. The engine provides a maximum torque of $T_E =$ 480 Nm. The ratio of the first gear is given by $i_G = 3.5$. The center differential distributes the driving torque with a ratio of 40:60 to the front and rear axles. The differentials at the front and rear axle provide an additional gear ratio of $i_D = 3.7$ and finally distribute the torque equally to the left and right wheels. The vehicle has mass $m = 2700$ kg. The position of the center of gravity is determined by $a_1 = 1.7$ m, $a_2 = 1.5$ m, $h = 0.9$ m, and the radius of each wheel is given by $r = 0.36$ m.

Determine the maximum acceleration of the vehicle.

Calculate the hereby required friction coefficients between tire and road at the front and rear axle.

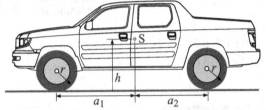

8.4 An SUV (vehicle I) of mass $m^I = 2500$ kg is towing a passenger car (vehicle II) with a mass of $m^{II} = 1800$ kg. The tow bar is connected at both vehicles in ball joints at the height of $k = 0.33\ m$. The distances

$$a_1^I = 1.25\ \text{m} \qquad a_2^I = 1.25\ \text{m} \qquad h^I = 0.70\ \text{m}$$
$$a_1^{II} = 1.40\ \text{m} \qquad a_1^{II} = 1.30\ \text{m} \qquad h^{II} = 0.55\ \text{m}$$

determine the centers of gravity at vehicle I and II.

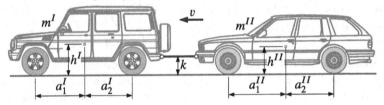

Calculate the maximum possible deceleration of the vehicle combination, when the tire/road coefficient of friction is given with $\mu = 1.2$ and vehicle II is not able to brake at all due to a defect.

Compute the force which is transmitted by the tow bar.

Determine the axle loads at both vehicles while braking with the maximum possible deceleration.

Specify the ratio F_{x2}^I/F_{x1}^I of the braking forces at vehicle I, which is required to achieve the maximum deceleration.

8.5 Extend the MATLAB-Script given in Listing 8.3 in order to produce the plots shown in Figs. 8.21 and 8.22.

Use part of the MATLAB-Script then to check the results of Exercise 8.1. Estimate suitable values for the axle masses, the suspension stiffness, and the tire stiffness. Try to reduce the brake pitch by an appropriate layout of the suspension kinematics.

Modify the MATLAB-Script to generate the results for vehicles with rear wheel and front-wheel drive.

9

Lateral Dynamics

CONTENTS

9.1 Kinematic Approach

9.1.1 Kinematic Tire Model

When a vehicle drives through a curve at low lateral acceleration, small lateral forces will be needed for course holding. Then, lateral slip hardly occurs at the wheels. In the ideal case at vanishing lateral slip, the wheels only move in a circumferential direction. The velocity component of the contact point in the lateral direction of the tire then vanishes

$$v_y = e_y^T v_{0P} = 0 \tag{9.1}$$

This constraint equation can be used as a "kinematic tire model" for course calculation of vehicles moving in the low lateral acceleration range.

9.1.2 Ackermann Geometry

Within the validity limits of the kinematic tire model, the necessary steering angle of the front wheels can be constructed via the given momentary pivot pole M, Figure 9.1. For slowly moving vehicles, the layout of the steering linkage is usually done according to Ackermann geometry. Then, the following relations apply,

$$\tan \delta_1 = \frac{a}{R} \quad \text{and} \quad \tan \delta_2 = \frac{a}{R + s} \tag{9.2}$$

where s is the track width and a denotes the wheel base. Eliminating the radius of curvature R, we get

$$\tan \delta_2 = \frac{a}{\dfrac{a}{\tan \delta_1} + s} \quad \text{or} \quad \tan \delta_2 = \frac{a \tan \delta_1}{a + s \tan \delta_1} \tag{9.3}$$

The deviations $\Delta \delta_2 = \delta_2^a - \delta_2^A$ of the actual steering angle δ_2^a from the Ackermann steering angle δ_2^A, which follows from Equation (9.3), are used, especially on commercial vehicles, to judge the quality of a steering system.

At a rotation around the momentary pivot pole M, the direction of the velocity is fixed for every point of the vehicle. The angle between the velocity vector v and

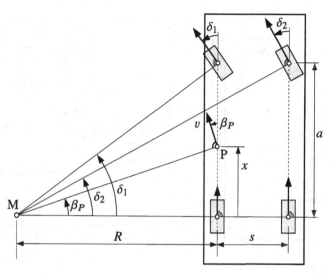

FIGURE 9.1
Ackermann steering geometry at a two-axled vehicle.

the longitudinal axis of the vehicle is called the side slip angle. The side slip angle is related to a specific point. At point P, which for the sake of simplicity is located on a line connecting the inner wheel centers, it is defined by

$$\tan \beta_P = \frac{x}{R} \quad \text{or} \quad \tan \beta_P = \frac{x}{a} \tan \delta_1 \tag{9.4}$$

where x defines the distance of point P to the inner rear wheel.

9.1.3 Space Requirement

The Ackermann approach can also be used to calculate the space requirement of a vehicle during cornering, left graph in Figure 9.2. If the front wheels of a two-axled vehicle are steered according to the Ackermann geometry, the outer point of the vehicle front will run on the maximum radius R_{max}, whereas a point on the inner side of the vehicle at the location of the rear axle will run on the minimum radius R_{min}. Within this simple approach, the outer contour of a vehicle is just approximated by a rectangular box. Then it holds,

$$R_{max}^2 = (R_{min} + b)^2 + (a + f)^2 \tag{9.5}$$

where a, b are the wheel base and the width of the vehicle, and f specifies the distance from the front of the vehicle to the front axle. Then, the space requirement $\Delta R = R_{max} - R_{min}$ can be specified as a function of the cornering radius R_{min} for a given vehicle dimension,

$$\Delta R = R_{max} - R_{min} = \sqrt{(R_{min} + b)^2 + (a + f)^2} - R_{min} \tag{9.6}$$

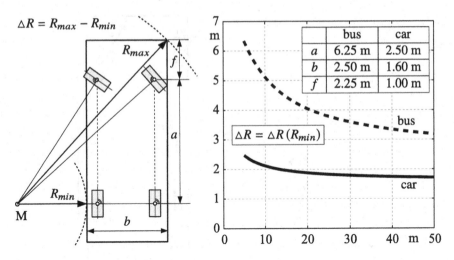

	bus	car
a	6.25 m	2.50 m
b	2.50 m	1.60 m
f	2.25 m	1.00 m

FIGURE 9.2
Definition of space requirement and typical example.

By combining Equation (9.5) with Equation (9.6), the space requirement of a typical passenger car and a bus is plotted on the right of Figure 9.2 versus the minimum cornering radius. In narrow curves, $R_{min} = 5.0$ m, a bus requires a space of 2.5 times the width, whereas a passenger car only needs 1.5 times the width.

9.1.4   Vehicle Model with Trailer

9.1.4.1   Kinematics

A simple model for a passenger car and a trailer is shown in Figure 9.3. Vehicle and trailer move on a horizontal track. The parameters a, b, and c describe the wheel base, the distance of the rear axle to the coupling point, and the distance from the coupling point to the axle of the trailer. The wheels at each axle are substituted by fictitious center wheels whose longitudinal direction is characterized by the axes x_1, x_2, and x_3. The rear axle of the vehicle and the trailer axle are not steered. Then, the position and the orientation of the vehicle with respect to the earth-fixed axis system x_0, y_0, z_0 is defined by the position vector from the origin 0 to the center of the rear axle

$$r_{02,0} = \begin{bmatrix} x \\ y \\ r_T \end{bmatrix} \tag{9.7}$$

and the rotation matrix

$$A_{02} = \begin{bmatrix} \cos\psi & -\sin\psi & 0 \\ \sin\psi & \cos\psi & 0 \\ 0 & 0 & 1 \end{bmatrix} \tag{9.8}$$

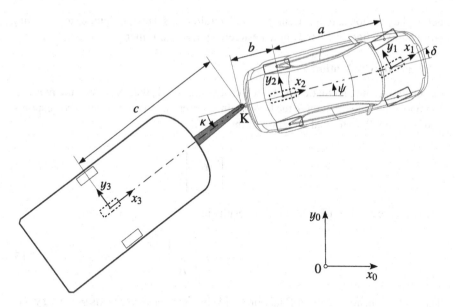

FIGURE 9.3
Vehicle and trailer with kinematic tire model.

Here, the tire radius r_T is considered constant, and x, y as well as the yaw angle ψ are used as generalized coordinates. The position vector

$$r_{01,0} = r_{02,0} + A_{02}\, r_{21,2} \quad \text{with} \quad r_{21,2} = \begin{bmatrix} a \\ 0 \\ 0 \end{bmatrix} \tag{9.9}$$

and the rotation matrix

$$A_{01} = A_{02}\, A_{21} \quad \text{with} \quad A_{21} = \begin{bmatrix} \cos\delta & -\sin\delta & 0 \\ \sin\delta & \cos\delta & 0 \\ 0 & 0 & 1 \end{bmatrix} \tag{9.10}$$

describe the position and the orientation of the front axle, where δ denotes the steering angle of the fictitious center wheel. Finally, the position vector

$$r_{03,0} = r_{02,0} + A_{02}\left(r_{2K,2} + A_{23}\, r_{K3,3} \right) \tag{9.11}$$

with

$$r_{2K,2} = \begin{bmatrix} -b \\ 0 \\ 0 \end{bmatrix} \quad \text{and} \quad r_{K3,2} = \begin{bmatrix} -c \\ 0 \\ 0 \end{bmatrix} \tag{9.12}$$

and the rotation matrix

$$A_{03} = A_{02}\, A_{23} \quad \text{with} \quad A_{23} = \begin{bmatrix} \cos\kappa & -\sin\kappa & 0 \\ \sin\kappa & \cos\kappa & 0 \\ 0 & 0 & 1 \end{bmatrix} \tag{9.13}$$

define the position and the orientation of the trailer axis. Here, K denotes the coupling point and κ defines the bend angle between vehicle and trailer.

9.1.4.2 Vehicle Motion

According to the kinematic tire model, cf. Section 9.1.1, the velocity at the rear axle can only have a component in the longitudinal direction of the tire, which corresponds here with the longitudinal direction of the vehicle,

$$v_{02,2} = \begin{bmatrix} v_{x2} \\ 0 \\ 0 \end{bmatrix} \tag{9.14}$$

The time derivative of Equation (9.7) results in

$$v_{02,0} = \dot{r}_{02,0} = \begin{bmatrix} \dot{x} \\ \dot{y} \\ 0 \end{bmatrix} \tag{9.15}$$

The vector transformation of Equation (9.14) into the earth-fixed axis system 0 yields

$$v_{02,0} = A_{02}\, v_{02,2} = A_{02} \begin{bmatrix} v_{x2} \\ 0 \\ 0 \end{bmatrix} = \begin{bmatrix} \cos\psi\, v_{x2} \\ \sin\psi\, v_{x2} \\ 0 \end{bmatrix} \tag{9.16}$$

Equating it with Equation (9.15) results in two first-order differential equations for the position coordinates x and y,

$$\dot{x} = v_{x2} \cos\psi \tag{9.17}$$

$$\dot{y} = v_{x2} \sin\psi \tag{9.18}$$

The time derivative of Equation (9.9) delivers the velocity at the front axle,

$$v_{01,0} = \dot{r}_{01,0} = \dot{r}_{02,0} + \omega_{02,0} \times A_{02}\, r_{21,2} \tag{9.19}$$

The transformation into the vehicle-fixed axis system x_2, y_2, z_2 results in

$$v_{01,2} = \underbrace{\begin{bmatrix} v_{x2} \\ 0 \\ 0 \end{bmatrix}}_{v_{02,2}} + \underbrace{\begin{bmatrix} 0 \\ 0 \\ \dot{\psi} \end{bmatrix}}_{\omega_{02,2}} \times \underbrace{\begin{bmatrix} a \\ 0 \\ 0 \end{bmatrix}}_{r_{21,2}} = \begin{bmatrix} v_{x2} \\ a\dot{\psi} \\ 0 \end{bmatrix} \tag{9.20}$$

The unit vectors

$$e_{x1,2} = \begin{bmatrix} \cos\delta \\ \sin\delta \\ 0 \end{bmatrix} \quad \text{and} \quad e_{y1,2} = \begin{bmatrix} -\sin\delta \\ \cos\delta \\ 0 \end{bmatrix} \tag{9.21}$$

define the longitudinal and lateral direction at the front center wheel. According to Equation (9.1), the velocity component lateral to the wheel must vanish,

$$e_{y1,2}^T v_{01,2} = -\sin\delta\, v_{x2} + \cos\delta\, a\,\dot\psi = 0 \qquad (9.22)$$

whereas in longitudinal direction the velocity

$$e_{x1,2}^T v_{01,2} = \cos\delta\, v_{x2} + \sin\delta\, a\,\dot\psi = v_{x1} \qquad (9.23)$$

will remain. Rearranging Equation (9.22) results in a first-order differential equation for the yaw angle,

$$\dot\psi = \frac{v_{x2}}{a}\tan\delta \qquad (9.24)$$

The momentary position $x = x(t)$, $y = y(t)$ and the orientation $\psi = \psi(t)$ of the vehicle are defined by three differential equations (9.17), (9.18), and (9.24), which are driven by the vehicle velocity v_{x2} and the steering angle δ.

9.1.4.3 Entering a Curve

In analogy with Equation (9.2), the steering angle δ can be related to the current track radius R or with $\varrho = 1/R$ to the current track curvature

$$\tan\delta = \frac{a}{R} = a\frac{1}{R} = a\varrho \qquad (9.25)$$

Then, the differential equation for the yaw angle reads as

$$\dot\psi = v_{x2}\,\varrho\,. \qquad (9.26)$$

With the curvature gradient

$$\varrho = \varrho(t) = \varrho_C\frac{t}{T} \qquad (9.27)$$

the entering of a curve is described as a continuous transition from a straight line with the curvature $\varrho = 0$ into a circle with the curvature $\varrho = \varrho_C$. Now the yaw angle of the vehicle can be calculated by simple integration,

$$\psi(t) = \frac{v_{x2}\,\varrho_C}{T}\frac{t^2}{2} \qquad (9.28)$$

where at time $t = 0$ a vanishing yaw angle, $\psi(t=0) = 0$, has been assumed. Then the position of the vehicle follows with Equation (9.28) from the differential equations defined by Equations (9.17) and (9.18)

$$x = v_{x2}\int_{t=0}^{t=T}\cos\left(\frac{v_{x2}\,\varrho_C}{T}\frac{t^2}{2}\right)dt \quad\text{and}\quad y = v_{x2}\int_{t=0}^{t=T}\sin\left(\frac{v_{x2}\,\varrho_C}{T}\frac{t^2}{2}\right)dt \qquad (9.29)$$

At constant vehicle speed, $v_{x2} = const.$, Equation (9.29) is the parameterized form of a clothoide. From Equation (9.25) the necessary steering angle can be calculated too.

If only small steering angles are necessary for driving through the curve, the tangent function can be approximated by its argument. Then, the continuous steer motion

$$\delta = \delta(t) \approx a\varrho = a\varrho c \frac{t}{T} \tag{9.30}$$

will drive the vehicle along a clothoide-like curve.

9.1.4.4   Trailer Motions

The time derivative of the position vector, defined in Equation (9.11), delivers the velocity of the trailer axis as

$$v_{03,0} = \dot{r}_{03,0} = \dot{r}_{02,0} + \omega_{02,0} \times A_{02} \, r_{23,2} + A_{02} \, \dot{r}_{23,2} \tag{9.31}$$

The velocity $\dot{r}_{02,0} = v_{02,0}$ and the angular velocity $\omega_{02,0}$ of the vehicle are defined in Equations (9.16) and (9.20). The vector from the rear axle to the axle of the trailer is given by

$$r_{23,2} = r_{2K,2} + A_{23} \, r_{K3,3} = \begin{bmatrix} -b - c \cos \kappa \\ -c \sin \kappa \\ 0 \end{bmatrix} \tag{9.32}$$

where $r_{2K,2}$ and $r_{K3,3}$ are defined in Equation (9.12). The time derivative of Equation (9.32) results in

$$\dot{r}_{23,2} = \underbrace{\begin{bmatrix} 0 \\ 0 \\ \dot{\kappa} \end{bmatrix}}_{\omega_{23,2}} \times \underbrace{\begin{bmatrix} -c \cos \kappa \\ -c \sin \kappa \\ 0 \end{bmatrix}}_{A_{23} \, r_{K3,3}} = \begin{bmatrix} c \sin \kappa \, \dot{\kappa} \\ -c \cos \kappa \, \dot{\kappa} \\ 0 \end{bmatrix} \tag{9.33}$$

The velocity of the trailer axle, defined in Equation (9.31), is transformed into the vehicle-fixed axis system x_2, y_2, z_2 now

$$v_{03,2} = \overbrace{\begin{bmatrix} v_{x2} \\ 0 \\ 0 \end{bmatrix}}^{v_{02,2}} + \overbrace{\begin{bmatrix} 0 \\ 0 \\ \dot{\psi} \end{bmatrix}}^{\omega_{02,2}} \times \overbrace{\begin{bmatrix} -b - c \cos \kappa \\ -c \sin \kappa \\ 0 \end{bmatrix}}^{r_{23,2}} + \overbrace{\begin{bmatrix} c \sin \kappa \, \dot{\kappa} \\ -c \cos \kappa \, \dot{\kappa} \\ 0 \end{bmatrix}}^{\dot{r}_{23,2}}$$

$$= \begin{bmatrix} v_{x2} + c \sin \kappa \, (\dot{\kappa} + \dot{\psi}) \\ -b\dot{\psi} - c \cos \kappa \, (\dot{\kappa} + \dot{\psi}) \\ 0 \end{bmatrix} \tag{9.34}$$

The longitudinal and lateral direction at the trailer axle are defined by the unit vectors

$$e_{x3,2} = \begin{bmatrix} \cos \kappa \\ \sin \kappa \\ 0 \end{bmatrix} \quad \text{and} \quad e_{y3,2} = \begin{bmatrix} -\sin \kappa \\ \cos \kappa \\ 0 \end{bmatrix} \tag{9.35}$$

At the trailer axle, the lateral velocity must vanish too

$$e^T_{y3,2}v_{03,2} = -\sin\kappa\left(v_{x2}+c\,\sin\kappa\,(\dot\kappa+\dot\psi)\right) + \cos\kappa\left(-b\dot\psi-c\,\cos\kappa\,(\dot\kappa+\dot\psi)\right) = 0 \quad (9.36)$$

whereas in longitudinal direction the velocity

$$e^T_{x3,2}v_{03,2} = \cos\kappa\left(v_{x2}+c\,\sin\kappa\,(\dot\kappa+\dot\psi)\right) + \sin\kappa\left(-b\dot\psi-c\,\cos\kappa\,(\dot\kappa+\dot\psi)\right) = v_{x3} \quad (9.37)$$

remains. If Equation (9.24) is inserted into Equation (9.36) now, one will get a first-order differential equation for the bend angle,

$$\dot\kappa = -\frac{v_{x2}}{a}\left(\frac{a}{c}\sin\kappa + \left(\frac{b}{c}\cos\kappa + 1\right)\tan\delta\right) \quad (9.38)$$

The differential equations provided by Equations (9.17), (9.18), and (9.24) describe the position and the orientation of the vehicle within the x_0, y_0 plane. The differential equation (9.38), which characterizes the motion of the trailer relative to the vehicle, depends nonlinearly on the bend angle κ and becomes unstable for small bend angles when vehicle and trailer are driven backward. In a steady state, the bend angle κ^{st} is defined by

$$0 = -\frac{v_{x2}}{a}\left(\frac{a}{c}\sin\kappa^{st} + \left(\frac{b}{c}\cos\kappa^{st} + 1\right)\tan\delta\right) \quad (9.39)$$

which represents a trigonometric equation of type (5.54) and according to Equation (5.56) is solved by

$$\kappa^{st} = \arcsin\frac{-c\,\tan\delta}{\sqrt{(b\,\tan\delta)^2 + a^2}} - \arctan\frac{b\,\tan\delta}{a} \quad (9.40)$$

9.1.4.5 Course Calculations

The function given in Listing 9.1 provides the state equation of the vehicle and trailer track model.

Listing 9.1
Function fun_09_track_model: vehicle and trailer track model dynamics

```
1 function xdot = fun_09_track_model( t, x, p )
2
3 psi=x(3); ka=x(4);              % get yaw and bend angle
4 d = interp1(p.t_d,p.d_i,t);    % steering angle
5
6 % vehicle and trailer dynamics
7 x2dot = p.v*cos(psi);
8 y2dot = p.v*sin(psi);
9 gadot = p.v/p.a*tan(d);
10 kadot =-p.v/p.a*(p.a/p.c*sin(ka)+(p.b/p.c*cos(ka)+1)*tan(d));
11
12 xdot = [ x2dot; y2dot; gadot; kadot ]; % state derivatives
13
14 end
```

The MATLAB-Function `interp1` uses a lookup table to compute the actual steering angle $\delta = \delta(t)$ via a linear interpolation. The velocity of the vehicle v provided in the structure element `p.v` is kept constant here. The MATLAB-Script in Listing 9.2 defines the parameter of the vehicle and trailer track model, performs a simulation, and plots some results. The parameter[1] corresponds with a typical passenger car and a camping trailer.

Listing 9.2

Script `app_09_track_model.m`: vehicle and trailer track model

```
1  p.a = 2.6;    % wheel base in m
2  p.b = 1.0;    % distance rear axle --> coupling point in m
3  p.c = 2.0;    % coupling point --> trailer axle in m
4  p.v = 5./3.6; % driving velocity, km/h --> m/s
5
6  % define parts of lookup table for steer input [time, steering angle]
7  p.t_d = [  0   20   30 ] ;        % in s
8  p.d_i = [  0   45   45 ]*pi/180 ; % deg --> rad
9
10 % get time interval
11 t0 = p.t_d(1);   te = p.t_d(length(p.t_d)) ;
12
13 % initial conditions (vehicle position and orientation)
14 x20 = 0;   y20 = 0;   ga0 = 0/180*pi ;  % yaw angle [deg --> rad]
15 % bending angle adjusted to steering angle (steady state solution)
16 d0  = interp1 ( p.t_d, p.d_i, t0 );
17 ka0 = asin(-p.c*tan(d0)/sqrt((p.b*tan(d0))^2+p.a^2)) ...
18       - atan2(p.b*tan(d0),p.a) ;
19 x0 = [ x20; y20;  ga0;   ka0 ];
20
21 % time simulation
22 [t,xout] = ode23(@(t,x) fun_09_track_model(t,x,p), [t0,te], x0 );
23
24 % post processing
25 d = interp1(p.t_d,p.d_i,t); % time history of steering angle
26 % steady-state bend angle
27 kas= asin(-p.c*tan(d)./sqrt((p.b*tan(d)).^2+p.a^2)) ...
28       -atan2(p.b*tan(d),p.a);
29 % axle positions
30 x1 = xout(:,1)+p.a*cos(xout(:,3));
31 y1 = xout(:,2)+p.a*sin(xout(:,3));
32 x2 = xout(:,1);
33 y2 = xout(:,2);
34 x3 = xout(:,1)-p.b*cos(xout(:,3)) - p.c*cos(xout(:,3)+xout(:,4));
35 y3 = xout(:,2)-p.b*sin(xout(:,3)) - p.c*sin(xout(:,3)+xout(:,4));
36
37 % plot some results
38 subplot(2,2,1)
39 plot(t,d*180/pi), grid on, title('\delta(t) in deg')
40 subplot(2,2,2)
41 plot(t,p.v^2/p.a*tan(d)/9.81), hold on, grid on, axis equal
42 title('a_y(t)/g')
43 subplot(2,2,3)
44 plot(x1,y1,'k'), hold on, plot(x2,y2,'--b'), plot(x3,y3,'r'), grid on
45 title('path of axles'), xlabel('m'), ylabel('m'), legend('1','2','3')
46 subplot(2,2,4)
47 plot(t,[xout(:,4),kas]*180/pi), grid on
48 title('\kappa(t) in deg'), legend('dyn','st')
```

---

[1]Note: Negative values for c will move the coupling point in front of the rear axle of the vehicle, which will match with the layout of a tractor semitrailer combination.

The value pairs defined by p.t_d and p.d_i describe a slowly increasing steering angle, which is kept constant at $t \geq 20\,s$ in this particular case. The resulting paths of the axle centers, plotted in Figure 9.4, show clothoide-like curves that end up in circles when the steering angle is finally kept constant. The dynamics of the trailer motion

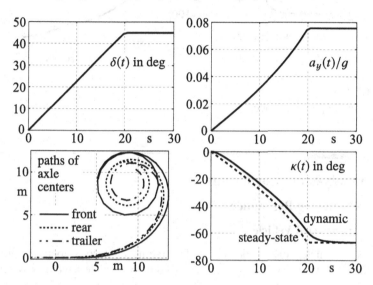

FIGURE 9.4
Entering a curve.

is visible in the lower right plot, where the time history of the dynamic bend angle κ is compared to its steady-state value provided in Equation (9.40). The kinematic tire model, on which this track model is based, is valid in the low range of lateral vehicle accelerations only. Here, the lateral acceleration on the rear axle center is simply given by $a_{y2} = v^2/R$, where the radius of curvature is provided as $R = a/\tan(\delta)$ via the Ackermann geometry.

Setting the driving vehicle velocity via p.v = -2.0/3.6; to $v = -0.5556$ m/s and replacing the code lines 7 and 8 in Listing 9.2 by

```
t_d = [ 0   3   9  12   15 ] ;        % in s
d_i = [ 0 -45  45   0    0 ]*pi/180 ; % deg --> rad
```

will provide the input for backing into a parking space. Some simulation results are plotted in Figure 9.5. The paths of the front and rear axle are as expected. The rear axle just moves on an S-shaped curve and ends up with a horizontal displacement of approximately 1.8 m. The movement of the front axle is more complicated but arrives at same lateral shift. However, the path of the trailer axle is quite different. When driving backward the differential equation (9.38) becomes unstable. As a result, the bend angle increases rapidly, which would cause severe damage to the vehicle and the trailer. The path of the trailer axle shows a bifurcation point. After this point, the trailer moves forward while the vehicle is still moving backward. Now, the trailer motion is stable and will tend toward $-180°$ finally.

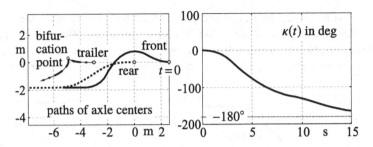

FIGURE 9.5
Backing into a parking space.

9.2 Steady-State Cornering

9.2.1 Cornering Resistance

9.2.1.1 Two-Axled Vehicle

A two-axled vehicle is driving on a circle of the radius R with the velocity v, Figure 9.6. Then, the velocity state of the vehicle expressed in the vehicle-fixed axis system V is

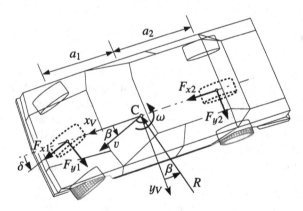

FIGURE 9.6
Two-axled vehicle in steady-state cornering.

defined by

$$v_{0C,V} = \begin{bmatrix} v\cos\beta \\ v\sin\beta \\ 0 \end{bmatrix} \quad \text{and} \quad \omega_{0F,V} = \begin{bmatrix} 0 \\ 0 \\ \omega \end{bmatrix} \tag{9.41}$$

where β denotes the side slip angle of the vehicle measured at the vehicle center of gravity. In steady-state cornering, the angular velocity ω is related via

$$\omega = \frac{v}{R} = \varrho\, v \tag{9.42}$$

to the vehicle velocity v and the circle radius R or the curvature ϱ, respectively.

As the vehicle is supposed to drive with constant speed ($v = const.$), the linear momentum applied in the direction of the x_V- and y_V-axis and the angular momentum about the vertical axis result in

$$m\left(-\varrho\, v^2\, \sin\beta\right) = F_{x1}\cos\delta - F_{y1}\sin\delta + F_{x2} \tag{9.43}$$

$$m\left(\varrho\, v^2\, \cos\beta\right) = F_{x1}\sin\delta + F_{y1}\cos\delta + F_{y2} \tag{9.44}$$

$$0 = a_1\left(F_{x1}\sin\delta + F_{y1}\cos\delta\right) - a_2\,F_{y2} \tag{9.45}$$

where m denotes the mass of the vehicle; F_{x1}, F_{x2}, F_{y1}, F_{y2} are the resulting forces in the longitudinal and vertical direction applied at the front and rear axle; and δ specifies the average steering angle at the front axle.

The engine torque is distributed by the center differential to the front and rear axle in standard drive trains. Then, under steady-state conditions

$$F_{x1} = \eta\, F_D \quad \text{and} \quad F_{x2} = (1-\eta)\, F_D \tag{9.46}$$

will hold, where F_D is the driving force. The dimensionless parameter η makes it possible to model different layouts of drive trains, Table 9.1. Inserting Equation (9.46)

TABLE 9.1
Different driving force distribution

$\eta = 0$	Rear-wheel drive	$F_{x1} = 0, \quad F_{x2} = F_D$
$0 < \eta < 1$	All-wheel drive	$\dfrac{F_{x1}}{F_{x2}} = \dfrac{\eta}{1-\eta}$
$\eta = 1$	Front-wheel drive	$F_{x1} = F_D, \quad F_{x2} = 0$

into Equation (9.43), one gets

$$\begin{aligned}
(\eta\cos\delta + (1-\eta))\, F_D - \sin\delta\, F_{y1} &= -\varrho\, m\, v^2\, \sin\beta \\
\eta\sin\delta\, F_D + \cos\delta\, F_{y1} + F_{y2} &= \varrho\, m\, v^2\, \cos\beta \\
a_1\eta\sin\delta\, F_D + a_1\cos\delta\, F_{y1} - a_2\, F_{y2} &= 0
\end{aligned} \tag{9.47}$$

These equations can be resolved for the driving force,

$$F_D = \frac{\dfrac{a_2}{a_1+a_2}\cos\beta\sin\delta - \sin\beta\cos\delta}{\eta + (1-\eta)\cos\delta}\, \varrho\, m\, v^2 \tag{9.48}$$

The driving force will vanish if

$$\frac{a_2}{a_1+a_2}\cos\beta\sin\delta = \sin\beta\cos\delta \quad \text{or} \quad \frac{a_2}{a_1+a_2}\tan\delta = \tan\beta \tag{9.49}$$

holds. As a matter of fact, this fully corresponds with the Ackermann geometry. However, the Ackermann geometry applies for small lateral accelerations only. In real driving situations, the side slip angle β of a vehicle at the center of gravity will always be smaller than the Ackermann side slip angle β_A. Then, due to $\tan\beta < \tan\beta_A = a_2/(a_1 + a_2)\tan\delta$, a driving force $F_D > 0$ will be needed to overcome the "cornering resistance" of the vehicle.

In [54] the vehicle model is extended to four wheels. But then, the slip angle, which occurs at each of the wheels when cornering, must be taken into account, in addition. It is shown there, that the cornering resistance of a vehicle can be reduced slightly further by completely shifting the driving torque to the outer wheels.

9.2.1.2 Four-Axled Vehicle

Most heavy-duty trucks have more than two axles. The special-purpose truck shown in Figure 9.7 is equipped with two steerable, single-tired axles in the front and two driven axles with twin-tires at the rear. Again, the wheels on each axle are summarized in a

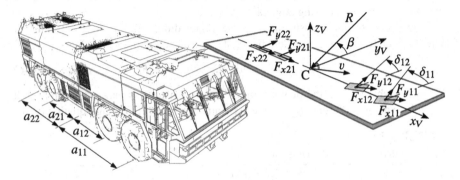

FIGURE 9.7
Four-axled vehicle.

fictitious center wheel and the vehicle is supposed to perform a steady-state cornering. As an extension to Equation (9.43), the linear and angular momentum will now read

$$-\varrho m v^2 \sin\beta = \begin{aligned}&F_{x11}\cos\delta_{11} - F_{y11}\sin\delta_{11} + F_{x12} + \\ &F_{x12}\cos\delta_{12} - F_{y12}\sin\delta_{12} + F_{x22}\end{aligned} \qquad (9.50)$$

$$\varrho m v^2 \cos\beta = \begin{aligned}&F_{x11}\sin\delta_{11} + F_{y11}\cos\delta_{11} + F_{y12} + \\ &F_{x12}\sin\delta_{12} + F_{y12}\cos\delta_{12} + F_{y22}\end{aligned} \qquad (9.51)$$

$$0 = \begin{aligned}&a_{11}(F_{x11}\sin\delta_{11} + F_{y11}\cos\delta_{11}) - a_{21}F_{y12} + \\ &a_{12}(F_{x12}\sin\delta_{12} + F_{y12}\cos\delta_{12}) - a_{22}F_{y22}\end{aligned} \qquad (9.52)$$

where δ_{11} and δ_{12} denote the average steering angles at the front axles. The parameter a_{11}, a_{12}, a_{21}, and a_{22} define the location of the axles with respect to the vehicle center. All axles may be driven in general. Assuming that the driving torque is distributed differently to the front and rear but equally to each front and rear axle, then the driving forces may be modeled by

$$F_{x11} = F_{x12} = F_{x1} = \eta\, F_D \quad \text{and} \quad F_{x21} = F_{x22} = F_{x2} = (1 - \eta)\, F_D \qquad (9.53)$$

where F_D denotes the overall driving force and the parameter η makes it possible to describe different driving scenarios according to Table 9.1. The lateral forces are modeled as linear functions of the corresponding lateral slips in a first approximation

$$F_{y11} = k_{SF} s_{y11}, \; F_{y12} = k_{SF} s_{y12}, \quad F_{y21} = k_{SR} s_{y21}, \; F_{y22} = k_{SR} s_{y22} \qquad (9.54)$$

where k_{SF} and k_{SR} summarize the cornering stiffness of all tires at each front and rear axle. In this simple approach, it is assumed that both front and rear axles are loaded equally and are equipped with the same tires. The vehicle is neither braked nor accelerated. Then, all wheels are close to a rolling situation, which will be characterized by nearly vanishing longitudinal slips. Then, $r_D \Omega \approx v_x$ will hold, and the lateral slips defined in Equation (3.72) are given by

$$s_{y11} = -\frac{-\sin\delta_{11}\, v\cos\beta + \cos\delta_{11}\, (v\sin\beta + a_{11}\omega)}{|\cos\delta_{11}\, v\cos\beta + \sin\delta_{11}\, (v\sin\beta + a_{11}\omega)|} \qquad (9.55)$$

$$s_{y12} = -\frac{-\sin\delta_{12}\, v\cos\beta + \cos\delta_{12}\, (v\sin\beta + a_{12}\omega)}{|\cos\delta_{12}\, v\cos\beta + \sin\delta_{12}\, (v\sin\beta + a_{12}\omega)|} \qquad (9.56)$$

$$s_{y21} = -\frac{v\sin\beta - a_{21}\omega}{|v\cos\beta|} \quad \text{and} \quad s_{y22} = -\frac{v\sin\beta - a_{22}\omega}{|v\cos\beta|} \qquad (9.57)$$

In steady-state cornering, the angular velocity ω is related to the vehicle velocity v and the curvature $\varrho = 1/R$ according to Equation (9.42).

For a given driving velocity v and various steering angles δ_{11}, δ_{12}, Equations (9.50) to (9.52) in combination with Equations (9.53) to (9.57) and Equation (9.42) represent a set of nonlinear equations for the side slip angle β, the curvature ϱ, and the driving force F_D or the driving resistance $f_D = F_D/(mg)$, respectively.

A standard MATLAB installation provides the function fzero to solve a nonlinear equation of type $f(x) = 0$ but the algorithm is restricted to one single equation. If the MATLAB Optimization Toolbox is available, the function fsolve will be the proper choice for solving a system of nonlinear equations. Here, the system of nonlinear equations, arranged in the form of a vector equation $f(x) = 0$, will be solved by minimizing the overall quadratic error of the vector function

$$f(x) = 0 \quad \Longrightarrow \quad \epsilon^2 = f(x)^T f(x) \to \text{Minimum} \qquad (9.58)$$

Then, the MATLAB standard function fminsearch can be used to solve the problem.

The function given in Listing 9.3 provides the overall quadratic error $\epsilon^2 = f(x)^T f(x)$ that characterizes the steady-state cornering of a four-axled truck. The vector function $f(x)$ incorporates here the force balances into the longitudinal and lateral directions as well as the torque balance about the vertical axis.

Listing 9.3

Function `fun_09_truck4a`: four-axled truck in steady-state cornering

```
1 function epsq = fun_09_truck4a(x,p)
2
3 rho = x(1); % curvature
4 be  = x(2); % side slip angle at cog
5 fd  = x(3); % normalized cornering resistance
6
7 % shortcuts
8 om = p.v*rho;        % angular vehicle velocity
9 FD = p.m*p.grav*fd; % driving force
10 vx = p.v*cos(be);   % velocity in longitudinal direction of vehicle
11 sd11=sin(p.d11); cd11=cos(p.d11); sd12=sin(p.d12); cd12=cos(p.d12);
12
13 % longitudinal forces
14 Fx11 = 0.5*p.eta*FD; Fx12=Fx11; Fx21 = 0.5*(1-p.eta)*FD; Fx22=Fx21;
15
16 % lateral slips (sx=0 assumed)
17 vy11 = p.v*sin(be)+p.a11*om;
18 sy11 = -(-sd11*vx+cd11*vy11)/(cd11*vx+sd11*vy11);
19 vy12 = p.v*sin(be)+p.a12*om;
20 sy12 = -(-sd12*vx+cd12*vy12)/(cd12*vx+sd12*vy12);
21 vy21 = p.v*sin(be)-p.a21*om;  sy21 = - vy21/vx ;
22 vy22 = p.v*sin(be)-p.a22*om;  sy22 = - vy22/vx ;
23
24 % lateral forces (linearized)
25 Fy11 = p.ksf*sy11; Fy12 = p.ksf*sy12;
26 Fy21 = p.ksr*sy21; Fy22 = p.ksr*sy22;
27
28 % force and torque balances
29 Fx = Fx11*cd11-Fy11*sd11 + Fx21 ...
30     + Fx12*cd12-Fy12*sd12 + Fx22  + p.m*p.v^2*rho*sin(be);
31 Fy = Fx11*sd11+Fy11*cd11 + Fy21 ...
32     + Fx12*sd12+Fy12*cd12 + Fy22  - p.m*p.v^2*rho*cos(be);
33 Tz = p.a11*(Fx11*sd11+Fy11*cd11) - p.a21*Fy21 ...
34     + p.a12*(Fx12*sd12+Fy12*cd12) - p.a22*Fy22;
35
36 % overall quadratic error
37 epsq = Fx^2 + Fy^2 + Tz^2;
38
39 end
```

The MATLAB-Script in Listing 9.4 provides the vehicle parameter, computes the cornering resistance for a variety of steering angles, and plots the results as a three-dimensional contour plot by using the MATLAB function `contour3`.

Listing 9.4

Script `app_09_truck4a.m`: cornering resistance of a four-axled truck

```
1 p.grav = 9.81;     % constant of gravity in m/s^2
2 p.m = 32000;       % mass of vehicle in kg
3 p.eta = 0.0;       % drive torque distribution (rear wheel drive)
4 p.a11 = 3.3750;    % cog --> midth of first front axle in m
5 p.a12 = 1.6250;    % cog --> midth of second front axle in m
6 p.a21 = 1.2500;    % cog --> midth of first rear axle in m
7 p.a22 = 2.7500;    % cog --> midth of second rear axle in m
8 p.ksf = 2*300000;  % front axle cornering stiffness in N/-
9 p.ksr = 4*280000;  % rear axle cornering stiffness in N/-
10 p.v = 5/3.6;       % vehicle velocity (low speed), km/h --> m/s
11
12 d11var=linspace(5,55,11)/180*pi; % range of steering angle 11
13 d12var=linspace(5,45, 9)/180*pi; % range of steering angle 12
14
```

```
15 % pre-allocate cornering resistance to speed up loops
16 n11=length(d11var); n12=length(d12var); crvar=zeros(n11,n12);
17
18 % solve nonlinear equations (trivial initial values)
19 for i=1:n11
20   p.d11=d11var(i);
21   for j=1:n12
22     p.d12=d12var(j);
23     x = fminsearch(@(x) fun_09_truck4a(x,p),[0; 0; 0]);
24     rho=x(1);   be=x(2);   fd=x(3);   crvar(i,j)=fd;
25   end
26 end
27
28 % plot cornering resistance as a function of d11 and d12
29 contour3(d12var*180/pi,d11var*180/pi,crvar,100), colormap('gray')
30 title('Normalized cornering resistance'), xlabel('d12'),ylabel('d11')
31
32 % compute and plot cornering resistance for the Ackermann geometry
33 d12a=atan2(((p.a21+p.a22)/2+p.a12)*tan(d11var),(p.a21+p.a22)/2+p.a11);
34 fd=zeros(size(d12a));
35 for i=1:length(d12a)
36   p.d11=d11var(i); p.d12=d12a(i);
37   x = fminsearch(@(x) fun_09_truck4a(x,p),[0; 0; 0]); fd(i)=x(3);
38 end
39 hold on, plot3(d12a*180/pi,d11var*180/pi,fd,'b','LineWidth',2)
```

The MATLAB script also computes and plots the cornering resistance for steering angles that are based on Ackermann geometry, Figure 9.8. A fictitious pivot point

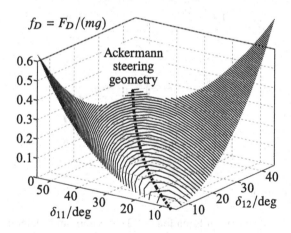

FIGURE 9.8
Normalized cornering resistance of a four-axled truck for different steering angles.

located at a line right between the two rear axles is assumed hereby. Similar to Equation (9.2), the steering angles at the front axles will then be given by

$$\tan \delta_{11} = \frac{\frac{1}{2}(a_{21}+a_{22}) + a_{11}}{R} \quad \text{and} \quad \tan \delta_{12} = \frac{\frac{1}{2}(a_{21}+a_{22}) + a_{12}}{R} \qquad (9.59)$$

Eliminating the curve radius R, one gets

$$\tan \delta_{12} = \frac{\frac{1}{2}(a_{21}+a_{22}) + a_{12}}{\frac{1}{2}(a_{21}+a_{22}) + a_{11}} \tan \delta_{11} \tag{9.60}$$

The broken line in Figure 9.8 represents the normalized cornering resistance of the four-axled truck, computed for Ackermann steering angles defined by Equation (9.60). Although the four-axled vehicle has no geometrically defined turning point, the Ackermann approach results even here in a driving performance that is characterized by a minimized or at least nearly minimized cornering resistance. That is why, most commercial trucks or special-purpose vehicle are equipped with steering systems that are based on the Ackermann geometry.

9.2.2 Overturning Limit

9.2.2.1 Static Stability Factor

If a vehicle of mass m is cornering with the lateral acceleration a_y, then the centrifugal force $m\,a_y$ will be generated, Figure 9.9. Similar to the maximum longitudinal

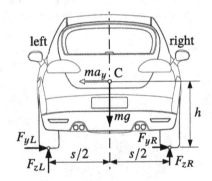

FIGURE 9.9
Vehicle in a right turn.

acceleration a_x^{max} provided in Equation (8.22), the maximum lateral acceleration is determined by the coefficient of friction

$$a_y^{max} = \mu g \tag{9.61}$$

However, the risk of rollover may reduce this limit further on. If the vehicle turns to the right, which is assumed in Figure 9.9, then a beginning rollover will be indicated by the liftoff of the tires on the right. For vanishing tire forces at the right $F_{yR} = 0$ and $F_{zR} = 0$, the torque balance about a line that is defined by the left contact points simply delivers

$$m g \frac{s}{2} - m a_y^T h = 0 \quad \text{or} \quad \frac{a_y^T}{g} = \frac{s/2}{h} \tag{9.62}$$

where h is the height of the center of gravity and $s/2$ denotes half of the track width. The vehicle will overturn rather than slide if

$$a_y^T < a_y^{max} \quad \text{or} \quad \frac{s/2}{h} < \mu \tag{9.63}$$

holds. As a consequence, the risk of rollover may be judged by the "static stability factor"

$$s_F = \frac{s/2}{h} \tag{9.64}$$

Values of $s_F \leq 1$ indicate a high rollover risk on paved dry roads, where $\mu = 1$ may be assumed, as done here.

9.2.2.2 Enhanced Rollover Model

However, the static stability factor serves as a rough indicator only. To determine the overturning limit, the tire deflection and the body roll have to be taken into account too, Figure 9.10. The balance of torques at the height of the track plane applied at the

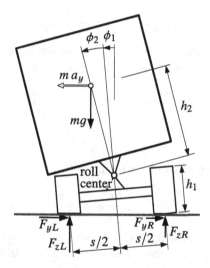

FIGURE 9.10
Overturning hazard on trucks.

already inclined vehicle results in

$$(F_{zL} - F_{zR})\frac{s}{2} = m\,a_y\,(h_1 + h_2) + m\,g\,[(h_1 + h_2)\phi_1 + h_2\phi_2] \tag{9.65}$$

where a_y describes the lateral acceleration, m is the vehicle mass, and small roll angles of the axle and the body were assumed, $\phi_1 \ll 1$, $\phi_2 \ll 1$. For a left-hand tilt, the tire forces at the right will vanish

$$F_{yR}^T = 0 \quad \text{and} \quad F_{zR}^T = 0 \tag{9.66}$$

whereas the left wheel carries the vehicle weight,

$$F_{zL}^T = m g \tag{9.67}$$

The superscript T indicates the tilting limit, hereby. Inserting Equations (9.66) and (9.67) into Equation (9.65) yields the lateral acceleration

$$\frac{a_y^T}{g} = \frac{\frac{s}{2}}{h_1 + h_2} - \phi_1^T - \frac{h_2}{h_1 + h_2} \phi_2^T \tag{9.68}$$

where the vehicle will start to roll over or tilt, respectively. Taking into consideration that $h = h_1 + h_2$ describes the height of the center of gravity above the road, the first term in Equation (9.68) corresponds with the static stability factor defined in Equation (9.64). However, the compliance of the tires and the suspension, which cause the axle and the body to roll ($\phi_1^T > 0$ and $\phi_2^T > 0$) will increase the hazard of overturning.

If the vehicle drives straight ahead, the weight of the vehicle is equally distributed to both sides

$$F_{zR}^{stat} = F_{zL}^{stat} = \frac{1}{2} m g \tag{9.69}$$

when a symmetric loading condition is assumed in this basic study. On a right turn, the wheel load at the left is increased and the one at the right is decreased by the same amount

$$F_{zL}^T = F_{zL}^{stat} + \Delta F_z \quad \text{and} \quad F_{zR}^T = F_{zR}^{stat} - \Delta F_z \tag{9.70}$$

At the tilting limit,

$$\Delta F_z^T = \frac{1}{2} m g \tag{9.71}$$

will hold, which corresponds with the tire deflection

$$\Delta r^T = \frac{\Delta F_z^T}{c_R} = \frac{1}{2} \frac{m g}{c_R} \tag{9.72}$$

where c_R denotes the radial tire stiffness. Because the right tire simultaneously rebounds by the same amount, the roll angle of the axle is defined by

$$2 \Delta r^T = s \phi_1^T \quad \text{or} \quad \phi_1^T = \frac{2 \Delta r}{s} = \frac{m g}{s c_R} \tag{9.73}$$

where s denotes the track width.

In analogy with Equation (9.65), the balance of torques at the body applied at its roll center yields

$$c_W \phi_2 = m a_y h_2 + m g h_2 (\phi_1 + \phi_2) \tag{9.74}$$

where c_W describes the roll stiffness of the body suspension. In particular, at the overturning limit $a_y = a_y^T$,

$$\phi_2^T = \frac{a_y^T}{g} \frac{m g h_2}{c_W - m g h_2} + \frac{m g h_2}{c_W - m g h_2} \phi_1^T \tag{9.75}$$

applies. At $c_W \to mgh_2$ the body roll angle tends to infinity ($\phi_2^T \to \infty$) which would cause the vehicle to overturn already at standstill. Hence, a minimum roll stiffness of $c_W > c_W^{min} = mgh_2$ is required at least to keep the body in an upright position. With Equations (9.73) and (9.75), the overturning condition in Equation (9.68) reads as

$$(h_1 + h_2)\frac{a_y^T}{g} = \frac{s}{2} - (h_1 + h_2)\frac{1}{c_R^*} - h_2\frac{a_y^T}{g}\frac{1}{c_W^* - 1} - h_2\frac{1}{c_W^* - 1}\frac{1}{c_R*} \quad (9.76)$$

where, for abbreviation purposes, the dimensionless quantities

$$c_R^* = \frac{s\,c_R}{m\,g} \quad \text{and} \quad c_W^* = \frac{c_W}{m\,g\,h_2} \quad (9.77)$$

have been introduced. Resolved for the normalized lateral acceleration,

$$\frac{a_y^T}{g} = \frac{\dfrac{s}{2}}{h_1 + h_2 + \dfrac{h_2}{c_W^* - 1}} - \frac{1}{c_R^*} \quad (9.78)$$

will finally remain.

For heavy trucks, a twin-tire axle may be loaded with $m = 13\,000$ kg. The radial stiffness of one tire is given by $c_R = 800\,000$ N/m, and the track width can be set to $s = 2$ m. The values $h_1 = 0.8$ m and $h_2 = 1.0$ m or $h = h_1 + h_2 = 1.8$ m define the height of the center of gravity. The corresponding results obtained from Equations (9.78), (9.73), and (9.75) are shown in Figure 9.11. Even with a rigid body suspension

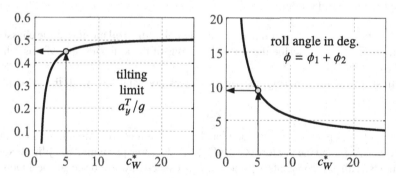

FIGURE 9.11

Tilting limit for a typical truck at steady-state cornering.

$c_W^* \to \infty$, the vehicle will turn over at a lateral acceleration of $a_y \approx 0.5g$. Then, the roll angle of the vehicle solely results from the tire deflection. At a normalized roll stiffness of $c_W^* = 5$, the overturning limit lies at $a_y \approx 0.45g$ and thus reaches 90% of the maximum, left plot in Figure 9.11. The vehicle will then turn over at a roll angle of $\phi = \phi_1 + \phi_2 \approx 9°$, right plot in Figure 9.11.

Depending on the shape of the tank, a liquid load may reduce the rollover threshold further on. In particular, in non-steady-state maneuvers, such as braking in a turn, dynamic sloshing forces affect vehicle behavior significantly [49].

9.2.3   Roll Support and Camber Compensation

When a vehicle drives in a steady state through a curve with the lateral acceleration a_y, centrifugal forces are applied to the single masses. The planar vehicle model, shown in Figure 9.12, consists of three bodies: the chassis of mass m_C and two knuckles with wheels attached to each of mass m_W. The generalized coordinates y_C,

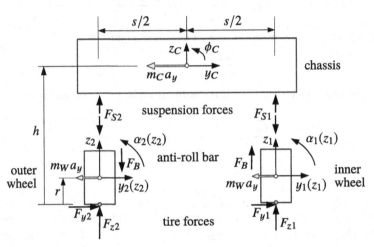

FIGURE 9.12
Simple vehicle roll model.

z_C, ϕ_C characterize the lateral and vertical as well as the roll motion of the chassis. In addition, z_1 and z_2 describe the vertical motions of the knuckles and wheels relative to the chassis.

In steady-state cornering, the vehicle forces are balanced. With the principle of virtual work

$$\delta W = 0 \tag{9.79}$$

the equilibrium position can be calculated. For the vehicle roll model, the suspension forces F_{S1}, F_{S2}, and the tire forces F_{y1}, F_{z1}, F_{y2}, F_{z2}, are approximated by linear spring elements with the constants c_S and c_y, c_z, respectively. As done in the kinematic approach, the lateral slip is neglected. The work W of these forces can be calculated directly or much more conveniently via the potential V, where $W = -V$ holds. For small rotations and deflections, linearized kinematics can be applied and one gets as a consequence,

$$
\begin{aligned}
W = &-m_C\, a_y\, y_C \\
&-m_W\, a_y\, (y_C + h_R\, \phi_C + y_1) - m_W\, a_y\, (y_C + h_R\, \phi_C + y_2) \\
&-\tfrac{1}{2}\, c_S\, z_1^2 - \tfrac{1}{2}\, c_S\, z_2^2 \\
&-\tfrac{1}{2}\, c_B\, (z_1 - z_2)^2 \\
&-\tfrac{1}{2}\, c_y\, (y_C + h\phi_C + y_1 + r\alpha_1)^2 - \tfrac{1}{2}\, c_y\, (y_C + h\phi_C + y_2 + r\alpha_2)^2 \\
&-\tfrac{1}{2}\, c_z\, (z_A + \tfrac{s}{2}\phi_C + z_1)^2 - \tfrac{1}{2}\, c_z\, (z_A - \tfrac{s}{2}\phi_C + z_2)^2
\end{aligned}
\tag{9.80}
$$

where the abbreviation $h_R = h - r$ was used, and c_B denotes the stiffness of the anti-roll bar, converted to the vertical displacement of the wheel centers. The kinematics of wheel suspensions are symmetrical in general. Then, the lateral motion of the knuckles and wheels and their rotation about the longitudinal axis (camber change), which results from the specific suspension kinematics, can be approximated by

$$y_1 = y_1(z_1) \approx \frac{\partial y}{\partial z} z_1, \quad \alpha_1 = \alpha_1(z_1) \approx \frac{\partial \alpha}{\partial z} z_1 \tag{9.81}$$

and

$$y_2 = y_2(z_2) \approx -\frac{\partial y}{\partial z} z_2, \quad \alpha_2 = \alpha_2(z_2) \approx -\frac{\partial \alpha}{\partial z} z_2 \tag{9.82}$$

As done in Section 8.5.1, partial derivatives were used here, to indicate that the constrained axle motions may depend not only on the vertical motion in general. Now, the work defined by Equation (9.80) can be described as a function of the vector of generalized coordinates

$$y = [\, y_C, \, z_C, \, \phi_C, \, z_1, \, z_2 \,]^T \tag{9.83}$$

Due to $W = W(y)$, the principle of virtual work provided in Equation (9.79) leads to

$$\delta W = \frac{\partial W(y)}{\partial y} \delta y = 0 \tag{9.84}$$

Because the virtual displacements will not vanish in general, $\delta y \neq 0$, Equation (9.84) delivers a system of linear equations in the form of

$$C\, y = b \tag{9.85}$$

The matrix C and the vector b are given by

$$C = \begin{bmatrix} 2c_y & 0 & 2c_y h & \frac{\partial \hat{y}}{\partial z} c_y & -\frac{\partial \hat{y}}{\partial z} c_y \\ 0 & 2c_z & 0 & c_z & c_z \\ 2c_y h & 0 & c_\phi & \frac{s}{2}c_z + h\frac{\partial \hat{y}}{\partial z} c_y & -\frac{s}{2}c_z - h\frac{\partial \hat{y}}{\partial z} c_y \\ \frac{\partial \hat{y}}{\partial z} c_y & c_z & \frac{s}{2}c_z + h\frac{\partial \hat{y}}{\partial z} c_y & c_S^* + c_B + c_z & -c_B \\ -\frac{\partial \hat{y}}{\partial z} c_y & c_z & -\frac{s}{2}c_z - h\frac{\partial \hat{y}}{\partial z} c_y & -c_B & c_S^* + c_B + c_z \end{bmatrix} \tag{9.86}$$

$$b = - \begin{bmatrix} m_C + 2m_W \\ 0 \\ (m_1 + m_2)\, h_R \\ m_W\, \partial y/\partial z \\ -m_W\, \partial y/\partial z \end{bmatrix} a_y \tag{9.87}$$

where the abbreviations

$$\frac{\partial \hat{y}}{\partial z} = \frac{\partial y}{\partial z} + r\frac{\partial \alpha}{\partial z}, \quad c_S^* = c_S + c_y \left(\frac{\partial y}{\partial z}\right)^2, \quad c_\phi = 2c_y h^2 + 2c_z \left(\frac{b}{2}\right)^2 \tag{9.88}$$

have been used. The MATLAB-Script in Listing 9.5 provides the data, solves the set of linear equations, computes the wheel loads, and displays the results.

Listing 9.5

Script `app_09_roll_model.m`: vehicle roll model

```
 1 g   = 9.81 ;  % constant of cravity in m/s^2
 2 ay  = 0.75*g; % set lateral acceleration in m/s^2
 3 mc  = 600;    % chassis mass (half vehicle) in kg
 4 mw  = 50;     % mass of each knuckle and wheel in kg
 5 s   = 1.5;    % track width in m
 6 h   = 0.6;    % height of center of gravity in m
 7 r   = 0.3;    % static tire radius in m
 8 cs  = 20000;  % suspension stiffness in N/m
 9 cb  = 10000;  % stiffness of anti roll bar in N/m
10 cy  =180000;  % tire lateral stiffness in N/m
11 cz  =200000;  % tire vertical stiffness in N/m
12
13 % axle kinematics
14 dydz = 0 ;       % lateral motion of wheel center caused by vertical
15 dadz = 0/180*pi; % camber change caused by vertical motion in Grad/m
16
17 hr = h-r ; sh = s/2;   dyqdz = dydz + r*dadz;   % abbreviations
18
19 % stiffness matrix (column by column)
20 C(:,1) = [          2*cy              ; ...
21                      0                ; ...
22                    2*cy*h             ; ...
23                  cy*dyqdz             ; ...
24                 -cy*dyqdz             ];
25 C(:,2) = [          0                 ; ...
26                    2*cz               ; ...
27                     0                 ; ...
28                     cz                ; ...
29                     cz                ];
30 C(:,3) = [        2*cy*h              ; ...
31                      0                ; ...
32            2*cz*sh^2+2*cy*h^2         ; ...
33           sh*cz+h*dyqdz*cy           ; ...
34          -sh*cz-h*dyqdz*cy           ];
35 C(:,4) = [       cy*dyqdz            ; ...
36                     cz               ; ...
37           sh*cz+h*dyqdz*cy           ; ...
38         cs+cb+cz+dyqdz^2*cy          ; ...
39                    -cb               ];
40 C(:,5) = [      -cy*dyqdz            ; ...
41                     cz               ; ...
42          -sh*cz-h*dyqdz*cy           ; ...
43                    -cb               ; ...
44         cs+cb+cz+dyqdz^2*cy          ];
45
46 % right hand side
47 b = -[ mc+2*mw;  0;  2*mw*hr;  mw*dydz; -mw*dydz ]*ay;
48
49 x = C\b; % solve set of linear equations and display results
50 disp(['lateral chassis displacement  y_C =',num2str(x(1))])
51 disp(['vertical chassis displacement z_C =',num2str(x(2))])
52 disp(['chassis roll motion         phi_C =',num2str(x(3)*180/pi)])
53 disp(['suspension travel inner wheel z_1 =',num2str(x(4))])
54 disp(['suspension travel outer wheel z_2 =',num2str(x(5))])
55 % compute and display wheel loads
56 Fz1=(mw+mc/2)*g+cz*(x(2)+s/2*x(3)+x(4)); disp(['Fz1=',num2str(Fz1)])
57 Fz2=(mw+mc/2)*g+cz*(x(2)-s/2*x(3)+x(5)); disp(['Fz2=',num2str(Fz2)])
```

The results for simple axle kinematics (dydz=0, dadz=0), which will be generated by a trailing arm suspension for example, are visualized in Figure 9.13. It can be seen

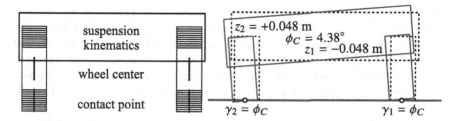

FIGURE 9.13
Vehicle roll for simple axle kinematics (dydz=0, dadz=0).

that due to simple kinematics, the roll angle of the chassis is transferred directly to the wheels, which will result in a tire camber angle of the same amount, $\gamma_1 = \gamma_2 = \phi_C$.

Of course, the anti-roll bar stiffness c_B affects the roll angle strongly. For example, omitting the anti-roll bar $c_B = 0$ will increase the roll angle up to $\phi_C = 8°$, which certainly is too much for a passenger car even when cornering hard with a lateral acceleration of $a_y = 0.75\,g$. On the other hand, the anti-roll bar couples the vertical motions of the left and right wheels, which affects the ride comfort in particular when crossing a bump on one side. That is why, controlled anti-roll bars equipped with actuators are used in high-level cars today.

Suspension kinematics, which move the wheels outward on jounce, will provide kinematically generated roll support, Figure 9.14. Now, the roll center of the chassis

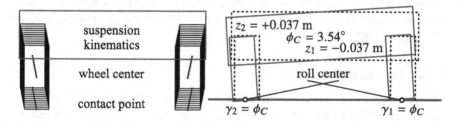

FIGURE 9.14
Roll support kinematically generated (dydz=0.2, dadz=0).

is located above the road, which reduces the roll angle. But, as the wheels do not rotate on jounce and rebound, the tire camber angle is still the same as the chassis roll angle, $\gamma_1 = \gamma_2 = \phi_C$. This kind of suspension kinematics will be generated, for instance, by a double wishbone suspension with inclined mounted wishbones of the same size. It is not really practical because the lateral movements of the contact points during jounce and rebound would cause enormous tire wear just when driving straight ahead on uneven roads.

Large tire camber angles result in unfavorable pressure distribution in the contact area, which leads to a reduction in maximally transmittable lateral forces. Thus, for more sportive vehicles, axle kinematics are employed, where the wheels are rotated around the longitudinal axis at jounce and rebound, $\alpha_1 = \alpha_1(z_1)$ and $\alpha_2 = \alpha_2(z_2)$. Even a full "camber compensation" can be achieved with $\gamma_1 \approx 0$ and $\gamma_2 \approx 0$. For simple wishbone suspension systems, the rotation of the wheels around the longitudinal move the contact points outward on jounce, which provides significant roll support in addition but will cause severe tire wear. In practice, most suspension systems generate partial camber compensation, Figure 9.15. If the rotation of the wheel around the

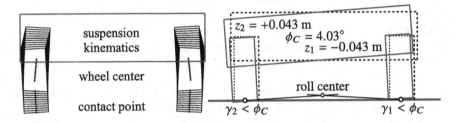

FIGURE 9.15
Partial camber compensation (dydz=-0.1, dadz=30/180*pi).

longitudinal axis (dadz>0) is combined with inward movement of the wheel center on jounce (dydz<0), the roll center can be placed on or slightly above the road at least. Double wishbone suspension systems with different-sized control arms make it possible to realize camber compensation and control the amount of roll support. The kinematic analysis of a typical double wishbone suspension performed in Section 5.4 shows a camber change (partial camber compensation) combined with slight outward movement of the reference point (contact point) on jounce that generates a moderate roll support, Figure 5.14.

9.2.4   Roll Center and Roll Axis

The roll center can be constructed from the lateral motion of the wheel contact points at each axle. The line through the roll center at the front and rear axle, is called the roll axis, Figure 9.16.

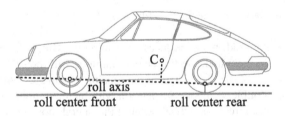

FIGURE 9.16
Roll axis.

    The chassis will roll about this axis when cornering. The distance from the roll axis to the center of gravity is responsible for the amount of body roll.

9.2.5 Wheel Load Transfer

The roll angle of a vehicle during cornering depends on the roll stiffness of the axle and on the position of the roll center. Different axle layouts at the front and rear axle may result in different roll angles of the front and rear part of the chassis, Figure 9.17. A chassis with a significant torsional compliance would allow its front and rear parts

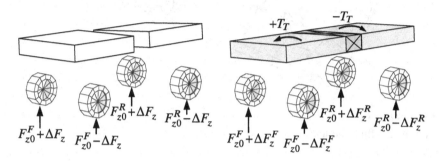

FIGURE 9.17
Wheel load transfer on a flexible and a rigid chassis.

to roll nearly independently. Then, the load transfer ΔF_z from the inner to the outer wheels will be the same[2] at both axles, if the front and rear parts have the same height of the center of gravity. On most passenger cars, the chassis is rather stiff, however. In this case, front and rear parts of the chassis are forced by an internal torque T_T to an overall chassis roll angle. This torque affects the wheel loads and thus generates a different load transfer at the front and rear wheels, $\Delta F_z^F \neq \Delta F_z^R$. Due to the degressive influence of the wheel load on longitudinal and lateral tire forces, the steering tendency of a vehicle can be affected.

9.3 Simple Handling Model

9.3.1 Modeling Concept

The main vehicle motions take place in a horizontal plane defined by the earth-fixed axis system 0, Figure 9.18. The yaw angle ψ describes the orientation of the vehicle-fixed longitudinal axis x_V, whereas β indicates the direction of travel by measuring the angle between the vehicle velocity v and the longitudinal axis x_V. Again, the wheels at each axle are summarized in one fictitious wheel, where δ denotes the mean steering angle at the front axle. Only the lateral tire forces F_{y1}, F_{y2} are taken into account because focus is laid on the lateral dynamics here. In addition, aerodynamic forces and torques, applied at the vehicle, are not taken into consideration.

---

[2]This can easily be checked by running the MATLAB-Script in Listing 9.5 with different model data and observing the wheel loads.

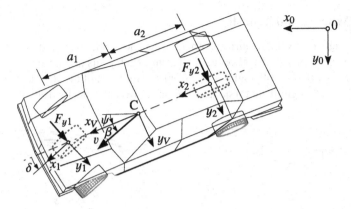

FIGURE 9.18
Simple handling model.

9.3.2 Kinematics

The vehicle velocity at the center of gravity can be expressed easily in the vehicle-fixed axis system x_V, y_V, z_V,

$$v_{C,V} = \begin{bmatrix} v \cos \beta \\ v \sin \beta \\ 0 \end{bmatrix} \tag{9.89}$$

where β is the side slip angle and v denotes the magnitude of the vehicle velocity. The velocity vectors and the unit vectors in the longitudinal and lateral directions of the axles are needed for computation of the lateral slips. One gets

$$e_{x_1,V} = \begin{bmatrix} \cos \delta \\ \sin \delta \\ 0 \end{bmatrix}, \quad e_{y_1,V} = \begin{bmatrix} -\sin \delta \\ \cos \delta \\ 0 \end{bmatrix}, \quad v_{01,V} = \begin{bmatrix} v \cos \beta \\ v \sin \beta + a_1 \dot{\psi} \\ 0 \end{bmatrix} \tag{9.90}$$

as well as

$$e_{x_2,V} = \begin{bmatrix} 1 \\ 0 \\ 0 \end{bmatrix}, \quad e_{y_2,V} = \begin{bmatrix} 0 \\ 1 \\ 0 \end{bmatrix}, \quad v_{02,V} = \begin{bmatrix} v \cos \beta \\ v \sin \beta - a_2 \dot{\psi} \\ 0 \end{bmatrix} \tag{9.91}$$

where a_1 and a_2 are the distances from the center of gravity C to the front and the rear axle. Finally, $\dot{\psi}$ denotes the yaw angular velocity of the vehicle.

9.3.3 Tire Forces

Unlike the kinematic tire model, now small lateral sliding motions in the contact points are permitted. For small lateral slips, the lateral force can be approximated by a linear approach

$$F_y = k_S s_y \tag{9.92}$$

where the cornering stiffness k_S depends on the wheel load F_z, and the lateral slip s_y is defined by Equation (3.72). Because the vehicle is neither accelerated nor braked, the rolling condition

$$r_D \, \Omega \;=\; e_x^T \, v_{0P} \tag{9.93}$$

is fulfilled at each wheel. Here, r_D is the dynamic tire radius, v_{0P} the contact point velocity, and e_x the unit vector in the longitudinal direction. With the lateral tire velocity

$$v_y \;=\; e_y^T \, v_{0P} \tag{9.94}$$

and the rolling condition defined in Equation (9.93), the lateral slip is given by

$$s_y \;=\; \frac{-e_y^T \, v_{0P}}{|\, e_x^T \, v_{0P} \,|} \tag{9.95}$$

where e_y defines the unit vector in the lateral direction of the tire. So, the lateral tire forces are simply modeled as

$$F_{y1} \;=\; k_{S1} \, s_{y1} \quad \text{and} \quad F_{y2} \;=\; k_{S2} \, s_{y2} \tag{9.96}$$

Even if the same tires are mounted at the front and rear axle, the cornering stiffness k_{S1} may vary from k_{S2} because of different wheel loads.

9.3.4 Lateral Slips

Using Equation (9.91), the lateral slip at the front axle defined by Equation (9.95) reads as

$$s_{y1} \;=\; \frac{+\sin\delta \, (v \, \cos\beta) - \cos\delta \, (v \, \sin\beta + a_1 \, \dot\psi)}{|\, \cos\delta \, (v \, \cos\beta) + \sin\delta \, (v \, \sin\beta + a_1 \, \dot\psi) \,|} \tag{9.97}$$

The lateral slip at the rear axle,

$$s_{y2} \;=\; -\frac{v \, \sin\beta - a_2 \, \dot\psi}{|\, v \, \cos\beta \,|} \tag{9.98}$$

is simpler because no steer angle is considered here. Now, the yaw velocity of the vehicle $\dot\psi$, the side slip angle β, and the steering angle δ are assumed to be small,

$$|a_1 \, \dot\psi| \ll |v|, \quad |a_2 \, \dot\psi| \ll |v| \tag{9.99}$$

$$|\beta| \ll 1, \quad \text{and} \quad |\delta| \ll 1 \tag{9.100}$$

Because the side slip angle β always describes the smaller angle between the velocity vector v and the vehicle longitudinal axis x_V, the straightforward linearization $v \, \sin\beta \approx v \, \beta$ is replaced by the more sophisticated one,

$$v \, \sin\beta \approx |v| \, \beta \tag{9.101}$$

Comparatively small values for the yaw velocity and the side slip angle correspond quite well with normal driving situations. However, small steering angles at the front axles will exclude the simulation of parking maneuvers, for example. Now, Equations (9.97) and (9.98) result in

$$s_{y1} = -\beta - \frac{a_1}{|v|} \dot{\psi} + \frac{v}{|v|} \delta \tag{9.102}$$

$$s_{y2} = -\beta + \frac{a_2}{|v|} \dot{\psi} \tag{9.103}$$

where the consequences of Equations (9.99), (9.100), and (9.101) were taken into consideration.

9.3.5   Equations of Motion

The velocities, angular velocities, and accelerations are needed to derive the equations of motion. For small side slip angles $\beta \ll 1$, Equation (9.89) can be approximated by

$$v_{C,V} = \begin{bmatrix} v \\ |v|\,\beta \\ 0 \end{bmatrix} \tag{9.104}$$

The angular velocity is given by

$$\omega_{0F,V} = \begin{bmatrix} 0 \\ 0 \\ \dot{\psi} \end{bmatrix} \tag{9.105}$$

The acceleration of the vehicle, expressed in the vehicle-fixed axis system x_V, y_V, z_V, is given by

$$a_{C,V} = \omega_{0F,V} \times v_{C,V} + \dot{v}_{C,V} = \begin{bmatrix} 0 \\ v\,\dot{\psi} + |v|\,\dot{\beta} \\ 0 \end{bmatrix} \tag{9.106}$$

where a constant vehicle velocity $v = const.$ was assumed and higher-order terms are neglected. The angular acceleration is simply defined by

$$\dot{\omega}_{0F,V} = \begin{bmatrix} 0 \\ 0 \\ \dot{\omega} \end{bmatrix} \tag{9.107}$$

where the yaw angular velocity

$$\omega = \dot{\psi} \tag{9.108}$$

substitutes the time derivative of the yaw angle. The linear momentum in the lateral direction of the vehicle reads as

$$m(v\,\omega + |v|\,\dot{\beta}) = F_{y1} + F_{y2} \tag{9.109}$$

where, due to the small steering angle, the term $F_{y1} \cos \delta$ has been approximated by F_{y1}, and m is the vehicle mass. With Equation (9.108), the angular momentum yields

$$\Theta \dot{\omega} = a_1 F_{y1} - a_2 F_{y2} \qquad (9.110)$$

where Θ denotes the inertia of vehicle about a vertical axis located at the center of gravity. Inserting the linear description of the lateral forces Equation (9.96) combined with the lateral slips Equations (9.102), (9.103) into Equations (9.109) and (9.110) yields[3]

$$\dot{\beta} = \frac{k_{S1}}{m\,|v|}\left(-\beta - \frac{a_1}{|v|}\omega + \frac{v}{|v|}\delta\right) + \frac{k_{S2}}{m\,|v|}\left(-\beta + \frac{a_2}{|v|}\omega\right) - \frac{v}{|v|}\omega \qquad (9.111)$$

$$\dot{\omega} = \frac{a_1\,k_{S1}}{\Theta}\left(-\beta - \frac{a_1}{|v|}\omega + \frac{v}{|v|}\delta\right) - \frac{a_2\,k_{S2}}{\Theta}\left(-\beta + \frac{a_2}{|v|}\omega\right) \qquad (9.112)$$

This represents two coupled, but linear first-order differential equations, which can be written in the form of a linear state equation,

$$\underbrace{\begin{bmatrix}\dot{\beta}\\\dot{\omega}\end{bmatrix}}_{\dot{x}} = \underbrace{\begin{bmatrix} -\dfrac{k_{S1}+k_{S2}}{m\,|v|} & \dfrac{a_2 k_{S2}-a_1 k_{S1}}{m\,|v||v|} - \dfrac{v}{|v|} \\[3mm] \dfrac{a_2 k_{S2}-a_1 k_{S1}}{\Theta} & -\dfrac{a_1^2 k_{S1}+a_2^2 k_{S2}}{\Theta\,|v|} \end{bmatrix}}_{A} \underbrace{\begin{bmatrix}\beta\\\omega\end{bmatrix}}_{x} + \underbrace{\begin{bmatrix} \dfrac{v}{|v|}\dfrac{k_{S1}}{m\,|v|} \\[3mm] \dfrac{v}{|v|}\dfrac{a_1 k_{S1}}{\Theta}\end{bmatrix}}_{B} \underbrace{\begin{bmatrix}\delta\end{bmatrix}}_{u} \qquad (9.113)$$

If a system can be at least approximately described by a linear state equation, then the stability, steady-state solutions, transient response, and optimal controlling can be calculated with classic methods of system dynamics.

9.3.6 Stability

9.3.6.1 Eigenvalues

The homogeneous state equation

$$\dot{x} = A\,x \qquad (9.114)$$

describes the eigendynamics. If the approach

$$x_h(t) = x_0\,e^{\lambda t} \qquad (9.115)$$

is inserted into Equation (9.114), the homogeneous equation will remain,

$$(\lambda E - A)\,x_0 = 0 \qquad (9.116)$$

One gets nontrivial solutions $x_0 \neq 0$ for

$$det\,|\lambda E - A| = 0 \qquad (9.117)$$

The eigenvalues λ provide information concerning the stability of the system.

---

[3]This simple planar model, often called the "bicycle model," was first published by P. Riekert and T. E. Schunck: Zur Fahrmechanik des gummibereiften Kraftfahrzeugs, Ingenieur-Archiv, 11, 1940, S. 210-224. It is still used for fundamental studies or the basic layout of control systems.

9.3.6.2   Low-Speed Approximation

The state matrix

$$
A^{v \to 0} = \begin{bmatrix} -\dfrac{k_{S1} + k_{S2}}{m \, |v|} & \dfrac{a_2 \, k_{S2} - a_1 \, k_{S1}}{m \, |v| |v|} - \dfrac{v}{|v|} \\ 0 & -\dfrac{a_1^2 \, k_{S1} + a_2^2 \, k_{S2}}{\Theta \, |v|} \end{bmatrix}
\tag{9.118}
$$

approximates the eigendynamics of vehicles at low speeds, $v \to 0$. The matrix in Equation (9.118) has the eigenvalues

$$
\lambda_1^{v \to 0} = -\frac{k_{S1} + k_{S2}}{m \, |v|} \quad \text{and} \quad \lambda_2^{v \to 0} = -\frac{a_1^2 \, k_{S1} + a_2^2 \, k_{S2}}{\Theta \, |v|}
\tag{9.119}
$$

Both eigenvalues are real, always negative, and independent of the model data and the driving direction. Thus, all vehicles will possess an asymptotically stable driving behavior at low speed!

9.3.6.3   High-Speed Approximation

At high driving velocities, $v \to \infty$, the state matrix can be approximated by

$$
A^{v \to \infty} = \begin{bmatrix} 0 & -\dfrac{v}{|v|} \\ \dfrac{a_2 \, k_{S2} - a_1 \, k_{S1}}{\Theta} & 0 \end{bmatrix}
\tag{9.120}
$$

Using Equation (9.120) one obtains from Equation (9.117) the relation

$$
(\lambda^{v \to \infty})^2 + \frac{v}{|v|} \frac{a_2 \, k_{S2} - a_1 \, k_{S1}}{\Theta} = 0
\tag{9.121}
$$

with the solutions

$$
\lambda_{1,2}^{v \to \infty} = \pm \sqrt{-\frac{v}{|v|} \frac{a_2 \, k_{S2} - a_1 \, k_{S1}}{\Theta}}
\tag{9.122}
$$

When driving forward with $v > 0$, the root argument will be positive if

$$
a_2 \, k_{S2} - a_1 \, k_{S1} < 0
\tag{9.123}
$$

holds. Then, one eigenvalue will be definitely positive, and the system is unstable. Two zero-eigenvalues $\lambda_1^{v \to \infty} = 0$ and $\lambda_2^{v \to \infty} = 0$ are obtained just for

$$
a_1 \, k_{S1} = a_2 \, k_{S2}
\tag{9.124}
$$

The driving behavior is indifferent then. Slight parameter variations may lead to unstable behavior, however. A set of parameters, resulting in

$$
a_2 \, k_{S2} - a_1 \, k_{S1} > 0 \quad \text{or} \quad a_1 \, k_{S1} < a_2 \, k_{S2}
\tag{9.125}
$$

will produce a negative root argument in Equation (9.122) on forward drive, $v > 0$. Then, the eigenvalues are imaginary, which characterizes undamped vibrations. To avoid instability on forward drive with high speed, vehicles have to satisfy the condition in Equation (9.125). However, the root argument in Equation (9.122) changes the sign on backward drive. As a consequence, all vehicles showing stable driving behavior at forward drive will definitely become unstable at fast backward drive!

9.3.6.4 Critical Speed

The condition for nontrivial solutions (9.117) results in a quadratic equation for the eigenvalues λ here,

$$\det |\lambda E - A| = \lambda^2 + v_1 \lambda + v_2 = 0 \tag{9.126}$$

which is solved by

$$\lambda_{1,2} = -\frac{v_1}{2} \pm \sqrt{\left(\frac{v_1}{2}\right)^2 - v_2} \tag{9.127}$$

Asymptotically stable solutions ($\lambda_{1,2}$ real and $\lambda_1 < 0$, $\lambda_2 < 0$) demand at least for

$$v_1 > 0 \quad \text{and} \quad v_2 > 0 \tag{9.128}$$

which exactly corresponds with the stability criteria of Stodola and Hurwitz [27]. According to Equation (9.113), the coefficients in Equation (9.126) can be derived from the vehicle data

$$v_1 = \frac{k_{S1} + k_{S2}}{m\,|v|} + \frac{a_1^2 k_{S1} + a_2^2 k_{S2}}{\Theta\,|v|} \tag{9.129}$$

$$\begin{aligned}
v_2 &= \frac{k_{S1} + k_{S2}}{m\,|v|} \frac{a_1^2 k_{S1} + a_2^2 k_{S2}}{\Theta\,|v|} - \frac{(a_2 k_{S2} - a_1 k_{S1})^2}{\Theta\,m\,|v|\,|v|} + \frac{v}{|v|} \frac{a_2 k_{S2} - a_1 k_{S1}}{\Theta} \\
&= \frac{k_{S1} k_{S2}\,(a_1 + a_2)^2}{m\,\Theta\,v^2} \left(1 + \frac{v}{|v|} \frac{a_2 k_{S2} - a_1 k_{S1}}{k_{S1} k_{S2}\,(a_1 + a_2)^2} m\,v^2\right)
\end{aligned} \tag{9.130}$$

The coefficient v_1 is always positive, whereas $v_2 > 0$ is fulfilled only if

$$1 + \frac{v}{|v|} \frac{a_2 k_{S2} - a_1 k_{S1}}{k_{S1} k_{S2}\,(a_1 + a_2)^2} m\,v^2 > 0 \tag{9.131}$$

holds. Hence, a vehicle designed to be stable for arbitrary velocities in the forward direction becomes unstable when it drives too fast backward. Because the coefficient v_2 is positive ($v_2 > 0$) for $a_2 k_{S2} - a_1 k_{S1} > 0$ and $v < 0$ only if $v > v_C^-$ holds, where according to Equation (9.131) the critical backward velocity is given by

$$v_C^- = -\sqrt{\frac{k_{S1} k_{S2}\,(a_1 + a_2)^2}{m\,(a_2 k_{S2} - a_1 k_{S1})}} \tag{9.132}$$

On the other hand, vehicle layouts with $a_2 k_{S2} - a_1 k_{S1} < 0$ are only stable while driving forward as long as $v < v_C^+$ holds. Here, Equation (9.131) yields the critical forward velocity of

$$v_C^+ = \sqrt{\frac{k_{S1} k_{S2} (a_1 + a_2)^2}{m (a_1 k_{S1} - a_2 k_{S2})}} \qquad (9.133)$$

Most vehicles are designed to be stable for fast forward drive. Then, the backward velocity must be limited in order to avoid stability problems. That is why, fast driving vehicles have four or more gears for forward drive but only one or two reverse gears.

9.3.6.5 Example

A passenger car is described by the data $m = 1600\ kg$, $\Theta = 2000\ kgm^2$, $a_1 = 1.1$ m, $a_2 = 1.4$ m, $k_{S1} = 124\,000$ N/-, $k_{S2} = 120\,000$ N/-. The vehicle is stable on forward drive because $a_2 * k_{S2} = 168\,000$ Nm is significantly larger than $a_1 * k_{S1} = 136\,400$ Nm here. Then, the vehicle is unstable at fast backward drive. According to Equation (9.132), the critical speed results in $v_C^- = -154.4$ km/h in this specific case. The eigenvalues are plotted in Figure 9.19 for different vehicle velocities, ranging from $v = -60$ km/h to $v = 180$ km/h. At higher velocities ($v > 24.1$ km/h) the eigenvalues become complex

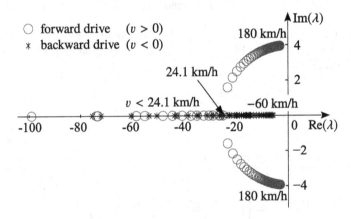

FIGURE 9.19
Stability analysis of a typical passenger car for different vehicle velocities.

conjugate, which characterizes damped vibrations. However, with increasing velocity, the real part $Re(\lambda)$, which indicates the amount of damping, becomes smaller and smaller. At $v \to \infty$ it will vanish completely. On backward drive, the vehicle is stable, which is indicated by real and negative eigenvalues, as long as $v < v_C^-$ holds.

9.3.7 Steady-State Solution

9.3.7.1 Steering Tendency

At a given steering angle $\delta = \delta_0$, a stable system reaches a steady state after a certain time. Then, the vehicle will drive on a circle with the radius R_{st}, which is determined by

$$\omega_{st} = \frac{v}{R_{st}} \tag{9.134}$$

where v is the velocity of the vehicle and ω_{st} denotes its steady-state angular velocity. With $x_{st} = const.$ or $\dot{x}_{st} = 0$, the state equation (9.113) simplifies to a system of linear equations,

$$A\,x_{st} = -B\,u \tag{9.135}$$

Using Equation (9.134), the state vector can be described in a steady state by

$$x_{st} = \begin{bmatrix} \beta_{st} \\ v/R_{st} \end{bmatrix} \tag{9.136}$$

where β_{st} denotes the steady-state side slip angle. With $u = [\delta_0]$, and the elements of the state matrix A and the vector B that are defined in Equation (9.113), the system of linear equations (9.135) yields

$$(k_{S1} + k_{S2})\beta_{st} + (m\,v\,|v| + a_1\,k_{S1} - a_2\,k_{S2})\frac{v}{|v|}\frac{1}{R_{st}} = \frac{v}{|v|}k_{S1}\,\delta_0 \tag{9.137}$$

$$(a_1\,k_{S1} - a_2\,k_{S2})\beta_{st} + (a_1^2\,k_{S1} + a_2^2\,k_{S2})\frac{v}{|v|}\frac{1}{R_{st}} = \frac{v}{|v|}a_1\,k_{S1}\,\delta_0 \tag{9.138}$$

where the first equation has been multiplied by $-m\,|v|$ and the second by $-\Theta$. Eliminating the steady-state side slip angle β_{st} results in

$$\left[m v |v| (a_1 k_{S1} - a_2 k_{S2}) + (a_1 k_{S1} - a_2 k_{S2})^2 - (k_{S1} + k_{S2})(a_1^2 k_{S1} + a_2^2 k_{S2}) \right] \frac{v}{|v|}\frac{1}{R_{st}}$$
$$= [a_1 k_{S1} - a_2 k_{S2} - a_1 (k_{S1} + k_{S2})] \frac{v}{|v|}k_{S1}\delta_0 \tag{9.139}$$

which can be simplified to

$$\left[m v |v| (a_1 k_{S1} - a_2 k_{S2}) - k_{S1} k_{S2} (a_1 + a_2)^2 \right] \frac{v}{|v|}\frac{1}{R_{st}} = -\frac{v}{|v|}k_{S1} k_{S2} (a_1 + a_2)\delta_0 \tag{9.140}$$

Hence, driving the vehicle at a certain radius requires a steering angle of

$$\delta_0 = \frac{a_1 + a_2}{R_{st}} + m\frac{v|v|}{R_{st}}\frac{a_2\,k_{S2} - a_1\,k_{S1}}{k_{S1}\,k_{S2}\,(a_1 + a_2)} \tag{9.141}$$

The first term is just the Ackermann steering angle, which follows from Equation (9.2) with the wheel base of $a = a_1 + a_2$ and the approximation for small steering angles

$\tan \delta_0 \approx \delta_0$. In fact, the Ackermann steering angle provides a good approximation for slowly moving vehicles, because the second expression in Equation (9.141) is negligibly small at $v \to 0$. Depending on the value of $a_2\, k_{S2} - a_1\, k_{S1}$ and the driving direction, forward $v > 0$ or backward $v < 0$, the necessary steering angle differs from the Ackermann steering angle at higher speeds. The difference is proportional to the lateral acceleration

$$a_y = \frac{v|v|}{R_{st}} = \pm \frac{v^2}{R_{st}} \qquad (9.142)$$

Now, Equation (9.141) can be written as

$$\delta_0 = \delta_A + p\, a_y \qquad (9.143)$$

where $\delta_A = (a_1 + a_2)/R_{st}$ is the Ackermann steering angle, p summarizes the relevant vehicle parameter, and $a_y = v^2/R_{st}$ denotes the lateral acceleration of the vehicle. In a diagram where the steering angle δ_0 is plotted versus the lateral acceleration a_y, Equation (9.143) will represent a straight line, Figure 9.20. On forward drive, $v > 0$,

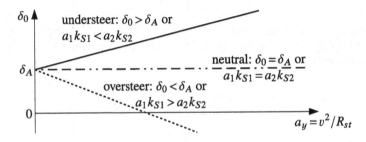

FIGURE 9.20
Different steering tendencies.

the inclination of this line, which is also called the steering gradient, is given by

$$\frac{d\,\delta_0}{d\,a_y} = p = \frac{m\,(a_2\, k_{S2} - a_1\, k_{S1})}{k_{S1}\, k_{S2}\,(a_1 + a_2)} \quad \text{in} \quad \frac{\text{rad}}{\text{m/s}^2} \qquad (9.144)$$

For steady-state cornering, the amount of the steering angle $\delta_0 \lesseqgtr \delta_A$ and hence, the steering tendency depends at increasing velocity on the sign of the stability condition $a_2 k_{S2} - a_1 k_{S1}$. The various steering tendencies are also arranged in Table 9.2.

TABLE 9.2
Steering tendencies of a vehicle at forward drive

Understeer	$\delta_0 > \delta_0^A$	or	$a_1 k_{S1} < a_2 k_{S2}$	or	$a_1 k_{S1} / a_2 k_{S2} < 1$
Neutral	$\delta_0 = \delta_0^A$	or	$a_1 k_{S1} = a_2 k_{S2}$	or	$a_1 k_{S1} / a_2 k_{S2} = 1$
Oversteer	$\delta_0 < \delta_0^A$	or	$a_1 k_{S1} > a_2 k_{S2}$	or	$a_1 k_{S1} / a_2 k_{S2} > 1$

9.3.7.2 Side Slip Angle

Resolving Equations (9.137) and (9.138) for the steady-state side slip angle yields

$$\beta_{st} = \frac{v}{|v|} \frac{a_2}{R_{st}} - \frac{m v |v|}{R_{st}} \frac{a_1}{k_{S2}(a_1 + a_2)} \tag{9.145}$$

As shown in Figure 9.21 the steady-state side slip angle starts with the kinematic value

$$\beta_{st}^{v \to 0} = \frac{v}{|v|} \frac{a_2}{R_{st}} \tag{9.146}$$

On forward drive ($v > 0$), it decreases with increasing velocity v or increasing lateral acceleration $a_y = v^2/R_{st}$, respectively. It changes sign at

$$_v \beta_{st}^{=0} = \sqrt{\frac{a_2\, k_{S2}\,(a_1 + a_2)}{m\, a_1}} \quad \text{or} \quad a_y^{\beta_{st}=0} = \frac{a_2\, k_{S2}\,(a_1 + a_2)}{m\, a_1\, R_{st}} \tag{9.147}$$

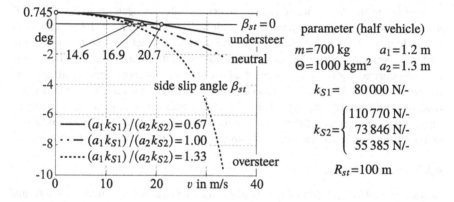

FIGURE 9.21
Steady-state side slip angle as a function of the vehicle velocity.

Note that the side slip angle defined in Equation (9.145) will decay on forward drive with increasing velocity or increasing lateral acceleration regardless of the steering tendency. As a consequence, the sign change defined by Equation (9.147) is just affected by the magnitude of the cornering stiffness k_{S2} at the rear axle.

9.3.7.3 Curve Radius

Usually a driver estimates the required steering angle $\delta = \delta_0$ when entering a curve of a given or rated radius R. However, the actual driven radius R_{st} depends on the vehicle speed v and the steering tendency of the vehicle. Rearranging Equation (9.141) yields

$$R_{st} = \frac{a_1 + a_2}{\delta_0} + m \frac{v |v|}{\delta_0} \frac{a_2\, k_{S2} - a_1\, k_{S1}}{k_{S1}\, k_{S2}\,(a_1 + a_2)} \tag{9.148}$$

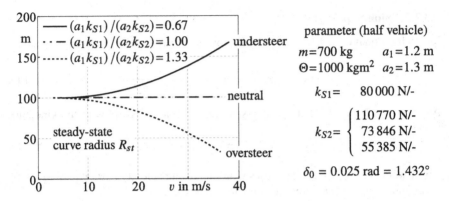

FIGURE 9.22
Curve radius at steady-state cornering with different velocities.

The actual driven curve radius R_{st} is plotted in Figure 9.22 versus the driving speed v for different steering tendencies. The steering angle has been set to $\delta_0 = 1.4321°$ here, in order to let the vehicle drive a circle with the radius $R_0 = 100\ m$ at $v \to 0$. Depending on the steering tendency of the vehicle, the actual driven curve radius will increase, remain constant, or even decrease with the vehicle velocity. In order to keep the desired path, the driver must adjust the steering input appropriately in the case of a vehicle with an under- or an oversteer tendency. If a vehicle with understeer tendency is cornering fast, the driver will usually over-estimate the required steering input. As a consequence, the vehicle will drive a curve with a smaller radius, which increases the lateral acceleration and may result in a critical driving situation.

9.3.7.4   Lateral Slips

At a steady state, $\dot{\beta}_{st} = 0$ and $\dot{\omega}_{st} = 0$ hold. Then, the equations of motion (9.109) and (9.110) can be resolved for the lateral forces

$$F_{y1_{st}} = \frac{a_2}{a_1 + a_2}\, m\, \frac{v^2}{R_{st}}$$
$$\text{or} \qquad \frac{a_1}{a_2} = \frac{F_{y2_{st}}}{F_{y1_{st}}} \qquad (9.149)$$
$$F_{y2_{st}} = \frac{a_1}{a_1 + a_2}\, m\, \frac{v^2}{R_{st}}$$

where Equation (9.134) was taken into account. With the linear tire model in Equation (9.92) one gets in addition

$$F_{y1}^{st} = k_{S1}\, s_{y1}^{st} \quad \text{and} \quad F_{y2}^{st} = k_{S2}\, s_{y2}^{st} \qquad (9.150)$$

where s_{yA1}^{st} and s_{yA2}^{st} denote the steady-state lateral slips at the front and rear axles. Now, the Equations (9.149) and (9.150) deliver

$$\frac{a_1}{a_2} = \frac{F_{y2}^{st}}{F_{y1}^{st}} = \frac{k_{S2}\, s_{y2}^{st}}{k_{S1}\, s_{y1}^{st}} \qquad \text{or} \qquad \frac{a_1\, k_{S1}}{a_2\, k_{S2}} = \frac{s_{y2}^{st}}{s_{y1}^{st}} \qquad (9.151)$$

Hence, the steering tendency of a vehicle is also indicated by the lateral slip ratio. According to Equation (3.72) the lateral slip is strongly related to the slip angle. As a consequence, the slip angles at the front axle will be larger than the slip angles at the rear axle ($s_{y1}^{st} > s_{y2}^{st}$) if a vehicle shows the tendency to understeer ($a_1 k_{S1} < a_2 k_{S2}$) at steady-state cornering.

9.3.8 Influence of Wheel Load on Cornering Stiffness

With identical tires at the front and rear axles, given a linear influence of wheel load on the increase of the lateral force over the lateral slip,

$$k_{S1}^{lin} = k_S F_{z1} \quad \text{and} \quad k_{S2}^{lin} = k_S F_{z2} \tag{9.152}$$

holds. The weight of the vehicle $G = mg$ is distributed over the axles according to the position of the center of gravity,

$$F_{z1} = \frac{a_2}{a_1 + a_2} G \quad \text{and} \quad F_{z2} = \frac{a_1}{a_1 + a_2} G \tag{9.153}$$

With Equation (9.152) and Equation (9.153) one obtains

$$a_1 k_{S1}^{lin} = a_1 k_S \frac{a_2}{a_1 + a_2} G \tag{9.154}$$

and

$$a_2 k_{S2}^{lin} = a_2 k_S \frac{a_1}{a_1 + a_2} G \tag{9.155}$$

Thus, a vehicle with identical tires would be steering neutrally with a linear influence of the wheel load on the cornering stiffness, because of

$$a_1 k_{S1}^{lin} = a_2 k_{S2}^{lin} \tag{9.156}$$

In practice, the lateral force is applied behind the center of the contact patch at the caster offset distance. Hence, the lever arms of the lateral forces will change to $a_1 \rightarrow a_1 - \frac{v}{|v|} n_{L_1}$ and $a_2 \rightarrow a_2 + \frac{v}{|v|} n_{L_1}$, which stabilizes the vehicle, independently from the driving direction.

For a real tire, a degressive influence of the wheel load on the tire forces is observed, Figure 9.23. According to Equation (9.110), the rotation of the vehicle is stable if the torque from the lateral forces F_{y1} and F_{y2} is aligning, i.e.,

$$a_1 F_{y1} - a_2 F_{y2} < 0 \tag{9.157}$$

holds. For a vehicle with the wheel base $a = 2.45$ m, the axle loads $F_{z1} = 4000$ N and $F_{z2} = 3000$ N yield the position of the center of gravity at $a_1 = 1.05$ m and $a_2 = 1.40$ m. For equal slip angles on the front and rear axles, one gets from the table in Figure 9.23, which holds for a specific side slip angle, the lateral forces $F_{y1} = 2576$ N and $F_{y2} = 2043$ N. With this, the condition in Equation (9.157) yields $1.05 * 2576 - 1.45 * 2043 = -257.55$. The value is significantly negative and thus stabilizing.

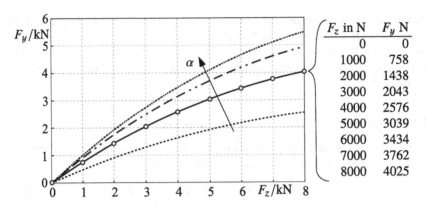

F_z in N	F_y N
0	0
1000	758
2000	1438
3000	2043
4000	2576
5000	3039
6000	3434
7000	3762
8000	4025

FIGURE 9.23
Lateral force F_y versus wheel load F_z at different slip angles α.

Vehicles with $a_1 < a_2$ have a stable, i.e., understeering driving behavior. If the axle load at the rear axle is larger than at the front axle ($a_1 > a_2$), generally a stable driving behavior can only be achieved with different tires.

At increasing lateral acceleration, the vehicle is more and more supported by the outer wheels. The wheel load differences can differ at a sufficiently rigid vehicle body, because of different kinematics (roll support) or different roll stiffness. Due to the degressive influence of wheel load, the lateral force at an axle decreases with increasing wheel load difference. If the wheel load is split more strongly at the front axle than at the rear axle, the lateral force potential at the front axle will decrease more than at the rear axle and the vehicle will become more stable with an increasing lateral force, i.e., more understeering.

9.4   Mechatronic Systems

9.4.1   Electronic Stability Program (ESP)

Electronic Stability Program (ESP) is the generic term for systems designed to improve a vehicle's handling, particularly at the limits where the driver might lose control of the vehicle. Robert Bosch GmbH was the first to deploy an ESP system, called the Electronic Stability Program, that was used by Mercedes-Benz. ESP compares the driver's intended direction in steering and braking inputs, to the vehicle's response, via lateral acceleration, rotation (yaw), and individual wheel speeds. ESP then brakes individual front or rear wheels and/or reduces excess engine power as needed to help correct understeer or oversteer, Figure 9.24. ESP also integrates all-speed traction control, which senses drive-wheel slip under acceleration and individually brakes

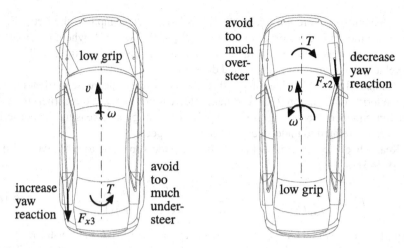

FIGURE 9.24
ESP braking concepts.

the slipping wheel or wheels, and/or reduces excess engine power, until control is regained. ESP combines anti-lock brakes, traction control, and yaw control.

9.4.2 Rear-Wheel Steering

The term Bu in the state equation (9.113) represents the steer input, where u just contains the steering angle δ at the fictitious front wheel and B is a 2×1 matrix. The straightforward extension

$$
Bu = \begin{bmatrix} \dfrac{v}{|v|} \dfrac{k_{S1}}{m_F \, |v|} & \dfrac{v}{|v|} \dfrac{k_{S2}}{m_F \, |v|} \\[2ex] \dfrac{v}{|v|} \dfrac{a_1 \, k_{S1}}{\Theta_F} & -\dfrac{v}{|v|} \dfrac{a_2 \, k_{S2}}{\Theta_F} \end{bmatrix} \begin{bmatrix} \delta_1 \\[1ex] \delta_2 \end{bmatrix} \tag{9.158}
$$

takes a steered rear axle into account, where δ_1 and δ_2 are describing the steering angles at the fictitious front and rear wheels. In a basic approach, the rear steering angle is made proportional to the front steering angle

$$
\delta_2 = q \, \delta_1 \tag{9.159}
$$

where q is a dimensionless parameter, yet to determine properly. Similar to Equation (9.148) a steer input of $\delta_1 = \delta_0$ and $\delta_2 = q\delta_0$ will result in a steady-state cornering radius of

$$
R_{st} = \frac{a_1 + a_2}{\delta_0 \, (1 - q)} + m \, \frac{v^2}{\delta_0 \, (1 - q)} \, \frac{a_2 \, k_{S2} - a_1 \, k_{S1}}{k_{S1} \, k_{S2} \, (a_1 + a_2)} \tag{9.160}
$$

where forward drive with $v \geq 0$ is assumed hereby.

Values of $q > 1$ are not practical, because then a steer input to the left ($\delta_0 > 0$) will cause the vehicle to corner to the right ($R_{st} < 0$). At $q = 1$, where the front and the rear wheels are steered in parallel and the cure radius tends to infinity $R_{st} \to \infty$, the vehicle is moving just laterally and does not corner at all.

Compared to $q = 0$, values of $q < 0$ or $q > 0$ increase or decrease the denominator in both terms of Equation 9.159 and hence decrease or increase the cornering radius as a consequence. According to Figure 9.22 increasing or decreasing the cornering radius R_{st} is related to understeer or oversteer, respectively.

Rear-wheel steering affects the steady-state side slip angle too. Similar to Equation (9.145) one gets

$$\beta^{st} = \frac{a_2 + a_1\, q}{R_{st}\,(1-q)} + m_F\, \frac{v^2}{R_{st}}\, \frac{a_2\, k_{S2}\, q - a_1\, k_{S1}}{k_{S1}\, k_{S2}\,(a_1 + a_2)\,(1-q)} \tag{9.161}$$

where forward drive with $v \geq 0$ is assumed again. Most drivers are disturbed by the changes in the side slip angle, which occur on slalom courses in particular. That is why, a control parameter q that keeps the steady-state side slip angle at zero value is a commonly used approach. Equation (9.161) delivers the corresponding control parameter as

$$q = q_v = \frac{k_{S1}\, a_1\, m_F\, v^2 - k_{S2}\, a_2\,(a_1 + a_2)}{k_{S2}\, a_2\, m_F\, v^2 + k_{S1}\, a_1\,(a_1 + a_2)} \tag{9.162}$$

The control parameter defined by Equation (9.162) depends on the layout of the vehicle and on the driving velocity, Figure 9.25. The position of the center of gravity

$m = 2000$ kg, $a_1 = 1.475$ m, $a_2 = 1.425$ m, $c_{S1} = 150000$ N/-, $c_{S2} = 180000$ N/-

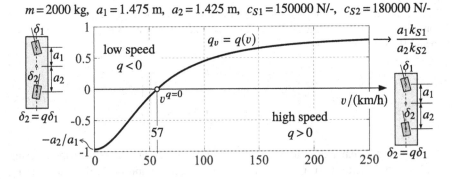

FIGURE 9.25
Control parameter $q_v = q(v)$ for a rear-wheel steering with zero side slip angle.

defined by a_1 and a_2 as well as the cornering stiffness at the front and rear axles k_{S1} and k_{S2} determine the low- and high-speed limit values

$$q_v^{v \to 0} = -\frac{a_2}{a_1} \quad \text{and} \quad q_v^{v \to \infty} = \frac{a_1\, k_{S1}}{a_2\, k_{S2}} \tag{9.163}$$

As stated above, values of $q > 1$ are not practical. Hence, this approach is limited to vehicles with an understeer tendency, where according to Table 9.2, $a_1 k_{S1} < a_2 k_{S2}$

will hold. The change from negative to positive values of the control parameter occurs at a velocity of

$$v^{q=0} = \sqrt{\frac{a_2 k_{S2}(a_1+a_2)}{a_1 m}} \qquad (9.164)$$

At negative values of the control parameter ($q < 0$) the rear wheels are steered in the opposite direction of the front wheels, thus reducing the steady-state curve radius R_{st} according to Equation (9.160) and enhancing the maneuverability significantly.

At positive values in the practical range of $0 < q < 1$, the wheels at the front and rear are steered in the same direction (in phase), thus increasing the steady-state cornering radius R_{st} according to Equation (9.160). This corresponds with an enhanced understeer tendency as demonstrated in Figure 9.22. But the in phase steering angle at the rear wheels simply amplifies the lateral and reduces the yaw response to a steering wheel input. As a result, the vehicle is easier to control by the driver, in particular at emergency like fast double lane change maneuvers.

In most passenger cars, the rear steering angle δ_2 is limited to a value δ_2^{max} of just a few degrees. Hence, the simple control law in Equation (9.159) with a parameter defined by Equation (9.162) cannot be realized in all driving situations. The restriction $|\delta_2| \leq \delta_2^{max}$ has an impact in particular in the low speed range because just here large steering angles will occur. The control will no longer nullify the steady-state side slip angle then, but will still enhance the maneuverability at low speeds.

Note, that the control law in Equation 9.159 will not affect the stability of the vehicle, because the state matrix A defined by Equation (9.113) remains unchanged.

A more detailed discussion of rear-wheel or all-wheel steering respectively, which includes also a proposal for a feedback control, is provided in [28].

9.4.3 Steer-by-Wire

Modern steer-by-wire systems can improve the handling properties of vehicles [66]. Usually an electronically controlled actuator is used to convert the rotation of the steering wheel into steer movements of the wheels. Steer-by-wire systems are based on mechanics, micro-controllers, electro-motors, power electronics, and digital sensors. At present, fail-safe systems with a mechanical backup system are under investigation. The potential of a modern active steering system can be demonstrated by the maneuver of braking on a μ-split [48]. The layout of a modern steering system and the different reactions of the vehicle are shown in Figure 9.26. The coefficient of friction on the left side of the vehicle is supposed to be 10% of the normal friction value on the right side. The vehicle speeds to $v = 130\,km/h$ and then the driver applies full brake pressure and fixes the steering wheel like he would do first in a panic reaction. During the whole maneuver, the anti-lock brake system was disabled. The different brake forces on the left and right tires make the car spin around the vertical axis.

Only skilled drivers will be able to stabilize the car by counter steering. The success of counter steering depends on the reactions in the first few seconds. A controller, that takes appropriate actions at the steering angle, can assist the driver's task.

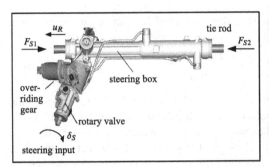
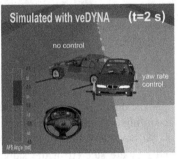

FIGURE 9.26
Steering system with over-riding gear and animated simulation screenshot of braking on μ-split with a standard and an active steering system.

Active front-wheel steering and direct yaw control as well as advanced driver assistance systems and autonomous driving are discussed in [28].

Exercises

9.1 A passenger car with a wheel base of $a = 2.6$ m is equipped with rear-wheel steering. The vehicle has a width of $b = 1.8$ m and the distance from the front axle to the front part of the vehicle is determined by $f = 0.8$ m. The track width at the front and the rear axles is given by $s = 1.5$ m.
In low-speed cornering, the rear wheels are steered opposite to the front wheels. The maximum steering angles are limited to $\delta_F \leq \pm 45°$ at the front and to $\delta_R \leq \pm 5°$ at the rear wheels.

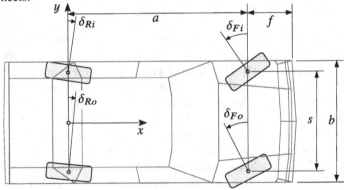

By assuming that the inner wheels are steered to their maximum values and based on the Ackermann geometry, the following quantities are to be calculated:

a) the coordinates x_M, y_M of the momentary turning center measured from the rear axle center,

b) the minimum turning circle diameter,

c) the maximum space requirement of the vehicle, and

d) the steering angles at the outer wheels.

9.2 A ready-mix truck with the given dimensions is equipped with two steerable front axles. The maximum steering angle of the inner wheel at the first front axle amounts to $\delta_{11}^{max} = 48°$.

To improve the cornering performance, the second rear axle is
designed to be steered too.

Compute the cornering radius R
measured from the turning pivot
point to wheel 11 and all steering
angles according to the Ackermann
geometry.

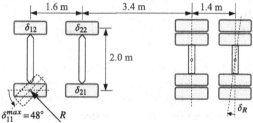

9.3 Compute the space requirement of a heavy-duty dumper with the given dimensions when cornering at a bend angle of 40°.

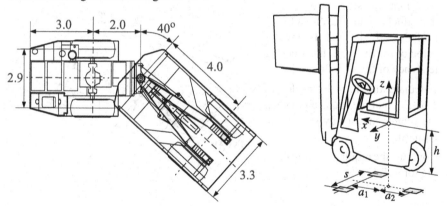

9.4 A three-wheeled forklift has a track width of $s = 1.0$ m at the front axle. The wheel base is defined by $a_1 = a_2 = 0.6$ m. The coefficient of friction between the tires and the ground is given by $\mu = 0.9$.

Compute the rollover threshold for the lateral acceleration when the center of gravity is located $h = 1.1$ m above the ground. Identify the potential line of rollover first and follow the approach in Section 9.2.2. Compute the maximum deceleration too.

9.5 Use the MATLAB-Script provided by Listing 9.2 and the MATLAB-Function in Listing 9.1 to investigate the cornering behavior of a tractor semi-trailer. Set $a = 3.5$ m, $b = -0.25$ m and $c = 6.0$ m.

Find a steering input that generates a double lane change with a lateral deviation of 5 m at a speed of 5 km/h.

9.6 The track width at the twin-tired rear axle of a bus is given by $s = 1.9$ m. The distances $h = 0.7$ m and $H = 2.1$ m define the height of the roll center and the height of the center of gravity.

The wheel loads $F_{z1} = F_{z2} = 60$ kN are measured at the rear axle while driving straight ahead.

The vertical stiffness of one single rear tire amounts to $c_R = 1000$ kN/m and the roll stiffness of the rear axle suspension is given by $c_W = 850$ kNm/rad.

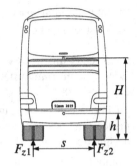

At what lateral acceleration will the bus will start to roll over?

How large is the roll angle of the bus?

9.7 A rear-wheel driven vehicle drives on a circle of radius $R = 40$ m with constant velocity. In the vehicle-fixed axis system, the acceleration components $a_{xV} = -0.133$ m/s^2 and $a_{yV} = 3.998$ m/s^2 are measured in the longitudinal and lateral direction. The average steering angle at the front axle is given by $\delta = 6.2°$. The vehicle has a mass $m = 1600$ kg. The distances $a_1 = 1.3$ m and $a_2 = 1.4$ m define the position of the center of gravity.

Determine the side slip angle β at the vehicle center C, the velocity v of the vehicle, and its angular velocity ω. Evaluate the steering tendency of the vehicle.

Calculate the lateral tire forces F_{y1}, F_{y2} that are required to keep the vehicle in a steady state when cornering.

Compute the required overall driving force F_D.

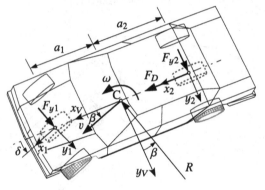

10

Driving Behavior of Single Vehicles

CONTENTS

10.1   Three-Dimensional Vehicle Model

10.1.1   Model Structure

As mentioned in Section 1.5, vehicles can be modeled by multibody systems. A full vehicle model consists of the vehicle framework supplemented by eventually separate modules for the steering system, the drive train, the tires, the load, passenger/seat models, an elastically suspended engine, and in the case of heavy trucks, by an elastically suspended driver's cab. The vehicle framework represents the kernel of such a full model. It includes at least the module chassis and modules for the wheel/axle suspension systems. Typical passenger cars are characterized by a sufficiently rigid chassis and independent wheel axle suspension systems. As an extension to the enhanced planar model, described in Section 8.5.1, the three-dimensional model for the vehicle framework now consists of $n = 9$ rigid bodies: four knuckles, four wheels, and the chassis, Figure 10.1. The vehicle-fixed axis system V is located in the middle of the

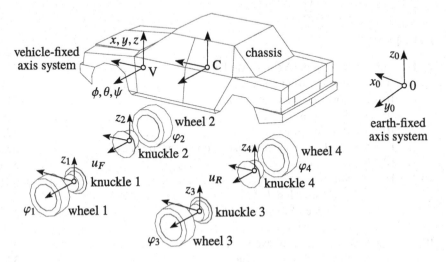

FIGURE 10.1
Bodies of the framework of a three-dimensional vehicle model.

front wheel centers; the x_V-axis points forward, the y_V-axis to the left, and the z_V-axis upward.

10.1.2   Position and Orientation

The position of the vehicle-fixed axis system V with respect to the earth-fixed axis system 0 is described by the components x, y, z of the position vector

$$r_{0V,0} = \begin{bmatrix} x & y & z \end{bmatrix}^T \tag{10.1}$$

Its orientation is defined by the rotation matrix

$$A_{0V} = \underbrace{\begin{bmatrix} \cos\psi & -\sin\psi & 0 \\ \sin\psi & \cos\psi & 0 \\ 0 & 0 & 1 \end{bmatrix}}_{A_\psi} \underbrace{\begin{bmatrix} \cos\theta & 0 & \sin\theta \\ 0 & 1 & 0 \\ -\sin\theta & 0 & \cos\theta \end{bmatrix}}_{A_\theta} \underbrace{\begin{bmatrix} 1 & 0 & 0 \\ 0 & \cos\phi & -\sin\phi \\ 0 & \sin\phi & \cos\phi \end{bmatrix}}_{A_\phi} \tag{10.2}$$

where ψ denotes the yaw angle and θ and ϕ characterize the pitch and roll motion. Now, the position and the orientation of the chassis are simply given by

$$r_{0C,0} = r_{0V,0} + A_{0V} \, r_{VC,V} \quad \text{and} \quad A_{0C} = A_{0V} \tag{10.3}$$

where the vector $r_{VC,V}$, which describes the position of the chassis center C relative to the origin V of the vehicle-fixed axis system, is constant and defined by the mass distribution of the chassis. Each wheel is supposed to be fully balanced. Then, its center is located on the rotation axis. In addition, it is assumed that the center of the corresponding knuckle will be sufficiently close by or will even coincide with the wheel center. As a consequence, the position of each knuckle, and simultaneously of each wheel, is defined by the vector

$$r_{0i,0} = r_{0V,0} + A_{0V} \, r_{Vi,V}, \quad i = 1(1)4 \tag{10.4}$$

If the rotation matrices A_{Vi}, $i = 1(1)4$ describe the orientation of each knuckle-fixed axis system relative to the vehicle-fixed axis system, then the rotation matrices

$$A_{0i} = A_{0V} \, A_{Vi}, \quad i = 1(1)4 \tag{10.5}$$

will define their orientation with respect to the earth-fixed axis system. A purely kinematic suspension model describes the position and orientation of each knuckle as a function of the jounce and rebound motion as well as the steering motion. The vertical motion z_i of each knuckle $i = 1(1)4$ relative to the chassis or, as done in Section 5.4, the rotation angle of the lower wishbone may be used to characterize the jounce and rebound motion. Assuming a rack-and-pinion steering system at both axles, the rack movements u_F and u_R at the front and rear axles will fully describe the steering motion at the corresponding axle. Then, the position and orientation of each knuckle and wheel center relative to the vehicle-fixed axis system is defined at the front ($i = 1, 2$) and at the rear axle ($i = 3, 4$) by

$$r_{Vi,V} = r_{Vi,V} \, (z_i, u_F), \quad A_{Vi} = A_{Vi} \, (z_i, u_F), \quad i = 1, 2 \tag{10.6}$$

$$r_{Vi,V} = r_{Vi,V} \, (z_i, u_R), \quad A_{Vi} = A_{Vi} \, (z_i, u_R), \quad i = 3, 4 \tag{10.7}$$

Finally, the angles φ_i, $i = 1(1)4$ describe the rotation of each wheel relative to the corresponding knuckle.

10.1.3 Velocities

Expressing the absolute velocity and the absolute angular velocity of the vehicle-fixed axis system in this axis system results in

$$v_{0V,V} = A_{0V}^T \, \dot{r}_{0V,0} = A_{0V}^T \begin{bmatrix} \dot{x} \\ \dot{y} \\ \dot{z} \end{bmatrix} \tag{10.8}$$

$$
\begin{aligned}
\omega_{0V,V} &= \begin{bmatrix} \dot{\phi} \\ 0 \\ 0 \end{bmatrix} + A_\phi^T \left\{ \begin{bmatrix} 0 \\ \dot{\theta} \\ 0 \end{bmatrix} + A_\theta^T \begin{bmatrix} 0 \\ 0 \\ \dot{\psi} \end{bmatrix} \right\} \\[2mm]
&= \begin{bmatrix} 1 & 0 & -\sin\theta \\ 0 & \cos\phi & \sin\phi\cos\theta \\ 0 & -\sin\phi & \cos\phi\cos\theta \end{bmatrix} \begin{bmatrix} \dot{\phi} \\ \dot{\theta} \\ \dot{\psi} \end{bmatrix} = K_R \begin{bmatrix} \dot{\phi} \\ \dot{\theta} \\ \dot{\psi} \end{bmatrix}
\end{aligned} \tag{10.9}
$$

where the components of $v_{0V,V}$ and $\omega_{0V,V}$ will be used as generalized velocities further on. Now, the absolute velocity and the angular velocity of the chassis are defined by

$$
v_{0C,V} = v_{0V,V} + \omega_{0V,V} \times r_{VC,V} \quad \text{and} \quad \omega_{0C,V} = \omega_{0V,V} \tag{10.10}
$$

At first, the time derivative of Equation (10.4) results in

$$
\dot{r}_{0i,0} = \underbrace{\dot{r}_{0V,0}}_{v_{0V,0}} + \omega_{0V,0} \times A_{0V}\, r_{Vi,V} + A_{0V}\, \dot{r}_{Vi,V}, \quad i = 1(1)4 \tag{10.11}
$$

Next, the transformation into the vehicle-fixed axis system yields

$$
v_{0i,V} = v_{0F,V} + \omega_{0V,V} \times r_{Vi,V} + \dot{r}_{Vi,V}, \quad i = 1(1)4 \tag{10.12}
$$

where the velocity state of the vehicle-fixed axis system, characterized by $v_{0V,V}$ and $\omega_{0V,V}$, is defined in Equations (10.8) and (10.9). The time derivatives of the position vectors, provided by Equations (10.6) and (10.7), result in

$$
\begin{aligned}
\dot{r}_{Vi,V} &= \frac{\partial r_{Vi,V}}{\partial z_i} \dot{z}_i + \frac{\partial r_{Vi,V}}{\partial u_F} \dot{u}_F = t_{zi,V}\, \dot{z}_i + t_{ui,V}\, \dot{u}_F, \quad i = 1, 2 \\[2mm]
\dot{r}_{Vi,V} &= \frac{\partial r_{Vi,V}}{\partial z_i} \dot{z}_i + \frac{\partial r_{Vi,V}}{\partial u_R} \dot{u}_R = t_{zi,V}\, \dot{z}_i + t_{ui,V}\, \dot{u}_R, \quad i = 3, 4
\end{aligned} \tag{10.13}
$$

where $t_{zi,V}$ and $t_{ui,V}$, $i = 1(1)4$ abbreviate the corresponding partial velocities. Similarly, the angular velocities of the knuckles may be written as

$$
\begin{aligned}
\omega_{0Ki,V} &= \omega_{0V,V} + d_{zi,V}\, \dot{z}_i + d_{ui,V}\, \dot{u}_F, \quad i = 1, 2 \\
\omega_{0Ki,V} &= \omega_{0V,V} + d_{zi,V}\, \dot{z}_i + d_{ui,V}\, \dot{u}_R, \quad i = 3, 4
\end{aligned} \tag{10.14}
$$

where $d_{zi,V}$ and $d_{ui,V}$, $i = 1(1)4$ abbreviate the corresponding partial angular velocities. Finally, the absolute angular velocity of each wheel is given by

$$
\begin{aligned}
\Omega_{0i,V} &= \omega_{0V,V} + d_{zi,V}\, \dot{z}_i + d_{ui,V}\, \dot{u}_F + A_{Vi}\, e_{yRi,i}\, \dot{\varphi}_i, \quad i = 1, 2 \\
\Omega_{0i,V} &= \omega_{0V,V} + d_{zi,V}\, \dot{z}_i + d_{ui,V}\, \dot{u}_R + A_{Vi}\, e_{yRi,i}\, \dot{\varphi}_i, \quad i = 3, 4
\end{aligned} \tag{10.15}
$$

where the unit vectors $e_{yRi,i}$, $i = 1(1)4$, describe the orientation of the wheel rotation axis in the design position. They are defined by a wheel alignment point or via the toe and camber angles as discussed in Section 1.3.5.

The partial velocities and the partial angular velocities of the vehicle model consisting of $n = 9$ model bodies are collected in Table 10.1 and Table 10.2.

TABLE 10.1

Partial velocities of a three-dimensional vehicle model

Body Name Mass	\multicolumn{12}{c}{Partial Velocities $\partial v_{0i,V}/\partial z_j$}
	\multicolumn{2}{c}{Vehicle}	\multicolumn{3}{c}{Front Susp.}	\multicolumn{3}{c}{Rear Susp.}	\multicolumn{4}{c}{Wheels}
	$v_{0V,V}$ v_x,v_y,v_z	$\omega_{0V,V}$ $\omega_x,\omega_y,\omega_z$	$\dot{z}_1$	$\dot{z}_2$	$\dot{u}_F$	$\dot{z}_3$	$\dot{z}_4$	$\dot{u}_R$	ω_1	ω_2	ω_3	ω_4
Knuckle + wheel 1 $m_1 = m_{K1} + m_{W1}$	$I_{3\times3}$	$\tilde{r}_{V1,V}^T$	t_{z1}	0	t_{u1}	0	0	0	0	0	0	0
Knuckle + wheel 2 $m_2 = m_{K2} + m_{W2}$	$I_{3\times3}$	$\tilde{r}_{V2,V}^T$	0	t_{z2}	t_{u2}	0	0	0	0	0	0	0
Knuckle + wheel 3 $m_3 = m_{K3} + m_{W3}$	$I_{3\times3}$	$\tilde{r}_{V3,V}^T$	0	0	0	t_{z3}	0	t_{u3}	0	0	0	0
Knuckle + wheel 4 $m_4 = m_{K4} + m_{W4}$	$I_{3\times3}$	$\tilde{r}_{V4,V}^T$	0	0	0	0	t_{z4}	t_{u4}	0	0	0	0
Chassis m_C	$I_{3\times3}$	$\tilde{r}_{VC,V}^T$	0	0	0	0	0	0	0	0	0	0

TABLE 10.2

Partial angular velocities of a three-dimensional vehicle model

Body name	Inertia	\multicolumn{12}{c}{Partial Angular Velocities $\partial \omega_{0i,V}/\partial z_j$}
		\multicolumn{2}{c}{Vehicle}	\multicolumn{3}{c}{Front Susp.}	\multicolumn{3}{c}{Rear Susp.}	\multicolumn{4}{c}{Wheels}
		$v_{0V,V}$ v_x,v_y,v_z	$\omega_{0V,V}$ $\omega_x,\omega_y,\omega_z$	$\dot{z}_1$	$\dot{z}_2$	$\dot{u}_F$	$\dot{z}_3$	$\dot{z}_4$	$\dot{u}_R$	ω_1	ω_2	ω_3	ω_4
Knuckle 1 Θ_{K1}		0	$I_{3\times3}$	d_{z1}	0	d_{u1}	0	0	0	0	0	0	0
Wheel 1 Θ_{W1}		0	$I_{3\times3}$	d_{z1}	0	d_{u1}	0	0	0	e_{yR1}	0	0	0
Knuckle 2 Θ_{K2}		0	$I_{3\times3}$	0	d_{z2}	d_{u2}	0	0	0	0	0	0	0
Wheel 2 Θ_{W2}		0	$I_{3\times3}$	0	d_{z2}	d_{u2}	0	0	0	0	e_{yR2}	0	0
Knuckle 3 Θ_{K3}		0	$I_{3\times3}$	0	0	0	d_{z3}	0	d_{u3}	0	0	0	0
Wheel 3 Θ_{W3}		0	$I_{3\times3}$	0	0	0	d_{z3}	0	d_{u3}	0	0	e_{yR3}	0
Knuckle 4 Θ_{K4}		0	$I_{3\times3}$	0	0	0	0	d_{z4}	d_{u4}	0	0	0	0
Wheel 4 Θ_{W4}		0	$I_{3\times3}$	0	0	0	0	d_{z4}	d_{u4}	0	0	0	e_{yR4}
Chassis Θ_C		0	$I_{3\times3}$	0	0	0	0	0	0	0	0	0	0

The cross products in the velocity equations between the angular velocity and the position vectors are substituted via $\omega \times r = -r \times \omega = -\tilde{r}\,\omega = \tilde{r}^T \omega$ by multiplication with the corresponding skew symmetric matrices and $I_{3\times3}$ denotes the 3×3 matrix of identity, hereby.

In this model approach it is assumed that the centers of each knuckle and wheel will coincide. That is why knuckle and wheel are summarized as one body in Table 10.1. The components v_x, v_y, v_z of the velocity $v_{0V,V}$ and the components $\omega_x, \omega_y, \omega_z$ of the angular velocity $\omega_{0V,V}$; the time derivatives of vertical wheel center displacements $\dot{z}_1$ to $\dot{z}_2$; the time derivatives of the lateral rack movements $\dot{u}_F$ and $\dot{u}_R$; as well as the wheel angular velocities $\omega_i = \dot{\varphi}_i$, $i = 1(1)4$, are used as generalized velocities here. The three-dimensional vehicle model therefore has $f = 3+3+4+2+4 = 16$ degrees of freedom.

All partial velocities and all inertia tensors are expressed in the vehicle-fixed reference frame V. In case of double wishbone suspension systems, the partial velocities and the partial angular velocities are provided in Section 5.4. For arbitrary suspension systems, the position and orientation of each knuckle and wheel as well as the required derivatives may be provided via the design kinematics discussed in Section 5.5.

10.1.4 Accelerations

Now, the absolute acceleration of the chassis, expressed in the vehicle-fixed axis system, is obtained as

$$a_{0C,V} = \dot{v}_{0V,V} + \dot{\omega}_{0V,V} \times r_{VC,V} + \omega_{0V,V} \times v_{0C,V} \qquad (10.16)$$

where the fact that the vector $r_{VC,V}$ is constant was already taken into account. The last term, which does not depend on the time derivatives of the generalized velocities $\dot{v}_{0V,V}$ or $\dot{\omega}_{0V,V}$ represents the remaining term $a_{0C,V}^R$ here. The absolute angular acceleration, expressed in the vehicle-fixed axis system, is given by

$$\alpha_{0C,V} = \dot{\omega}_{0C,V} + \omega_{0V,V} \times \omega_{0C,V} = \dot{\omega}_{0V,V} + \omega_{0V,V} \times \omega_{0V,V} = \dot{\omega}_{0V,V} \quad (10.17)$$

and will contain no remaining acceleration terms, $\alpha_{0C,V}^R = 0$. The absolute acceleration of the knuckles and wheel centers, $i = 1(1)4$, is obtained by

$$\begin{aligned} a_{0i,V} = {} & \dot{v}_{0V,V} + \dot{\omega}_{0V,V} \times r_{Vi,V} + t_{zi,V}\ddot{z}_i + t_{ui,V}\ddot{u}_j \\ & + \omega_{0V,V} \times \dot{r}_{Vi,V} + \dot{t}_{zi,V}\dot{z}_i + \dot{t}_{ui,V}\ddot{u}_j + \omega_{0V,V} \times v_{0i,V} \end{aligned} \qquad (10.18)$$

The absolute angular acceleration of the knuckles, $i = 1(1)4$, is given by

$$\begin{aligned} \alpha_{0Ki,V} = {} & \dot{\omega}_{0V,V} + d_{zi,V}\ddot{z}_i + d_{ui,V}\ddot{u}_j \\ & + \dot{d}_{zi,V}\dot{z}_i + \dot{d}_{ui,V}\ddot{u}_j + \omega_{0V,V} \times \omega_{0Ki,V} \end{aligned} \qquad (10.19)$$

and the absolute angular acceleration of the wheels, $i = 1(1)4$, reads as

$$\begin{aligned} \alpha_{0Wi,V} = {} & \dot{\omega}_{0V,V} + d_{zi,V}\ddot{z}_i + d_{ui,V}\ddot{u}_j + A_{Vi}e_{yRi,i}\ddot{\varphi}_{Wi} \\ & + \dot{d}_{zi,V}\dot{z}_i + \dot{d}_{ui,V}\ddot{u}_j + \omega_{Vi,V} \times A_{Vi}e_{yRi,i}\dot{\varphi}_i + \omega_{0V,V} \times \omega_{0Wi,V} \end{aligned} \qquad (10.20)$$

The lateral rack movements were abbreviated by u_j, where $u_j = u_F$ holds at the front axle ($i = 1, 2$) and $u_j = u_R$ at the rear axle ($i = 3, 4$). The last lines in Equations (10.18) to (10.20) are representing the remaining terms $a_{0Ki,V}^R$, $a_{0Wi,V}^R$, and $\alpha_{0Ki,V}^R$, $\alpha_{0Wi,V}^R$, respectively, which in the case of the wheel angular acceleration include the gyroscopic torques generated by the wheel rotation. The parts in the remaining accelerations, which are generated by the time derivatives of the partial velocities and partial angular velocities, are small compared to the other parts in vehicle dynamics and may thus be neglected, c.f. [42].

10.1.5 Applied and Generalized Forces and Torques

Similar to the procedure used in Section 8.5.1, the forces and torques applied to each body may be specified in an additional column in Tables 10.1 and 10.2 and transformed via the partial velocities and the partial angular velocities to the corresponding generalized forces and torques. This method is applied here to the weight and inertia forces and to the gyroscopic torques only. According to Equation (1.31), the contribution of body i to the vector of generalized forces and torques is given by

$$q_{(i)} = \frac{\partial v_{0i,V}^T}{\partial z} \left(m_i g_{,V} - m_i a_{0i,V}^R \right)$$

$$+ \frac{\partial \omega_{0i,V}^T}{\partial z} \left(-\Theta_{i,V} \alpha_{0i,V}^R - \omega_{0i,V} \times \Theta_{i,V} \, \omega_{0i,V} \right)$$

$$i = 1(1)n \qquad (10.21)$$

The three-dimensional model for the vehicle framework consists here of $n = 9$ bodies, which in the case of translatory motions are reduced to five bodies by combining each knuckle and wheel into one body. All terms are expressed in the vehicle-fixed axis system now. The applied torque $T_{i,V}^a$ is omitted, and the applied force $F_{i,0}^a$ is just replaced by the body weight $G_{i,V} = m_i g_{,V}$, where $g_{,V}$ denotes the gravity vector expressed in the vehicle-fixed axis system.

All other applied forces and torques are transformed according to Equation (6.23) to generalized ones by just applying the principle of virtual power. This was already practiced for point-to-point force elements in Section 6.1.5 and for the tire forces and torques in Section 5.3.6.

The general layout of the drive train generates braking torques between knuckle and wheel and transmits driving torques via half-shafts from the chassis-mounted differentials to the wheels. In this model approach, the wheels are described relative to the knuckle and those relative to the vehicle (chassis). Then, similar to the results achieved in Equation (8.68) for the enhanced planar model, the driving and braking torques T_{Di}, T_{Bi}, $i = 1(1)4$, will act directly as generalized torques here in the components $q(13)$ to $q(16)$. Beyond that, the driving torques T_{Di}, $i = 1(1)4$, will generate the additional terms $d_{zi}^T e_{yRi} T_{Di}$, $i = 1(1)4$, in the components $q(7)$, $q(8)$, $q(10)$, $q(11)$, which are related to the jounce and rebound motions z_1 to z_4 here.

10.1.6 Equations of Motion

The equations of motion can be generated now via Jordain's principle of virtual power. According to Section 1.5.3, the equations of motion will be provided in two sets of first-order differential equations:

$$K(y)\, \dot{y} = z \qquad (10.22)$$

$$M(y)\, \dot{z} = q(y, z) \qquad (10.23)$$

Here, the vector of the generalized coordinates and generalized velocities are determined by

$$y = \begin{bmatrix} x, \ y, \ z, \ \phi, \ \theta, \ \psi, \ z_1, \ z_2, \ u_F, \ z_3, \ z_4, \ u_R, \ \varphi_1, \ \varphi_2, \ \varphi_3, \ \varphi_4 \end{bmatrix}^T \qquad (10.24)$$

$$z = \begin{bmatrix} v_x, v_y, v_z, \omega_x, \omega_y, \omega_z, \dot{z}_1, \dot{z}_2, \dot{u}_F, \dot{z}_3, \dot{z}_4, \dot{u}_R, \omega_1, \omega_2, \omega_3, \omega_4 \end{bmatrix}^T \qquad (10.25)$$

The components of the velocity v_x, v_y, v_z and angular velocity ω_x, ω_y, ω_z are related via Equations (10.8) and (10.9) to the corresponding time derivatives $\dot{x}$, $\dot{y}$, $\dot{z}$ and $\dot{\phi}$, $\dot{\theta}$, $\dot{\psi}$ of the generalized coordinates. In addition, the wheel angular velocities $\omega_i = \dot{\varphi}_i$, $i = 1(1)4$ just abbreviate the time derivatives of the corresponding angles. That is why the kinematical differential equation (10.22) can easily be resolved for the time derivative of the vector of generalized coordinates.

According to Equation (1.30), the elements of the 16×16-mass-matrix are defined by the masses and the inertias of each body as well as the partial velocities and the partial angular velocities, which are provided in the Tables 10.1 and 10.2 for the three-dimensional vehicle model.

The three-dimensional model for the vehicle framework must be supplemented with separate models for the tire, the drive train, and the steering system. The sophisticated handling tire model TMeasy was presented in Chapter 3. A model for a standard drive train including differentials, the gear box, the clutch, and the engine is discussed in Section 4.5. However, the drive train with its model and vibrational complexity can be bypassed by simply applying given or controlled driving torques directly to the wheels. A quite simple model approach of a steering system will just take the torsional compliance of the steering column and the ratio of the steering box into account, cf. Section 5.3.6. Then, the rotation of the steering wheel provided as simple time history or by a driver model makes it possible to simulate different driving maneuvers.

10.2   Driver Model

10.2.1   Standard Model

Many driving maneuvers require inputs by the driver at the steering wheel and the gas pedal that depend on the actual state of the vehicle. A real driver takes a lot of information provided by the vehicle and the environment into account. His actions are anticipatory and adapts his reactions to the dynamics of the particular vehicle. The modeling of human actions and reactions is a challenging task. That is why driving simulators operate with real drivers instead of driver models. However, offline simulations will require a suitable driver model.

Usually, driver models are based on simple, mostly linear vehicle models where the motion of the vehicle is reduced to horizontal movements and the wheels on each axle are lumped together [65]. Standard driver models, like the one shown in Figure 10.2, usually consist of two levels: anticipatory feed-forward (open-loop) and compensatory (closed-loop) control. The properties of the vehicle model and the capability of the driver are used to design appropriate transfer functions for the open- and closed-loop control. The model includes a path prediction and takes the reaction time of the driver into account.

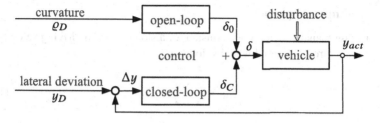

FIGURE 10.2
Two-level control driver model [26].

10.2.2 Enhanced Model

Different from technical controllers, a human driver normally does not simply follow a given trajectory, but sets the target course within given constraints (i.e., road width or lane width), Figure 10.3. At the anticipation level, the optimal trajectory for the

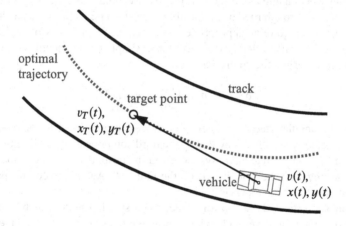

FIGURE 10.3
Enhanced driver model.

vehicle is predicted by repeatedly solving optimal control problems for a nonlinear bicycle model whereas on the stabilization level, a position control algorithm precisely guides the vehicle along the optimal trajectory [64]. The result is a virtual driver who is able to guide the virtual vehicle on a virtual road at high speeds as well as in limit situations where skidding and sliding effects take place. A broad variety of drivers ranging from unskilled to skilled or aggressive to non-aggressive can be described by this driver model [18].

10.2.3   Simple Approach

Many driving maneuvers require a constant or a slowly varying driving velocity. The average of all circumferential wheel velocities,

$$v_W = \frac{1}{4} \sum_{i=1}^{4} r_{Di} \Omega_i \qquad (10.26)$$

approximates the driving velocity v_D quite well. Here, r_{Di} is the dynamic tire radius and Ω_i describes the absolute angular velocity of each wheel, $i = 1(1)4$. Then, a simple *p*-controller,

$$T_D = p_D (v_D - v_W) \qquad (10.27)$$

may be used to generate an appropriate driving torque, which will be distributed to the wheels according to the layout of the driving system. Of course, the control parameter p_D must be adjusted to the specific vehicle properties.

Open-loop maneuvers, like a steering step input, are operated simply by providing the steering wheel angle as a function of time. Keeping a vehicle on a straight line when driving on a rough road or performing a steady-state cornering on a circle with given radius will require an appropriate controller. Again, a P-control applied to the steering wheel angular velocity may serve as a simple driver model. In the case of steady-state cornering, the controller

$$\dot{\delta}_{SW} = p_{SW} \left(\frac{v_D}{R} - \omega_{0F,0}(3) \right) \qquad (10.28)$$

generates the angular steering wheel velocity $\dot{\delta}_{SW}$ proportional to the deviation of actual yaw velocity, which is provided by the third component $\omega_{0F,0}(3)$ of the angular velocity of the vehicle-fixed axis system, from the desired yaw velocity $\omega_z = v_D/R$ when cornering with the velocity v_D on the radius R. Again, the control parameter p_{SW} must be chosen appropriately.

A more sophisticated PD-control applied to the steering angle, which incorporates the reaction time of a driver and makes it possible to perform a double lane change, is presented in [6].

10.3   Standard Driving Maneuvers

10.3.1   Steady-State Cornering

The steering tendency of a real vehicle is determined by the driving maneuver called steady-state cornering. The maneuver is performed quasi-static. The driver tries to keep the vehicle on a circle with the given radius R. He slowly increases the driving speed v and, due to $a_y = v^2/R$, also the lateral acceleration until reaching the limit. Whereas in the simulation a perfectly flat road is easily realized, a field test will usually be characterized by slight disturbances induced by a nonperfect road surface,

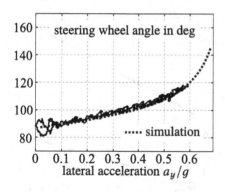 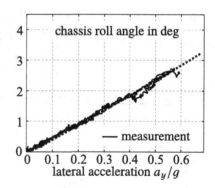

FIGURE 10.4
Steady-state cornering: Front-wheel driven passenger car.

Figure 10.4. The simulation results, obtained with a sophisticated three-dimensional vehicle model [61], match quite well with the measurements. The passenger car has a mass of $m = 1450$ kg and a wheel base of $a = 2.7$ m. The center of gravity is located closer to the front axle, $a = 1.1$ m, which is typical for front-wheel driven cars. The maneuver was performed on a circle of radius $R = 30$ m here. To drive this radius with low lateral acceleration, $a_y \to 0$, a steering angle at the front wheel of $\delta_A = \arctan(2.7/30) = 0.0898 \equiv 5.143°$ will be required. Inspecting Figure 10.4 yields a steering wheel angle of $\delta_{SW}^{a_y \to 0} \approx 87°$, which results in a overall steering ratio of $i_S = 87°/5.143° = 17$. The inclination in the diagram steering angle versus lateral velocity decides about the steering tendency and stability behavior. The tendency to understeer is significant in the whole acceleration range here. When approaching the limit range, which may be estimated by $a_y^{max} \approx 0.8g$, the understeer tendency and, as a consequence, the stability of the vehicle too, increase more and more. The roll angle of the vehicle increases in proportion to the lateral acceleration with a ratio of approximately $4.8°/g$.

Typical results for a rear-wheel driven car are displayed in Figure 10.5. The passenger car has mass of $m = 1300$ kg and a wheel base $a = 2.57$ m. The center of gravity is located nearly in the middle, at $a = 1.25$ m, which is quite typical for rear-wheel driven cars. The vehicle is driven on a radius of $R = 100$ m now, which means that the potential limit range $a_y^{limit} = 0.8g$ is reached at velocity

$$v^{limit} = \sqrt{a_y^{limit} R} = \sqrt{0.8 * 9.81 * 100} = 28 \text{ m/s} \quad \text{or} \quad v^{limit} = 100.8 \text{ km/h}$$
(10.29)

This corresponds quite well with the typical travel velocity on country roads in Europe. Estimating the overall steering ratio by inspecting the upper-left plot in Figure 10.5 and computing the Ackermann angle results in $i_S = \delta_{SW}/\delta_A = 30°/1.4722° = 20.4$. In the low acceleration range, the steering tendency is nearly neutral, which is the result of nearly equal loads at the front and rear axles. The nonlinear influence of the wheel load on the tire performance is used here to design a vehicle that is weakly stable but sensitive to steer input in the lower range of lateral acceleration, and is very

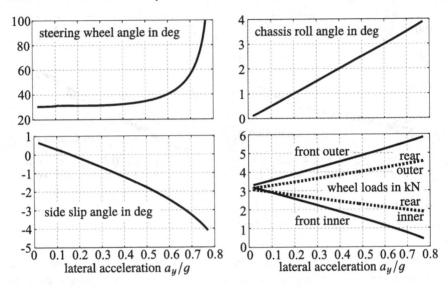

FIGURE 10.5
Steady-state cornering: Rear-wheel driven car.

stable but less sensitive to steer input in the limit range. The roll angle of the vehicle increases in proportion to the lateral acceleration with a ratio of approximately $5°/g$ here.

The side slip angle, measured at the vehicle center, starts with the value $\beta^{a_y \to 0} \approx 0.75°$, which coincides with the Ackermann value $\beta^{st} = \arctan(a_2/R) = \arctan(2.57-1.25)/100 = 0.0132 \equiv 0.756°$. It changes sign at $a_y^{\beta_{st}=0} \approx 0.16g$. Inserting this value into Equation (9.147) delivers cornering stiffness at the rear axle of

$$c_{S2} = \frac{a_1 R m \, a_y^{\beta_{st}=0}}{a_2\,(a_1+a_2)} = \frac{1.25 * 100 * 1300 * 0.16 * 9.81}{(2.57-1.25) * 2.57} = 75\,186 \text{ N/m} \qquad (10.30)$$

which indicates rather soft tires but corresponds to the fact that the vehicle under consideration is an older one. The steering gradient, which was estimated here with $k \approx 0$, is defined in Equation (9.144). It can also be used to determine the cornering stiffness c_{S1} of the front axle.

The overturning torque is intercepted by the wheel load differences between the outer and inner wheels. With a sufficiently rigid frame, the use of an anti-roll bar at the front axle allows for increasing the wheel load difference there and for decreasing it at the rear axle accordingly. Thus, the degressive influence of the wheel load on tire properties, cornering stiffness, and maximum possible lateral forces, is stressed more strongly at the front axle, and the vehicle becomes more under-steering and stable at increasing lateral acceleration, until it drifts out of the curve over the front axle in the limit situation. Problems occur with front-wheel driven vehicles, because due to the demand for traction, the front axle cannot be relieved at will.

Having a sufficiently large test site, the steady-state cornering maneuver can also be carried out at constant speed. There, the steering wheel is slowly turned until the vehicle reaches the limit range. That way, weakly motorized vehicles can also be tested at high lateral accelerations.

10.3.2 Step Steer Input

The dynamic response of a vehicle is often tested with a step steer input. Methods for the calculation and evaluation of an ideal response, as used in system theory or control techniques, cannot be used with a real car, as a step input at the steering wheel is not possible in practice. A real steering angle gradient and typical results are displayed in Figure 10.6 and Figure 10.7.

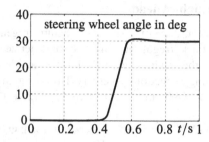

FIGURE 10.6
Time history of a typical step steer input.

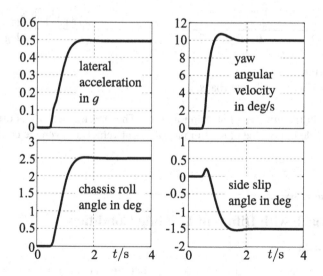

FIGURE 10.7
Step steer: Passenger car at $v = 100$ km/h.

Note that the angle at the steering wheel is the decisive factor for the driving behavior, but the steering angle at the wheels can differ from the steering wheel angle because of elasticities, friction influences, and a servo-support. For very fast steering movements, the dynamics of the tire forces also play an important role.

In practice, a step steer input is usually only used to judge vehicles subjectively. Exceedances in yaw velocity, roll angle, and especially side slip angle are felt as annoying. The vehicle under consideration behaves dynamically very well. Almost no overshoots occur in the time history of the roll angle and the lateral acceleration. However, small overshoots can be noticed at the yaw velocity and the side slip angle. As indicated in Section 9.4.2, an active rear-wheel steering can reduce the variations in the side slip angle in particular.

10.3.3   Driving Straight Ahead

The irregularities of a track are of a stochastic nature in general. As discussed in Section 2.3, stochastic track irregularities may be characterized by a reference power spectral density Φ_0 and the waviness w. A straightforward drive on an uneven track makes continuous steering corrections necessary. The histograms of the steering angle δ at a driving speed of $v = 90$ km/h are displayed in Figure 10.8. The track quality

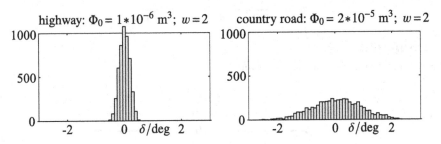

FIGURE 10.8
Steering activity on different roads.

is reflected in the number of steering actions. The steering activity is often used to judge a vehicle in practice or to compare different vehicles or vehicle concepts.

10.4   Coach with Different Loading Conditions

10.4.1   Data

The difference between empty and laden is sometimes very large at trucks and coaches. All relevant data of a travel coach under fully laden and empty conditions are listed in Table 10.3.

TABLE 10.3

Data for a laden and empty coach

Vehicle	Mass in kg	Center of Gravity in m	Inertias in kg m^2		
Empty	12 500	−3.800 | 0.000 | 1.500	12 500	0	0
			0	155 000	0
			0	0	155 000
Fully laden	18 000	−3.860 | 0.000 | 1.600	15 400	0	250
			0	200 550	0
			250	0	202 160

The coach has a wheel base of $a = 6.25$ m. The front axle with the track width $s_v = 2.046$ m has a double wishbone, single-wheel suspension. The twin-tire rear axle with track widths $s_h^o = 2.152$ m and $s_h^i = 1.492$ m is guided by two longitudinal links and an A-arm. The air springs are fitted to load variations via a level control.

10.4.2 Roll Steering

While the kinematics at the front axle hardly cause steering movements at roll motions, the kinematics at the rear axle are tuned in a way to cause a notable roll steering effect, Figure 10.9. This is achieved by moving the wheel center on jounce to the front and

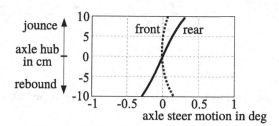

FIGURE 10.9

Roll steer motion of a solid axle.

on rebound to the rear. As a solid axle rigidly connects both wheels, this will cause the axle to rotate about the vertical axis, thus performing a steer motion when the chassis rolls and the outer wheel is forced to a jounce and the inner to a rebound motion.

10.4.3 Steady-State Cornering

Some results of a steady-state cornering on a radius of $R = 100$ m are plotted in Figure 10.10. The fully laden vehicle is slightly more understeering than the empty one. The roll steer behavior affects the steering tendency differently, because the laden vehicle exhibits a larger roll motion when cornering. In general, vehicles with a

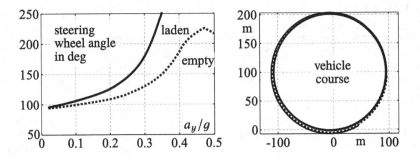

FIGURE 10.10
Coach at steady-state cornering.

twin-tired solid axle at the rear exhibit a basic understeer tendency. When cornering to the left, the forward velocities of the tire centers at the rear axle will increase from the inner left to the outer right, Figure 10.11. Not taking driving or braking maneuvers

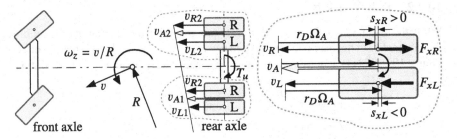

FIGURE 10.11
Vehicle with a twin-tired solid rear axle while cornering to the left.

into account, each rear wheel will rotate with an average angular velocity Ω_A, which results from the rolling condition $v_A = r_D \, \Omega_A$, where r_D is the dynamic rolling radius and $v_A = \frac{1}{2} (v_R + v_L)$ represents the average forward velocity of the right (R) and the left (L) tire center. As both twin tires are forced to rotate with this average angular velocity Ω_A, positive $s_{xR} > 0$ and negative $s_{xL} < 0$ longitudinal slips will result at the right and left tire as a consequence. The thereby caused couple of longitudinal forces F_{xR}, F_{xL} generates a torque about the vertical axes, which acts in the opposite direction of the yaw velocity ω_z of the vehicle. The same effect, which is similar to the tire bore torque mechanism, occurs on both wheels and adds to a significant torque T_u, thus producing the understeer tendency of those vehicles. The higher wheel loads of the laden coach will amplify this understeer effect, in addition.

The simulation was performed with P-controllers for the driving torque and the steering wheel angular velocity, which are provided in Equations (10.27) and (10.28). Despite the simple driver model, the coach is kept on a circled path quite well, right plot in Figure 10.10. When approaching the limit range, the understeer tendency of the laden coach is increasing rapidly, which results in a slight increase in the curve radius.

Due to the relatively high position of the center of gravity, the maximal lateral acceleration is limited by the overturning hazard. For the empty vehicle, the inner

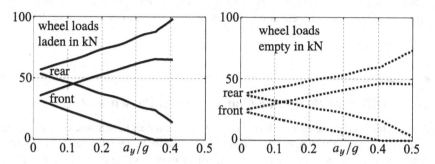

FIGURE 10.12
Wheel loads of a coach at steady-state cornering.

front wheel lifts off at a lateral acceleration of $a_y \approx 0.4\,g$, right plot in Figure 10.12. If the vehicle is fully laden, this effect will occur at a lateral acceleration of $a_y \approx 0.35\,g$, left plot in Figure 10.12.

10.4.4 Step Steer Input

The results of a step steer input at the driving velocity of $v = 80$ km/h can be seen in Figure 10.13. To achieve comparable acceleration values in steady-state condition,

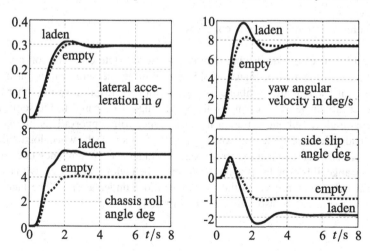

FIGURE 10.13
Step steer input to laden and empty coach.

the step steer input was done for the empty vehicle with $\delta = 90°$ and for the fully laden one with $\delta = 135°$. The steady-state roll angle is 50% larger at the fully laden coach than at the empty one. By the level control, the air spring stiffness increases with the load. Because the damper effect remains unchanged, the fully laden vehicle is not damped as well as the empty one. This results in larger overshoots in the time histories of the lateral acceleration, the yaw angular velocity, and the side slip angle.

10.5   Different Rear Axle Concepts for a Passenger Car

A medium-sized passenger car is equipped in standard design with a semi-trailing rear axle. By accordingly changed data, this axle can easily be transformed into a trailing arm or a single wishbone suspension. According to the roll support, the semi-trailing axle realized in serial production represents a compromise between the trailing arm and the single wishbone suspension, Figure 10.14.

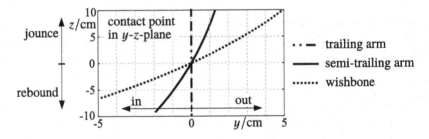

FIGURE 10.14
Kinematics of different rear axle concepts.

The influences on the driving behavior at steady-state cornering on a radius of $R = 100$ m are shown in Figure 10.15. Substituting the semi-trailing arm on the standard car by a single wishbone, one gets, without adaption of the other system parameter, a vehicle oversteering in the limit range. Compared to the semi-trailing arm, the single wishbone causes a notably higher roll support. This increases the wheel load difference at the rear axle, Figure 10.15. Because the wheel load difference simultaneously decreases at the front axle, the understeering tendency decreases. In the limit range, this even leads to an oversteering behavior. The vehicle with a trailing arm rear axle is, compared to the serial car, more understeering. The lack of roll support at the rear axle also causes a larger roll angle.

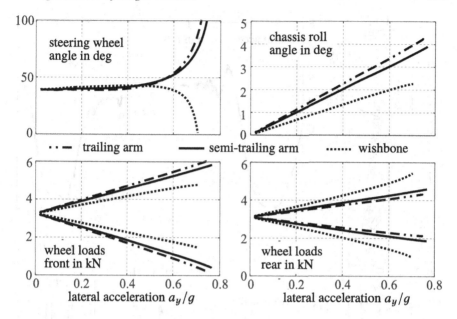

FIGURE 10.15
Steady-state cornering with different rear axle concepts.

10.6 Obstacle Avoidance and Off-Road Scenario

Mechatronic systems like the electronic stability program (ESP) or active rear-wheel steering (RWS) assist the driver in critical situations. An obstacle avoidance maneuver, where the vehicle even comes off-road, may serve as an example for such a critical situation.

At first, the vehicle drives straight ahead with a speed of $v = 90$ km/h. All of a sudden, an obstacle blocks the road. The driver takes the risk to go off-road by passing the obstacle on the right and tries to steer the vehicle back on the road again. The off-road scenario is simply modeled here by a reduced coefficient of friction ($\mu = 1 \rightarrow \mu = 0.4$) and a slightly lowered profile ($z = 0 \rightarrow z = -0.05$ m).

The maneuver is simulated with a fully non-linear and three-dimensional vehicle model, as presented in Section 10.1, with ESP switched on or off and with active rear-wheel steering in addition. The results are characterized in Figures 10.16 and 10.17 by top view multi-shots, the time history of the lateral acceleration, and the brake triggers or the wheel steering angles, respectively.

The multi-shots in Figure 10.16 demonstrate that without mechatronic assistance (ESP off) a normal driver, modeled by a PD-control here, is not able to cope with this critical avoidance maneuver. The vehicle with the ESP switched off, passes the obstacle and comes back to the road again. But the reduced friction in the off-road

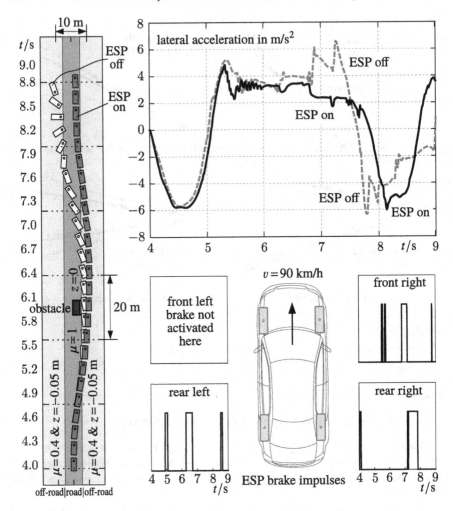

FIGURE 10.16
Obstacle avoidance with ESP on and off.

environment requires a comparatively large steering angle to generate sufficient lateral forces in order to bring the vehicle back to the road. The change from reduced friction to normal friction, which occurs at $t \approx 7$ s when the vehicle comes back to the road, affects at first the front wheels and then results in a distinct yaw reaction. A normal driver will counter steer immediately, which in this particular case causes the vehicle to go off-road on the left and spin finally.

The electronic stability program (ESP) is activated if too much over- or understeer is detected, as explained in Section 9.4. A simple linear handling model, characterized by the parameter in Table 10.4, is used to estimate the normal or expected vehicle response to a steer input and to activate the ESP control appropriately.

The first steer input at $t = 4$ s is already assisted by a brake impulse at the right rear wheel, which turns the vehicle to the right, thus enhancing the impact of this steer input to the vehicle. As can be seen from the corresponding time histories in Figure 10.16, the brakes at the rear wheels and on the right front wheels are triggered at certain events. The result is a vehicle trajectory, which corresponds to a nearly perfect double lane change and guides the vehicle with a comparatively smooth time history of the lateral acceleration past the obstacle.

The results of a vehicle without ESP but equipped with active rear-wheel steering (RWS) compared to the results of a vehicle just with ESP on are shown in Figure 10.17.

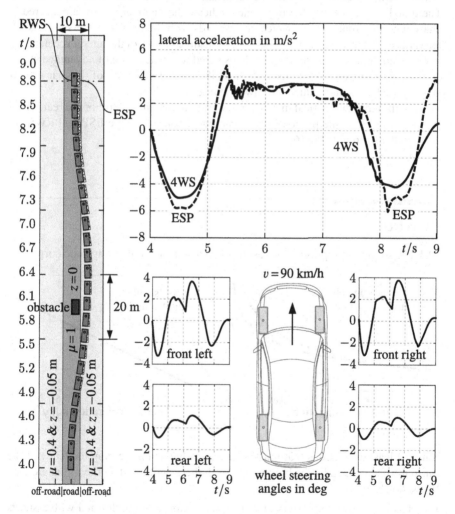

FIGURE 10.17
Obstacle avoidance with active wheel steering (RWS) compared to ESP.

TABLE 10.4

Parameter of a simple linear handling model characterizing a fullsize vehicle

Vehicle mass and inertia	Position of CoG	Cornering stiffness
m = 2127.8 kg	a_1 = 1.5 m	c_{S1} = 150 kN/-
Θ = 3358.3 kgm²	a_2 = 1.4 m	c_{S2} = 200 kN/-

The multi-shots demonstrate that the vehicle with active rear-wheel steering (RWS) is slightly closer to a perfect double lane change than the vehicle with ESP on. The time histories of the steering angles indicate that the rear wheels are steered in phase with the front wheels, which corresponds with the results of the simple feed-forward control discussed in Section 9.4.2. The time history of the lateral acceleration is very smooth now. This underlines that steered rear wheels are of advantage at fast double lane changes in particular. Because the in-phase steering of the rear wheels amplify the lateral and reduce the yaw response of a vehicle.

Details about this critical maneuver, the ESP model, the control of the rear-wheel steering, as well as the proposal of an integrated control, where ESP and RWS are combined, are provided in [6].

Exercises

10.1 The solid lines represent the results of a passenger car at steady-state cornering.

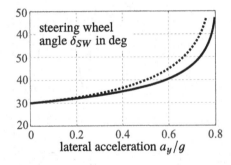 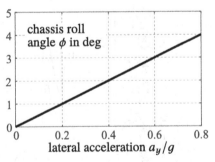

Name at least two actions that must be taken to change the steering tendency of the car to the one defined by the broken line. Will these actions affect the roll angle too? If yes, how?

10.2 The plot shows the results of a steady-state cornering performed with a sports car on a radius of $R = 100$ m, where the side slip angle was measured at the vehicle's center of gravity. The sports car has a wheelbase of $a = 2.35$ m and the wheel loads at the front and rear axles are specified by 5600 N and 7000 N.

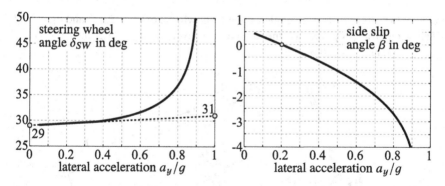

Compute the mass m and the location (a_1, a_2) of the center of gravity with respect to the front and the rear axle.

Calculate the average steering angle δ_A at the front wheels and the side slip angle β_A at the vehicle's center of gravity that hold according to Ackermann when driving with nearly vanishing lateral acceleration $a_y \to 0$ on the given radius.

Compute the overall steering ratio defined by $i_S = \delta_{SW}/\delta_A$, where δ_{SW} denotes the steering wheel angle at $a_y \to 0$ and δ_A is the Ackermann steering angle.

Calculate the cornering stiffness c_{S2} at the rear axle, the steering gradient, which according to Equation (9.144) will be defined here by

$$k = \frac{1}{i_S} \frac{d\,\delta_{SW}}{d\,a_y},$$

and the cornering stiffness c_{S1} at the front axle, which will all be valid in the low acceleration range.

10.3 A passenger car performs a step steer maneuver at the driving velocity of $v = 120$ km/h.

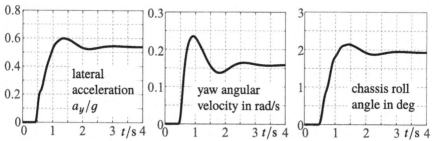

Judge the results.

At what radius is the vehicle driving at the end of the maneuver ($t = 4$ s)?

How will the plots change if the anti-roll bar stiffness at the rear axle is doubled and a rigid chassis as well as a degressive wheel load influence on the tire forces can be taken for granted?

10.4 A sensor located in point M at the center of the rear bumper measures the velocity components $v_{Mx} = 20$ m/s and $v_{My} = -0.50$ m/s into the longitudinal and lateral directions of the vehicle. Additional sensors provide the yaw angular velocity

$\omega = 0.25$ rad/s and steering angles at the front wheels of $\delta_1 = 2.07°$ and $\delta_1 = 2.03°$. The geometric properties of the vehicle are given by $a_1 = 1.3$ m, $a_2 = 1.4$ m, $b = 1.2$ m, and $s = 1.6$ m.

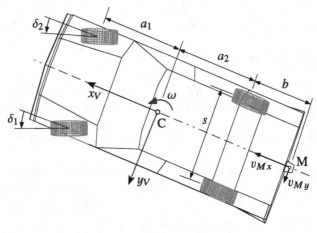

Is the steering system designed according to Ackermann? Compute the side slip angle at point M and at the center C. Compute the slip angles at all wheels by assuming that the longitudinal slips are negligibly small.

Bibliography

[1]   D. Bestle and M. Beffinger. "Design of Progressive Automotive Shock Absorbers." In: *Proceedings of Multibody Dynamics*. Madrid, 2005.

[2]   M. Blundell and D. Harty. *The Multibody System Approach to Vehicle Dynamics*. Elsevier Butterworth-Heinemann Publications, 2004.

[3]   *Bosch Automotive Handbook*. 7th ed. Robert Bosch GmbH, 2007.

[4]   H. Braun. "Untersuchung von Fahrbahnunebenheiten und Anwendung der Ergebnisse." PhD thesis. TU Braunschweig, 1969.

[5]   T. Butz et al. "A Realistic Road Model for Real-Time Vehicle Dynamics Simulation." In: *Society of Automotive Engineers*, SAE Paper 2004-01-1068 (2004).

[6]   A. A. Castro. "Development of a Robust and Fault Tolerant Integrated Control System to Improve the Stability of Road Vehicles in Critical Driving Scenarios." PhD thesis. Pontifical Catholic University of Rio de Janeiro, Brasil, 2017. URL: https://www.maxwell.vrac.puc-rio.br/colecao.php?strSecao= resultado&nrSeq=33168@2 (visited on 11/29/2019).

[7]   *CDTire – Scalable Tire Model*. URL: https://www.itwm.fraunhofer.de/ en/departments/mdf/cdtire.html (visited on 07/21/2019).

[8]   C. J. Dodds and J. D. Robson. "The Description of Road Surface Roughness." In: *Journal of Sound and Vibration* 31.2 (1973), pp. 175–183.

[9]   P. Dorato et al. *Linear-Quadratic Control. An Introduction*. Englewood Cliffs, New Jersey: Prentice-Hall, 1995.

[10] M. Eichler et al. *Dynamik von Luftfedersystemen mit Zusatzvolumen: Modellbildung, Fahrzeugsimulationen und Potenzial*. Bericht 1791. VDI, 2003.

[11] *enDYNA – Real-time Simulation of Combustion Engines*. URL: https://www. tesis.de/endyna/?r=1 (visited on 08/24/2019).

[12] P. Fickers and B. Richter. "Incorporating FEA-Techniques into MSA Illustrated by Several Rear Suspension Concepts." In: *9th European ADAMS User Conference*. Frankfurt, 1994.

[13] M. Gipser. "FTire, a New Fast Tire Model for Ride Comfort Simulations." In: *International ADAMS User Conference*. Berlin, 1999.

[14] P. Gruber and R. S. Sharp, eds. *Proceedings of the 4th International Tyre Colloquium*. 2015. URL: http://epubs.surrey.ac.uk/807823 (visited on 07/13/2019).

[15] A. Hackl et al. "Experimental Validation of the Maxwell Model for Description of Transient Tyre Forces." In: *Proceedings of the 16. Internationales Stuttgarter Symposium Automobil- und Motorentechnik*. FKFS. Stuttgart, Germany, 2016.

[16] W. Hirschberg et al. "Ermittlung der Potenziale zur LKW-Stabilisierung durch Fahrdynamiksimulation." In: *Berechnung und Simulation im Fahrzeugbau*. VDI, Sept. 2000.

[17] W. Hirschberg et al. "User-Appropriate Tyre-Modeling for Vehicle Dynamics in Standard and Limit Situations." In: *Vehicle System Dynamics* 38.2 (2002), pp. 103–125.

[18] M. Irmscher and M. Ehmann. "Driver Classification using ve-DYNA Advanced Driver." In: *SAE International* 01-0451 (2004).

[19] ISO 2631. "Mechanical Vibration and Shock Evaluation of Human Exposure to Whole-Body Vibration." In: *International Standard (ISO)* (1997).

[20] ISO 8608. "Mechanical Vibration - Road Surface Profiles - Reporting of Measured Data." In: *International Standard (ISO)* (1995).

[21] P. van der Jagt. "The Road to Virtual Vehicle Prototyping; New CAE-Models for Accelerated Vehicle Dynamics Development." ISBN 90-386-2552-9 NUGI 834. Tech. Univ. Eindhoven, 2000.

[22] R. T. Kane and D. A. Levinson. "Formulation of Equations of Motion for Complex Spacecraft." In: *Journal of Guidance and Control* 3(2) (1980), pp. 99–112.

[23] U. Kiencke and L. Nielsen. *Automotive Control Systems*. Springer, 2000.

[24] W. Kortüm and P. Lugner. *Systemdynamik und Regelung von Fahrzeugen*. Springer, 1993.

[25] P. Lugner and M. Plöchl. *Tire Model Performance Test (TMPT)*. Taylor & Francis, 2007.

[26] M. Mitschke and H. Wallentowitz. *Dynamik der Kraftfahrzeuge, 4. Auflage*. Springer, 2004.

[27] P.C. Müller and W.O. Schiehlen. *Lineare Schwingungen*. Wiesbaden: Akad. Verlagsgesellschaft, 1976.

[28] M. Nagai and P. Raksincharoensak. "Advanced Chassis Control and Automated Driving." In: *Vehicle Dynamics of Modern Passenger Cars*. Ed. by Peter Lugner. 582nd ed. CISM International Centre for Mechanical Sciences. Springer International Publishing, 2019, pp. 247–307.

[29] U. Neureder. *Untersuchungen zur Übertragung von Radlastschwankungen auf die Lenkung von Pkw mit Federbeinvorderachse und Zahnstangenlenkung*. Vol. 12(518). Fortschritt-Berichte VDI. Düsseldorf: VDI-Verlag, 2002.

[30] P. E. Nikravesh. *Computer-Aided Analysis of Mechanical Systems*. Englewood Cliffs, New Jersey: Prentice Hall, 1988.

[31]   Ch. Oertel and A. Fandre. "Ride Comfort Simulations and Steps Towards Life Time Calculations; RMOD-K and ADAMS." In: *International ADAMS User Conference*. Berlin, 1999.

[32]   oponeo. *How Much Does a Tyre Weigh*. URL: https://www.oponeo.co.uk/ tyre-article/how-much-does-a-tyre-weigh (visited on 07/13/2019).

[33]   H.B. Pacejka. *Tyre and Vehicle Dynamics*. Oxford: Butterworth-Heinemann, 2002.

[34]   H.B. Pacejka and E. Bakker. "The Magic Formula Tyre Model." In: *1st Int. Colloquium on Tyre Models for Vehicle Dynamic Analysis*. Lisse: Swets & Zeitlinger, 1993.

[35]   K. Popp and W. Schiehlen. *Fahrzeugdynamik*. Stuttgart: Teubner, 1993.

[36]   K. Popp and W.O. Schiehlen. *Ground Vehicle Dynamics*. Berlin: Springer, 2010.

[37]   J. Rauh. "Virtual Development of Ride and Handling Characteristics for Advanced Passenger Cars." In: *Vehicle System Dynamics* 40.1–3 (2003), pp. 135–155.

[38]   A. Riepl et al. "Rough Road Simulation with Tire Model RMOD-K and FTire." In: *Proceedings of the 18th IAVSD Symposium on the Dynamics of Vehicles on Roads and on Tracks. Kanagawa, Japan*. London: Taylor & Francis, 2003.

[39]   G. Rill. "A Modified Implicit Euler Algorithm for Solving Vehicle Dynamic Equations." In: *Multibody System Dynamics* 15.1 (2006), pp. 1–24.

[40]   G. Rill. "Demands on Vehicle Modeling." In: *The Dynamics of Vehicles on Road and on Tracks*. Ed. by R.J. Anderson. Lisse: Swets-Zeitlinger, 1990.

[41]   G. Rill. "First Order Tire Dynamics." In: *Proceedings of the III European Conference on Computational Mechanics Solids, Structures and Coupled Problems in Engineering*. Lisbon, Portugal, 2006.

[42]   G. Rill. *Simulation von Kraftfahrzeugen*. Braunschweig: Vieweg, 1994. URL: https://www.researchgate.net/publication/317037037_Simulation_ von_Kraftfahrzeugen (visited on 11/29/2019).

[43]   G. Rill. "The Influence of Correlated Random Road Excitation Processes on Vehicle Vibration." In: *The Dynamics of Vehicles on Road and on Tracks*. Ed. by K. Hedrik. Lisse: Swets-Zeitlinger, 1984.

[44]   G. Rill. "Vehicle Modeling by Subsystems." In: *Journal of the Brazilian Society of Mechanical Sciences & Engineering - ABCM* XXVIII.4 (2006), pp. 431–443.

[45]   G. Rill. "Vehicle Modeling for Real Time Applications." In: *Journal of the Brazilian Society of Mechanical Sciences - RBCM* XIX.2 (1997), pp. 192–206.

[46]   G. Rill. "Wheel Dynamics." In: *Proceedings of the XII International Symposium on Dynamic Problems of Mechanics (DINAME 2007)*. 2007.

[47]   G. Rill and C. Chucholowski. "A Modeling Technique for Fast Computer Sim-
       ulations of Configurable Vehicle Systems." In: *Proceedings of the 21st Inter-
       national Congress of Theoretical and Applied Mechanics (ICTAM)*. Warsaw,
       Poland, 2004.

[48]   G. Rill and C. Chucholowski. "Modeling Concepts for Modern Steering Sys-
       tems." In: *ECCOMAS Multibody Dynamics*. Madrid, Spain, 2005.

[49]   G. Rill and J. Rauh. "Simulation von Tankfahrzeugen." In: *Berechnung im
       Automobilbau, VDI-Bericht 1007*. Düsseldorf: VDI, 1992.

[50]   G. Rill and W. Schiehlen. "Performance Assessment of Time Integration Meth-
       ods for Vehicle Dynamics Simulations." In: *Multibody Dynamics 2009 (ECCO-
       MAS Thematic Conference, Warsaw, Poland, June 29 to July 2, 2009)*. Ed. by
       K. Arczewski and J. Fraczek. Warsaw: Faculty of Power and Aeronautical
       Engineering, Warsaw University of Technology, 2009.

[51]   G. Rill et al. "Leaf Spring Modeling for Real Time Applications." In: *The
       Dynamics of Vehicles on Road and on Tracks - Extensive Summaries, IAVSD
       03*. 2003.

[52]   G. Rill. "An Engineer's Guess on Tyre Parameter Made Possible with
       TMeasy." In: *Proceedings of the 4th International Tyre Colloquium in:
       http://epubs.surrey.ac.uk/807823*. Ed. by Patrick Gruber and Robin S. Sharp.
       University of Surrey, GB, 2015.

[53]   G. Rill. "Multibody Systems and Simulation Techniques." In: *Vehicle Dynam-
       ics of Modern Passenger Cars*. Ed. by Peter Lugner. 582nd ed. CISM Inter-
       national Centre for Mechanical Sciences. Springer International Publishing,
       2019, pp. 309–375.

[54]   G. Rill. "Reducing the Cornering Resistance by Torque Vectoring." In: *X
       International Conference on Structural Dynamics, EURODYN 2017, Rome,
       Italy*. Procedia Engineering, ELSEVIER, 2017.

[55]   G. Rill. "Sophisticated but Quite Simple Contact Calculation for Handling Tire
       Models." In: *Multibody System Dynamics* 45.2 (2019), pp. 131–153.

[56]   G. Rill. "TMeasy 6.0 – A Handling Tire Model that incorporates the first
       Two Belt Eigenmodes." In: *to appear in Proceedings of the XI International
       Conference on Structural Dynamics (EURODYN 2020)*. 2020.

[57]   G. Rill and A. A. Castro. "A Novel Approach for Parametrization of Suspension
       Kinematics." In: *Proceedings of the 26th IAVSD Symposium on the Dynamics
       of Vehicles on Roads and on Tracks. Gothenburg, Sweden*. Taylor & Francis,
       2019.

[58]   G. Rill and A. A. Castro. "The Influence of Axle Kinematics on Vehicle
       Dynamics." In: *Interdisciplinary Applications of Kinematics: Proceedings of
       the Third International Conference (IAK) Interdisciplinary Applications of
       Kinematics: Proceedings of the Third International Conference (IAK)*. Ed. by
       Kecskeméthy et al. Mechanisms and Machine Science. Springer, 2019.

[59] G. Rill et al. *Grundlagen und Methodik der Mehrköpersimulation*. 4th ed. Wiesbaden: Springer Vieweg, 2020.

[60] G. Rill et al. "VTT – A Virtual Test Truck for Modern Simulation Tasks." In: *Vehicle System Dynamics, DOI: 10.1080/00423114.2019.1705356* (2019).

[61] Th. Seibert and G. Rill. "Fahrkomfortberechnungen unter Einbeziehung der Motorschwingungen." In: *Berechnung und Simulation im Fahrzeugbau, VDI-Bericht 1411*. Düsseldorf: VDI, 1998.

[62] VIRES Simulationstechnologie. *Open File Formats and Open Source Tools for the Detailed Description, Creation and Evaluation of Road Surfaces*. URL: www.opencrg.org (visited on 07/04/2019).

[63] G. Smith and M. Blundell. "A New Efficient Free-Rolling Tyre-Testing Procedure for the Parameterisation of Vehicle Dynamics Tyre Models." In: *Journal of Automobile Engineering (Proc IMechE Part D)* 231.10 (2016), pp. 1435–1448.

[64] M. Vögel et al. "An Optimal Control Approach to Real-Time Vehicle Guidance." In: *Mathematics – Key Technology for the Future*. Ed. by W. Jäger and H.-J. Krebs. Berlin: Springer, 2003, pp. 84–102.

[65] M. Weigel et al. "A Driver Model for a Truck-Semitrailer Combination." In: *Vehicle System Dynamics Supplement* 41 (2004), pp. 321–331.

[66] H. Weinfurter et al. "Entwicklung einer Störgrößenkompensation für Nutzfahrzeuge mittels Steer-by-Wire durch Simulation." In: *Berechnung und Simulation im Fahrzeugbau, VDI-Berichte 1846*. Düsseldorf: VDI, 2004, pp. 923–941.

Glossary

ABS: The anti-lock braking system (ABS) is an automated system that imitates skilled drivers by controlled threshold and cadence braking. It maintains the maneuverability of the vehicle, in particular.

Ackermann steering: If the layout of a steering system is such that the wheels at one axle rotate about different axles, then the steering angle at the inner wheel must be larger than the one at the outer wheel in order to achieve a smooth and proper cornering of the vehicle. Georg Lankensperger applied this steering mechanism in 1816 at horse-drawn carriages and Rudolph Ackermann patented it later on.

Anti-squat: Special feature of suspension system that reduces the pitch angle of the chassis when the vehicle accelerates.

Brake pitch: Pitch angle of chassis induced by braking the vehicle.

Camber: The wheel camber describes the angle between the wheel rotation axis and a horizontal plane. It usually is measured at kinematics and compliance (K&C) test rigs. The angle between the wheel rotation axis and the local or effective road plane defines the tire camber, which affects the lateral tire force and the turn torque.

Contact point: The geometric contact point is the point on the intersection line between the rim center plane and the local or effective road plane that has the shortest distance to the wheel center.

Cornering stiffness: The cornering stiffness of a tire describes the linearized ability of a tire to generate lateral forces when cornering. It corresponds to the initial inclination of the lateral force plotted versus the lateral slip or slip angle, respectively. The cornering stiffness of an axle is not just the sum of the cornering stiffnesses of the tire. Because, the cornering stiffness of a tire strongly depends on the wheel load and the left and right wheel loads differ when cornering.

ESP: The electronic stability program assists the driver in critical situations. It triggers a brake impulse at a single front or rear wheel and counteracts a too distinct over- or understeer tendency of the vehicle.

Handling: Handling characterizes how a vehicle responds and reacts to the inputs of a driver or a driving machine, as well as how it moves along a road.

Oversteer: Vehicle tends to turn in or even skids while cornering.

Rear-wheel steering: Steering the rear wheels in addition to the front wheels enhances the handling attitude and improves cornering. Some suspension systems provide a passive rear-wheel steering by well-tuned bushing compliances. A controlled rear-wheel steering is usually performed such that at low speeds the rear wheels are steered in opposite and at higher speeds in the same direction of the front wheels.

Roll center: The roll center of each axle is defined by the kinematics of the suspension. The imaginary line, called roll axis, between the front and rear roll centers determine the axis around the chassis rolls.

Roll stiffness: The overall stiffness related to axle roll motions.

Side slip: Angle between the velocity vector measured at a specific vehicle point and the longitudinal direction of the vehicle.

Slip: Dimensionless ratio of contact sliding velocity versus transport velocity.

Slip angle: Angle between the contact point sliding velocity and the circumferential tire direction.

Toe: An inward (toe-in) or outward (toe-out) alignment of the forward edge of the wheels at an axle.

Track width: Distance of the contact points at an axle.
Note: twin tired axles are characterized by an inner and an outer track width.

Understeer: Vehicle tends to move straight ahead instead of cornering.

Wheel alignment point: A point located on the wheel rotation axis which does not coincide with the wheel center, thus defining the momentary direction of the wheel rotation axis.

Wheel base: Distance from the centers of the front and rear wheels or the front axle and the center point of the rear axles respectively.

Wheel load: Component of the tire force applied normal to the local road plane.
Note: if a vehicle is placed on a horizontal road, the sum of all wheel loads amounts to the weight of the vehicle.

Index

Printed in the United States
by Baker & Taylor Publisher Services